Case Files™

Biochemis

D1552778

Second Edition

EUGENE C. TOY, MD
The John S. Dunn Senior Academic Chief and Program Director
Obstetrics and Gynecology Residency Program
The Methodist Hospital, Houston
Clerkship Director and Clinical Associate Professor
Department of Obstetrics and Gynecology
University of Texas Medical School at Houston
Houston, Texas

WILLIAM E. SEIFERT, JR., PHD
Senior Lecturer and Medical Biochemistry Course Director
Department of Biochemistry and Molecular Biology
University of Texas Medical School at Houston
Houston, Texas

HENRY W. STROBEL, PHD
Professor
Department of Biochemistry and Molecular Biology
Assistant Dean For Student Affairs
Associate Dean for Faculty Affairs
University of Texas Medical School at Houston
Houston, Texas

KONRAD P. HARMS, MD
Clinical Assistant Professor
Weill Cornell College of Medicine
Associate Program Director
Obstetrics and Gynecology Residency
The Methodist Hospital, Houston
Houston, Texas

McGraw Hill Medical

New York Chicago San Francisco
Lisbon London Madrid Mexico City
Milan New Delhi San Juan Seoul
Singapore Sydney Toronto

The McGraw-Hill Companies

Case Files™: Biochemistry, Second Edition

Copyright © 2008 by The McGraw-Hill Companies, Inc. All rights reserved. Printed in the United States of America. Except as permitted under the United States Copyright Act of 1976, no part of this publication may be reproduced or distributed in any form or by any means, or stored in a data base or retrieval system, without the prior written permission of the publisher.

Previous edition copyright © 2005 by The McGraw-Hill Companies, Inc.

Case Files™ is a trademark of The McGraw-Hill Companies, Inc.

1 2 3 4 5 6 7 8 9 0 DOC/DOC 0 9 8

ISBN 978-0-07-148665-1
MHID 0-07-148665-8

Notice

Medicine is an ever-changing science. As new research and clinical experience broaden our knowledge, changes in treatment and drug therapy are required. The authors and the publisher of this work have checked with sources believed to be reliable in their efforts to provide information that is complete and generally in accord with the standard accepted at the time of publication. However, in view of the possibility of human error or changes in medical sciences, neither the editors nor the publisher nor any other party who has been involved in the preparation or publication of this work warrants that the information contained herein is in every respect accurate or complete, and they disclaim all responsibility for any errors or omissions or for the results obtained from use of the information contained in this work. Readers are encouraged to confirm the information contained herein with other sources. For example and in particular, readers are advised to check the product information sheet included in the package of each drug they plan to administer to be certain that the information contained in this work is accurate and that changes have not been made in the recommended dose or in the contraindications for administration. This recommendation is of particular importance in connection with new or infrequently used drugs.

This book was set in Times Roman by International Typesetting and Composition.
The editor was Catherine A. Johnson.
The production supervisor was Catherine Saggese.
Project management was provided by Gita Raman, International Typesetting and Composition.
The cover designer was Aimee Nordin.
RR Donnelley was printer and binder.

This book is printed on acid-free paper.

Library of Congress Cataloging-in-Publication Data

Case files. Biochemistry / Eugene C. Toy ... [et al.]—2nd ed.
p. ; cm.
Includes bibliographical references and index.
ISBN-13: 978-0-07-148665-1 (pbk. : alk. paper)
ISBN-10: 0-07-148665-8 (pbk. : alk. paper)
1. Clinical biochemistry—Case studies. I. Toy, Eugene C. II. Title: Biochemistry.
[DNLM: 1. Metabolism—Case Reports. 2. Metabolism—Examination Questions.
3. Biochemistry—Case Reports. 4. Biochemistry—Examination Questions.
QU 18.2 C337 2008]
RB112.5.C38 2008
612'.015—dc22 2008005740

❖ DEDICATION

To my friend, mentor, and role model, Dr. Benton Baker III,
who taught me by his example, the importance of integrity, unselfishness,
and teamwork.

—ECT

To my wife, Heidi, and to Koen and Kort,
for their unwavering love and support and
for reminding me daily of what is really important in life.

—KPH

To Dr. John A. DeMoss, Founding Chair of the Department of Biochemistry and Molecular Biology of the University of Texas Medical School at Houston. Dr. DeMoss served as Department Chair from 1971 until he stepped down from that office in 1992, serving as Professor until 2001 when he became Emeritus Professor, his present position. Dr. DeMoss is that unique kind of scholar who excels at both research and teaching. He loves teaching and he is gifted at it. Dr. DeMoss not only has taught and continues to teach generations of students, he also teaches faculty, especially young faculty, how to teach and to love teaching. It is therefore in gratitude for his teaching of us that we dedicate this book to Dr. John A. DeMoss.

—WES & HWS

CONTENTS

CONTRIBUTORS

Sayeepriyadarshini Anakk, PhD
Postdoctoral Fellow
Department of Molecular and Cell Biology
Baylor College of Medicine
Houston, Texas
β-Thalassemia
Human Immunodeficiency Virus (HIV)

Leposava Antonovic, PhD
Clinical Pharmacy Fellow
Department of Pharmacy
University of Texas–MD Anderson Cancer Center
Houston, Texas
Cystic Fibrosis
Polymerase Chain Reaction (PCR)

Michael R. Blackburn, PhD
Professor
Department of Biochemistry
University of Texas Medical School at Houston
Houston, Texas
Gout
Vegetarian–Essential Amino Acids

Phillip B. Carpenter, PhD
Associate Professor
Department of Biochemistry and Molecular Biology
University of Texas Medical School at Houston
Houston, Texas
Tumor Suppressor Genes and Cancer

Rebecca L. Cox, PhD
Medical Microbiology Course Director
Department of Microbiology and Molecular Genetics
University of Texas Medical School at Houston
Houston, Texas
Herpes Simplex Virus (HSV)

Jade M. Hatley
MD/PhD Candidate
Department of Biochemistry and Molecular Biology
University of Texas Medical School at Houston
Houston, Texas
Porphyria

Raegan D. Hunt, MD, PhD
Resident in Pediatrics
The Johns Hopkins Hospital
Baltimore, Maryland
Cholestasis of Pregnancy
Folate Deficiency

Avinash Kalsotra, PhD
Postdoctoral Associate
Department of Pathology
Baylor College of Medicine
Houston, Texas
Erythromycin
Quinolones

Richard J. Kulmacz, PhD
Professor
Departments of Internal Medicine and Biochemistry and Molecular Biology
University of Texas Medical School at Houston
Houston, Texas
NSAIDs

Julia E. Lever, PhD
Professor
Department of Biochemistry and Molecular Biology
University of Texas Medical School at Houston
Houston, Texas
Acromegaly
Addison Disease
Diabetes Insipidus
Hyperparathyroid Disease
Hypothyroidism
Menopause

Alan E. Levine, PhD, MEd
Associate Professor
Department of Biochemistry and Molecular Biology
University of Texas Medical School at Houston
Houston, Texas
Hypercholesterolemia
Methotrexate
Statin Medications
Tay-Sachs Disease

John A. Putkey, PhD
Professor
Department of Biochemistry and Molecular Biology
University of Texas Medical School at Houston
Houston, Texas
Sickle Cell Disease

Daniel J. Ryder
PhD Candidate
Department of Biochemistry and Molecular Biology
University of Texas Medical School at Houston
Houston, Texas
Cushing Syndrome

Ann-Bin Shyu, PhD
Professor and Jesse H. Jones Chair in Molecular Biology
Department of Biochemistry and Molecular Biology
University of Texas Medical School at Houston
Houston, Texas
Ribavirin Use in Influenza

Michael D. Spears, MD
Resident in Pathology
University of Texas Southwestern Medical Center
Dallas, Texas
Approach to Learning Biochemistry

Cheri M. Turman, PhD
Director
Chemistry Department
Analytical Food Laboratories
Grand Prairie, Texas
Fragile X Syndrome
Gallstones

Martin E. Young, DPhil
Assistant Professor
Department of Pediatrics
Children's Nutrition Research Center
Baylor College of Medicine
Houston, Texas
Anorexia Nervosa
Somogyi Effect
Type II Diabetes

ACKNOWLEDGMENTS

The inspiration for this basic science series occurred at an educational retreat led by Dr. Maximilian Buja, who at the time was the dean of the medical school. It has been such a joy to work together with Drs. William Seifert and Henry Strobel, who are both accomplished scientists and teachers, as well as the other excellent authors and contributors. It has been rewarding to collaborate with Dr. Konrad Harms, whom I have watched mature from a medical student to resident and now a brilliant faculty member. I would like to thank McGraw-Hill for believing in the concept of teaching by clinical cases. I owe a great debt to Catherine Johnson, who has been a fantastically encouraging and enthusiastic editor.

At the University of Texas Medical School at Houston, we would like to recognize Dr. Rodney E. Kellems, Chair of the Department of Biochemistry and Molecular Biology, for his encouragement and delight in this project; Johnna Kincaid, former Director of Management Operations for her support; Nina Smith and Bonnie Martinez for their help in preparing the manuscript; and Amy Gilbert for her assistance with the figures. We also would like to thank the many students that have allowed us to teach them over the years and who have in the process taught us. Dr. Seifert thanks his wife Margie for her encouragement, patience, and support.

At Methodist Hospital, I appreciate Drs. Mark Boom, Karin Pollock-Larsen, H. Dirk Sostman, and Judy Paukert, and Mr. John Lyle and Mr. Reggie Abraham. At St. Joseph Medical Center, I would like to recognize our outstanding administrators: Phil Robinson, Pat Mathews, Laura Fortin, Dori Upton, Cecile Reynolds, and Drs. John Bertini and Thomas V. Taylor. I appreciate Marla Buffington's advice and assistance. Without the help from my colleagues, Drs. Konrad Harms, Jeané Simmons Holmes, and Priti Schachel, this book could not have been written. Most important, I am humbled by the love, affection, and encouragement from my lovely wife, Terri and our four children, Andy, Michael, Allison, and Christina.

Eugene C. Toy

❖ INTRODUCTION

Often, the medical student will cringe at the "drudgery" of the basic science courses and see little connection between a field such as biochemistry and clinical problems. Clinicians, however, often wish they knew more about the basic sciences, because it is through the science that we can begin to understand the complexities of the human body and thus have rational methods of diagnosis and treatment.

Mastering the knowledge in a discipline such as biochemistry is a formidable task. It is even more difficult to retain this information and to recall it when the clinical setting is encountered. To accomplish this synthesis, biochemistry is optimally taught in the context of medical situations, and this is reinforced later during the clinical rotations. The gulf between the basic sciences and the patient arena is wide. Perhaps one way to bridge this gulf is with carefully constructed clinical cases that ask basic science-oriented questions. In an attempt to achieve this goal, we have designed a collection of patient cases to teach biochemistry related points. More importantly, the explanations for these cases emphasize the underlying mechanisms and relate the clinical setting to the basic science data. We explore the principles rather than emphasize rote memorization.

This book is organized for versatility: to allow the student "in a rush" to go quickly through the scenarios and check the corresponding answers and to provide more detailed information for the student who wants thought-provoking explanations. The answers are arranged from simple to complex: a summary of the pertinent points, the bare answers, a clinical correlation, an approach to the biochemistry topic, a comprehension test at the end to reinforcement or emphasis, and a list of references for further reading. The clinical cases are arranged by system to better reflect the organization within the basic science. Finally, to encourage thinking about mechanisms and relationships, we intentionally did not primarily use a multiple-choice format. Nevertheless, several multiple-choice questions are included at the end of each scenario to reinforce concepts or introduce related topics.

HOW TO GET THE MOST OUT OF THIS BOOK

Each case is designed to introduce a clinically related issue and includes open-ended questions usually asking a basic science question, but at times, to break up the monotony, there will be a clinical question. The answers are organized into four different parts:

PART I

1. **Summary**
2. A **straightforward answer** is given for each open-ended question.
3. **Clinical Correlation**—A discussion of the relevant points relating the basic science to the clinical manifestations, and perhaps introducing the student to issues such as diagnosis and treatment.

PART II

An **approach to the basic science concept** consisting of three parts

1. **Objectives**—A listing of the two to four main principles that are critical for understanding the underlying biochemistry to answer the question and relate to the clinical situation
2. **Definitions of basic terminology**
3. **Discussion of topic**

PART III

Comprehension Questions—Each case includes several multiple-choice questions that reinforce the material or introduces new and related concepts. Questions about the material not found in the text are explained in the answers.

PART IV

Biochemistry Pearls—A listing of several important points, many clinically relevant reiterated as a summation of the text and to allow for easy review, such as before an examination.

SECTION I

Applying the Basic Sciences to Clinical Medicine

Part 1. Approach to Learning Biochemistry

Part 2. Approach to Disease

Part 3. Approach to Reading

PART 1. APPROACH TO LEARNING BIOCHEMISTRY

Biochemistry is best learned by a systematic approach, first by learning the **language** of the discipline and then by understanding the **function** of the various processes. Increasingly, cellular and molecular biology play an important role in the understanding of disease processes and also in the treatment of disease. Initially, some of the terminology must be memorized in the same way that the alphabet must be learned by rote; however, the appreciation of the way that the biochemical words are constructed requires an understanding of mechanisms and a manipulation of the information.

PART 2. APPROACH TO DISEASE

Physicians usually tackle clinical situations by taking a history (asking questions), performing a physical examination, obtaining selective laboratory and imaging tests, and then formulating a diagnosis. The conglomeration of the history, physician examination, and laboratory tests is called the **clinical database.** After reaching a diagnosis, a treatment plan is usually initiated, and the patient is followed for a clinical response. Rational understanding of disease and plans for treatment are best acquired by learning about the normal human processes on a basic science level; likewise, being aware of how disease alters the normal physiologic processes is understood on a basic science level.

PART 3. APPROACH TO READING

There are six key questions that help to stimulate the application of basic science information to the clinical setting. These are:

1. **What is the most likely biochemical mechanism for the disease causing the patient's symptom or physical examination finding?**
2. **Which biochemical marker will be affected by treating a certain disease, and why?**
3. **Looking at graphical data, what is the most likely biochemical explanation for the results?**
4. **Based on the deoxyribonucleic acid (DNA) sequence, what is the most likely amino acid or protein result, and how will it be manifest in a clinical setting?**
5. **What hormone–receptor interaction is likely?**
6. **How does the presence or absence of enzyme activity affect the biochemical (molecular) conditions, and how does that in turn affect the patient's symptoms?**

1. **What is the most likely biochemical mechanism for the disease causing the patient's symptom or physical examination finding?**

   This is the fundamental question that basic scientists strive to answer—the underlying cause of a certain disease or symptom. Once this underlying mechanism is discovered, then progress can be made regarding methods of diagnosis and treatment. Otherwise, our attempts are only *empiric,* in other words, only by trial and error and observation of association. Students are encouraged to think about the mechanisms and underlying cause rather than just memorizing by rote. For example, in sickle cell disease, students should connect the various facts together, setting the foundation for understanding disease throughout their life:

In sickle cell disease, valine (a hydrophobic amino acid) is substituted for glutamate (a charged, hydrophilic amino acid) in the sixth position in the β-globin chain of hemoglobin (Hb). This decreases the solubility of hemoglobin when it is in the deoxygenated state, resulting in its precipitation into elongated fibers in the red blood cell.

This causes the red blood cell to have less distensibility and thus to *sickle,* leading to rupture of the red blood cell (hemolysis) and blockage in small capillaries. The *sludging* in small capillaries leads to poor oxygen delivery, ischemia, and pain.

2. **Which biochemical marker will be affected by therapy?**

After a diagnosis has been made and therapy initiated, then the patient response should be monitored to assure improvement. Ideally, the patient response should be obtained in a scientific manner: unbiased, precise, and consistent. Although more than one physician or nurse may be measuring the response, it should be as carefully performed with little inter- (one person to the next) or intravariation (one person measuring) as possible. One of the therapeutic measures includes serum or imaging markers; for example, in diabetic ketoacidosis, the serum glucose and pH would be measured to confirm improvement with therapy. Another example would be to follow the volume of a pulmonary mass imaged by CT scan after chemotherapy. The student must know enough about the disease process to know which marker to measure and the expected response over time.

3. **Looking at graphical data, what is the most likely biochemical explanation for the results?**

Medicine is art and science. The **art** aspect consists of the way that the physician deals with the human aspect of the patient, expressing empathy, compassion, establishing a therapeutic relationship, and dealing with uncertainty; the **science** is the attempts to understand disease processes, making rational treatment plans, and being objective in observations. The physician as scientist must be precise about how to elicit data and then carefully make sense of the information, using up-to-date evidence. Exercises to develop the skills of data analysis require interpretation of data in various representations, such as in tables or on graphs.

4. **Based on the DNA sequence, what amino acid or protein would be produced, and how would the protein be manifested in a clinical setting?**

The clinician–basic scientist collaboration requires each party to "speak the same language" and translate forward and backward from science to clinical, and vice versa. Biochemical thinking is very stepwise, for example, the relationship among DNA, ribonucleic acid (RNA), proteins, and clinical findings. Since the genomic information (DNA) codes for proteins that affect physiologic or pathologic changes, it is of fundamental importance that the student becomes very comfortable thinking about these relationships:

Forward: DNA → proteins → Clinical Manifestations
Backward: Clinical Findings → proteins' effects → DNA

5. **What hormone–receptor interaction is likely?**

A **hormone** is a substance, usually a peptide or steroid, produced by one tissue and conveyed by the bloodstream to another part of the body to effect physiological activity, such as growth or metabolism. A **receptor** is a cellular structure that mediates between a chemical agent (hormone) and the physiologic response. The way that the hormone causes its effect is vital to understand, because many diseases occur as a result of abnormal hormone production, abnormal hormone receptor interaction, or abnormal cellular response to the hormone–receptor complex. For example, diabetes mellitus is manifest clinically by high blood glucose levels. However, in type I diabetes (usually juvenile onset), the etiology is insufficient insulin secreted by the pancreas. (Insulin acts to put serum glucose into cells or store it as glycogen.) In contrast, the mechanism in type II diabetes (usually adult onset) is a defect of the insulin receptor messenger; in fact, the insulin levels in these individuals are usually higher than normal. Understanding the difference between the two mechanisms allows the scientist to approach individualized therapy, and it allows the clinician to understand the differences in these patients, such as the reason that type I diabetics are much more prone to diabetic ketoacidosis (because of insulin deficiency).

6. **How does the presence or absence of enzyme activity affect the biochemical (molecular) conditions, and how does that in turn affect patient symptoms?**

Enzymes are proteins that act as catalysts, speeding the rate at which biochemical reactions proceed but not altering the direction or nature of the reactions. The presence or absence of these important substances affects the biochemical conditions, which then influence the other physiologic processes in the body. Enzyme deficiencies are often inherited as autosomal recessive conditions and may be passed from parent to child. Clearly, when students begin to understand the role of the enzyme and the chemical reaction that it governs, they begin to understand the intricacies of the human biological processes.

BIOCHEMISTRY PEARLS

- ❖ There are six key questions to stimulate the application of basic science information to the clinical arena.
- ❖ Medicine consists of both art and science.
- ❖ The scientific aspect of medicine seeks to gather data in an objective manner, understand physiologic and pathologic processes in light of scientific information, and propose rational explanations.
- ❖ The skilled clinician must be able to translate back and forth between the basic sciences and the clinical sciences.

REFERENCE

Braunwald E, Fauci AS, Kasper KL, et al., eds. Harrison's Principles of Internal Medicine, 15th ed. New York: McGraw-Hill, 2001.

SECTION II

Clinical Cases

CASE 1

A 15-year-old African-American female presents to the emergency room with complaints of bilateral thigh and hip pain. The pain has been present for 1 day and is steadily increasing in severity. Acetaminophen and ibuprofen have not relieved her symptoms. She denies any recent trauma or excessive exercise. She does report feeling fatigued and has been having burning with urination along with urinating frequently. She reports having similar pain episodes in the past, sometimes requiring hospitalization. On examination, she is afebrile (without fever) and in no acute distress. No one in her family has similar episodes. Her conjunctiva and mucosal membranes are slightly pale in coloration. She has nonspecific bilateral anterior thigh pain with no abnormalities appreciated. The remainder of her examination is completely normal. Her white blood cell count is elevated at 17,000/mm^3, and her hemoglobin (Hb) level is decreased at 7.1 g/dL. The urinalysis demonstrated an abnormal number of numerous bacteria.

- **What is the most likely diagnosis?**
- **What is the molecular genetics behind this disorder?**
- **What is the pathophysiologic mechanism of her symptoms?**

ANSWERS TO CASE 1: SICKLE CELL DISEASE

Summary: A 15-year-old African-American female with recurrent bilateral thigh and hip pain, anemia, and symptoms and laboratory evidence of a urinary tract infection.

- **Most likely diagnosis:** Sickle cell disease (pain crisis).
- **Biochemical mechanism of disease:** Single amino acid substitution on hemoglobin beta chain, inherited in an autosomal recessive fashion (1 of 12 African Americans in United States are carriers of the trait).
- **Pathophysiologic mechanism of symptoms:** The sickled red blood cells cause infarction of bone, lung, kidney, and other tissue from vasoocclusion.

CLINICAL CORRELATION

This 15-year-old female's description of her pain is typical of a sickle cell pain crisis. Many times, infection is a trigger, most commonly pneumonia or a urinary tract infection. This case is consistent with a urinary tract infection, indicated by her symptoms of urinary frequency, and burning with urination (dysuria). Her white blood cell count is elevated in response to the infection. The low hemoglobin level is consistent with sickle cell anemia. Since she is homozygous (both genes coding for sickle hemoglobin), both her parents have sickle cell trait (heterozygous) and thus do not have symptoms. The diagnosis can be established with hemoglobin electrophoresis. Treatment includes searching for an underlying cause of crisis (infection, hypoxia, fever, excessive exercise, and extreme changes in temperature), administration of oxygen, intravenous fluids for hydration, pain management, and consideration of a blood transfusion.

APPROACH TO SICKLE CELL DISEASE

Objectives

1. Understand the primary, secondary, tertiary, and quaternary levels of protein structure.
2. Be able to describe the structure of hemoglobin and its role in oxygen binding and dissociation.
3. Be able to describe the mechanism that amino acid substitution results in sickle cell hemoglobin.

Definitions

Allosteric effectors: Molecules that bind to enzymes or protein carriers at sites other than the active or ligand binding site. On binding, allosteric effectors either positively or negatively affect the enzymatic activity or capability of the protein to bind its ligand.

Globin: The globular proteins that are the polypeptide components of myoglobin and hemoglobin. They contain a hydrophobic pocket that holds the heme prosthetic group.

Globin fold: The three-dimensional structure of the proteins that is common to myoglobin and the subunits of hemoglobin.

Heme: A porphyrin ring that has a Fe^{+2} ion coordinately bound in the center of the molecule. Heme binds oxygen in hemoglobin and myoglobin and serves as an electron carrier in the cytochromes.

Hemoglobin: The tetrameric protein in high concentration in red blood cells that binds oxygen in the capillaries of the lungs and delivers it to peripheral tissues. Each globin subunit contains a heme group that binds oxygen when the iron atom is in the ferrous (+2) oxidation state.

Myoglobin: A protein having a single globin polypeptide with a bound heme group. It is located primarily in muscle cells and stores oxygen for times when there is a high demand for energy.

Protein primary (I°) structure: The amino acid sequence of the protein, listed from the amino-terminal amino acid to the carboxy-terminal amino acid.

Protein secondary (II°) structure: The local three-dimensional spatial arrangement of amino acids close to one another in the primary sequence. α-Helices and β-sheets are the predominant secondary structures in proteins.

Protein tertiary (III°) structure: The overall three-dimensional structure of a single polypeptide chain, including positions of disulfide bonds. Noncovalent forces such as hydrogen bonding, electrostatic forces, and hydrophobic effects are also important.

Protein quaternary (IV°) structure: The overall three-dimensional arrangement of polypeptide subunits in a multi-subunit protein.

DISCUSSION

The **activity** of a given protein is dependent on **proper folding** of its polypeptide chain to assume a defined **three-dimensional structure.** The importance of protein folding to molecular medicine is emphasized by the fact that many disease-causing mutations do not directly affect the active or ligand binding site of proteins, but instead cause **local or global alterations in protein structure** or **disrupt the folding pathway** such that the native protein fold is not achieved, or undesirable interactions with other proteins are promoted. The molecular defect that changes adult hemoglobin (HbA) to sickle hemoglobin

(HbS), leading to sickle cell anemia, is a classic example of mutations that affect protein structure.

Protein structure is typically classified as consisting of four levels: **primary (I°)**, **secondary (II°)**, **tertiary (III°)**, and **quaternary (IV°)**. **Primary structure** is the **sequence of amino acids** in the protein. **Secondary structure** is the **local three-dimensional spatial arrangement of amino acids that are close to one another** in the primary sequence. **α-Helices and β-sheets** compose the majority of secondary structures in all known proteins. **Tertiary structure** is the **spatial arrangement of amino acid residues that are far apart** in the linear primary sequence of a single polypeptide chain, and it includes **disulfide bonds** and **noncovalent forces.** These noncovalent forces include **hydrogen bonding,** which is also the primary stabilization force for the formation of α-helices and β-sheets, **electrostatic interactions, van der Waals forces,** and **hydrophobic effects. Quaternary structure** is the manner in which **subunits of a multi-subunit protein are arranged** with respect to one another.

Normal **hemoglobin has four subunits called globins.** Adult hemoglobin has **two α (α_1 and α_2) and two β (β_1 and β_2) globin chains.** Each globin chain has an associated heme prosthetic group, which is the site of oxygen binding and release. All globin chains have similar primary sequences. The secondary structure of globin chains consists of approximately **75 percent** α-helix. The similar primary sequence promotes a similar tertiary structure in all globins that is called the **globin fold,** which is compact and globular in overall conformation. The **quaternary structure of HbA can be described as a dimer of $\alpha_1\beta_1$ and $\alpha_2\alpha_2$ dimers.** The αβ dimers move relative to one another during the binding and release of oxygen.

Hemoglobin must remain soluble at high concentrations within the red blood cell to support normal oxygen binding and release properties. This is made possible by a distribution of **amino acid side chains** in which **hydrophobic residues are sequestered in the interior core** of the folded globin subunits, while **hydrophilic residues** dominate the **water-exposed surface** of the globin fold. The disk-shaped heme prosthetic group is inserted into a hydrophobic pocket formed by the globin.

Hemoglobin, and the homologous monomeric protein myoglobin (Mb), both bind and release oxygen as a function of the surrounding concentration, or partial pressure of oxygen. A plot of percent saturation of Hb or Mb with oxygen versus the partial pressure of oxygen is called the oxygen dissociation curve. Unlike Mb, which has a simple hyperbolic **oxygen dissociation curve** typical of ligand binding to a monomeric protein, the **quaternary structure of HbA** allows it to bind oxygen with **positive cooperativity** giving a **sigmoidal** oxygen dissociation curve (Figure 1-1). Essentially, binding of oxygen

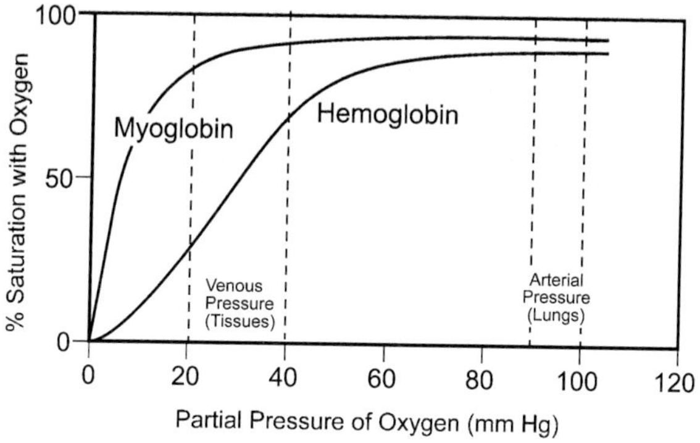

Figure 1-1. Oxygen saturation curves for hemoglobin and myoglobin.

to one HbA subunit increases the affinity of binding to other subunits in the tetramer, thereby shifting the equilibrium between oxy and deoxy forms. The net effect of this cooperativity is that HbA is able to release oxygen, whereas Mb globin would be most saturated with oxygen at the partial pressure of oxygen normally found in resting peripheral tissues. The quaternary structure of HbA also allows it to respond to **2,3-bisphosphoglycerate, carbon dioxide, and hydrogen ion, all of which are heterotropic negative allosteric effectors of oxygen binding.**

Sickle cell anemia results from the nonconservative substitution of **valine for glutamate at residue 6 (Val-6) in the β-chain of hemoglobin.** The mutated hemoglobin is called HbS. The intrinsic oxygen binding properties of HbA and HbS are the same, however, the **solubility of deoxy HbS is reduced** because of exposure of **Val-6** at the **surface of the β-chain.** Since hemoglobin is present at very high concentrations in the red blood cell, **deoxy HbS will precipitate inside the cell.** The precipitate takes the form of elongated fibers because of the association of complementary hydrophobic surfaces on the β- and α-chains of deoxy HbS. At oxygen saturations found in arterial blood, the oxy HbS predominates and HbS does not precipitate because Val-6 of the β-chain is not exposed to the surface.

The tendency for deoxy HbS to precipitate is why clinical manifestations of sickle cell anemia are brought on by **exertion** and why **treatment includes administration of oxygen.** The stiff fibrous precipitate causes the red blood cell to deform into the characteristic sickle shape and makes the normally malleable cell susceptible to hemolysis.

COMPREHENSION QUESTIONS

[1.1] A newly married African-American couple, both from families having histories of good health, are about to have a child. If the incidence of the sickle cell trait is approximately 1/12 among persons of African descent in the United States, what is the chance that they will give birth to a child that is affected by sickle cell disease?

A. 1/12
B. 1/24
C. 1/96
D. 1/288
E. 1/576

[1.2] A 25-year-old African-American male with sickle cell anemia, who has been hospitalized several times for painful sickle cell crises, has successfully been free of these crises since he has been on hydroxyurea therapy. Treatment with hydroxyurea results in which of the following?

A. An increase in the oxygen affinity of HbS
B. An increase in the levels of hemoglobin F (HbF) in red blood cells
C. A decreased cooperativity in oxygen binding by HbS
D. A posttranslational modification of HbS that prevents polymerization
E. A decreased ability of HbS to bind 2,3-bisphosphoglycerate (2,3-BPG)

[1.3] A pregnant woman is able to transfer oxygen to her fetus because fetal hemoglobin has a greater affinity for oxygen than does adult hemoglobin. Why is the affinity of fetal hemoglobin for oxygen higher?

A. The tense form of hemoglobin is more prevalent in the circulation of the fetus.
B. There is less 2,3-BPG in the fetal circulation as compared to maternal circulation.
C. Fetal hemoglobin binds 2,3-BPG with fewer ionic bonds than the adult form.
D. The Bohr effect is enhanced in the fetus.
E. The oxygen-binding curve of fetal hemoglobin is shifted to the right.

Answers

[1.1] **E.** Since sickle cell anemia has an autosomal recessive inheritance pattern and the incidence in the United States to persons of African descent is approximately 1/12, then each parent has a 1/12 chance of being a carrier. Since it is an autosomal recessive gene, then their offspring would have a 1/4 chance of being homozygous if both of their parents were also carriers. Thus, the chances of having a child with the sickle cell trait is (1/12)(1/12)(1/4) or 1/576.

[1.2] **B.** By inhibiting the enzyme ribonucleotide reductase, hydroxyurea has been shown to increase the levels of fetal Hb (HbF, $\alpha_2\gamma_2$) by mechanisms not fully understood. The increase in HbF concentrations has the effect of decreasing HbS levels in the red blood cell. The increased concentration of HbF disrupts the polymerization of HbS and decreases the incidence of sickle cell crises. Hydroxyurea does not affect either the oxygen affinity or cooperativity of oxygen binding of HbS, nor does it react with HbS to cause a posttranslational modification or affect 2,3-BPG binding.

[1.3] **C.** In adult hemoglobin 2,3-BPG binds to ionized residues in the interface between the two β-chains, thus decreasing the oxygen affinity. The γ-chains of fetal hemoglobin have fewer ionized residues in the corresponding interface and bind 2,3-BPG with less affinity, which allows for a greater binding affinity for oxygen.

BIOCHEMISTRY PEARLS

- The quaternary structure of HbA allows it to bind oxygen with positive cooperativity giving a sigmoidal oxygen dissociation curve.
- Sickle cell anemia results from the nonconservative substitution of valine for glutamate at the sixth residue of the β-chain of hemoglobin.
- Precipitation of HbS is more likely to occur from exertion and deoxygenation. Treatment consists therefore of oxygen and hydration.
- The stiff fibrous precipitate of HbS causes the red blood cell to deform into the characteristic sickle shape and makes the normally malleable cell susceptible to hemolysis.

REFERENCES

Schultz RM, Liebman MN. Proteins II: structure-function relationships in protein families. In: Devlin TM, ed. Textbook of Biochemistry with Clinical Correlations, 5th ed. New York: Wiley-Liss, 2002.

Weatherall DJ, et al. The hemoglobinopathies. In: Scriver CR, Beaudet AL, Sly W, et al., eds. The Metabolic & Molecular Basis of Inherited Disease, 8th ed. New York: McGraw-Hill, 2001.

❖ CASE 2

A 21-year-old college student presents to the clinic complaining of a sudden onset of chills and fever, muscle aches, headache, fatigue, sore throat, and painful nonproductive cough 3 days prior to fall final exams. Numerous friends of the patient in the dormitory reported similar symptoms and were given the diagnosis of influenza. He said that some of them were given a prescription for ribavirin. On examination, he appears ill with temperature 39.4°C (103°F). His skin is warm to the touch, but no rashes are appreciated. The patient has mild cervical lymph node enlargement but otherwise has a normal examination.

- **What is the most likely diagnosis?**
- **What is the biochemical mechanism of action of ribavirin?**
- **What is the genetic make up of this infectious organism?**

ANSWERS TO CASE 2: RIBAVIRIN AND INFLUENZA

Summary: A college student complains of the sudden onset of fever, chills, malaise, nonproductive cough, and numerous sick contacts in the fall season.

- **Likely diagnosis:** Acute influenza infection
- **Biochemical mechanism of action of ribavirin:** A nucleoside analogue with activity against a variety of viral infections
- **Genetic makeup of organism:** Ribonucleic acid (RNA) respiratory virus

CLINICAL CORRELATION

This 21-year-old college student has the clinical clues suggestive of acute influenza. Typically, the illness occurs in the winter months with an acute onset of fever, myalgias (muscle aches), headache, cough, and sore throat. Usually, there are outbreaks with many individuals with the same symptoms. This patient is young and healthy, and antiviral therapy is not mandatory. The best way to prevent the infection is by influenza vaccination, usually given in October or November of each year. Because of the antigenic changes of the virus, a new vaccine must be given each year. Patients who are at especially high risk for severe complications or death should receive the vaccine each year. These include the elderly and people with asthma, chronic lung disease, human immunodeficiency virus (HIV) infection, diabetes, or chronic renal insufficiency.

APPROACH TO THE USE OF RIBAVIRIN IN INFLUENZA

Objectives

1. Know the structure of deoxyribonucleic acid (DNA) and RNA.
2. Understand the processes of denaturation, renaturation, and hybridization of DNA.
3. Know the differences between RNA and DNA.
4. Be familiar with the differences between human and viral/bacterial DNA and RNA.

Definitions

Base pairing: The hydrogen bonds formed between complementary bases that are part of the polynucleotide chains of nucleic acids. The base pairing is specific in that adenine will base pair with thymine (uracil in RNA) and guanine will pair with cytosine.

Chargaff rule: the amount of adenine and thymine in DNA is equal; the amount of cytosine and guanine are equal. (A = T, C = G). The amount of purines equals the amount of pyrimidines.

Helicase: An enzyme that will catalyze the separation of the strands of the DNA double helix during replication.

Nucleoside: A nitrogenous base (purines such as adenine or guanine; pyrimidines such as uracil, thymine, or cytosine) in an *N*-β-glycosidic linkage to a pentose sugar (deoxyribose in DNA and ribose in RNA).

Nucleotide: A nucleoside that has a phosphoester bond to one of the hydroxyl groups of the pentose sugar.

Polymerase: An enzyme that will add nucleotides to a growing nucleic acid chain using a template strand to determine which nucleotide is added. A nucleoside triphosphate condenses with the growing strand releasing pyrophosphate.

Ribavirin: 1-β-D-ribofuranosyl-1*H*-1,2,4-triazole-3-carboxamide; a purine nucleoside analog that exhibits antiviral activity against a broad spectrum of DNA and RNA viruses.

DISCUSSION

DNA and RNA are both polymers of nucleosides joined by 3′,5′-phosphodiester linkages. Each **nucleoside** consists of a **heterocyclic nitrogenous base in a glycosidic link with a pentose sugar.** The **backbone** of both DNA and RNA is formed by the **phosphate bridges between the 3′-hydroxyl group of one pentose and the 5′-hydroxyl group of another.** The nitrogenous bases form the "side chains." **DNA contains the bases adenine (A), guanine (G), cytosine (C),** and **thymine (T),** whereas **RNA contains A, G, and C but has uracil (U) instead of T.** The **pentose sugar in DNA is 2′-deoxyribose,** whereas in **RNA it is ribose.**

The **most stable DNA structure** is formed when **two polynucleotide chains are joined by hydrogen bonding between the side chain bases.** The base pairing is specific in that **adenine forms hydrogen bonds with thymine,** whereas **guanine forms hydrogen bonds with cytosine.** The result is an **antiparallel double helix in which one polynucleotide strand runs in the 5′ to 3′** direction, **while the other runs in the 3′ to 5′** direction. The **phosphate groups are located on the outside of the double helix** with the **base pairs forming the "stair steps" in the center of the spiral. RNA,** on the other hand, is usually **single stranded** (the exception is certain RNA viruses), but the **strand can loop back on itself and form regions of base pairing** (A with U and G with C). The presence of the **hydroxyl group at the 2′-position makes RNA much more susceptible to hydrolysis and decreases its stability.**

The **double helical structure of DNA must be disrupted** during almost all **biological processes** in which it participates, including **DNA replication and repair, as well as transcription** of the DNA sequence information to RNA. Experimentally, the **double helix can be separated, or denatured, by increasing the temperature** to well above 50°C (122°F). If the temperature is carefully decreased, renaturation occurs when the base pairs reform. Under these conditions, hybridization can be induced by allowing the single strands of DNA form base pairs with complementary base sequences on another strand of DNA or RNA. During replication, repair and transcription, complex proteins (helicases during DNA replication, RNA polymerase during transcription) cause the separation of the two strands.

The **cellular DNA and RNA polymerases** of higher organisms are much more accurate than those of viruses because of their high specificities and, in the case of DNA polymerases, proofreading capabilities. This accounts for the high degree of mutation in viruses. However, this can have therapeutic advantages.

Ribavirin (1-β-D-ribofuranosyl-1*H*-1,2,4-triazole-3-carboxamide) is a **purine nucleoside analog** exhibiting in vitro antiviral activity against a broad spectrum of DNA and RNA viruses. Ribavirin has shown clinical efficacy against both influenza A and B viruses. This antiviral activity is a result of the **resemblance of this compound to nucleosides.** Studies showed that ribavirin most closely **resembles guanosine,** as determined by x-ray crystallography (Figure 2-1). Once in cells, ribavirin is converted to its 5′-phosphate derivatives by cellular enzymes. The major metabolite is ribavirin-5′-triphosphate (RTP), and the intracellular concentration of the mono-, di-, and triphosphate derivatives probably is similar to that of other cellular nucleotides. Although not all polymerase/replication systems use ribavirin as they do guanosine, it has been shown that many of the systems that are inhibited by ribavirin are reversed by the addition of guanosine.

Until 2001, it was thought that the mechanism of action of ribavirin involved a decrease in cellular guanosine triphosphate (GTP) pools resulting from inhibition of inosine monophosphate dehydrogenase by ribavirin monophosphate. More recently, the mechanism of action for ribavirin has been expanded to include lethal mutagenesis of the viral genome as a result of ribavirin triphosphate utilization by the error-prone viral RNA-dependent RNA polymerase, and incorporation of ribavirin into viral RNA.

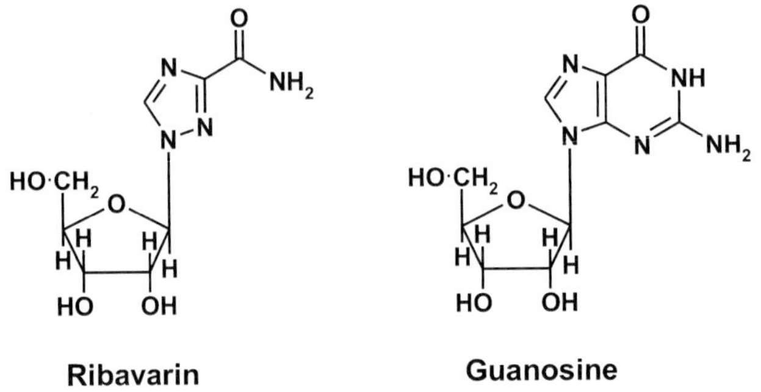

Figure 2-1. Comparison of the structures of ribavirin (1-β-D-ribofuranosyl-1*H*-1,2,4-triazole-3-carboxamide) with the purine nucleoside guanosine.

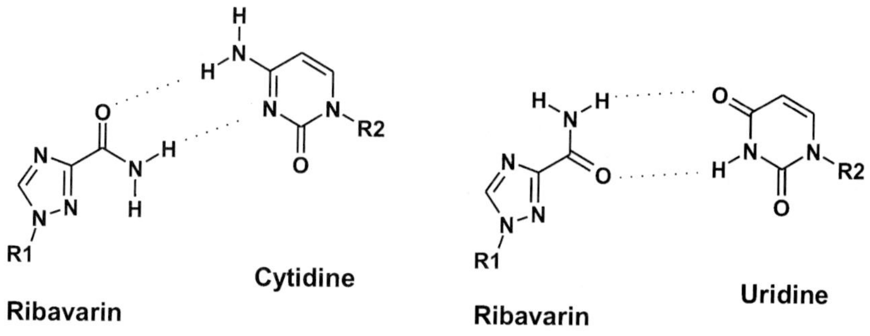

Figure 2-2. The pseudo base (1,2,4-triazole 3-carboxamide) of ribavirin pairs equivalently with cytosine and uracil. R denotes the polyribonucleotide strand.

In 2001 a critical study published in *Nature Medicine* showed in vitro use of ribavirin triphosphate by a model viral RNA polymerase, poliovirus 3D *pol.* **Ribavirin incorporation is mutagenic,** because it serves as a **template for incorporation of cytidine and uridine** with equal efficiency (Figure 2-2). Ribavirin reduces infectious poliovirus production to as little as 0.00001 percent in cell culture. The **antiviral activity of ribavirin correlates directly with its mutagenic activity.** These data indicate that ribavirin forces the virus into "error catastrophe." This study reveals a new concept about nucleoside analogs, that is, mutagenic ribonucleosides may represent an important class of anti-RNA virus agents.

RNA **viruses like HIV and the influenza virus use a naturally high mutation rate** of their RNA polymerases with low fidelity for replication to avoid and escape most treatments and vaccines. A **RNA virus tends to mistakenly insert ribavirin** into newly synthesized copies of their RNA genome. But ribavirin adds so many extra mutations to the virus during the replication of its genetic materials, it induces a flood of mutations and thus pushes the virus into a kind of **genetic meltdown**.

COMPREHENSION QUESTIONS

[2.1] Influenza virus is a class Vb virus, which means that it has a single (–)-stranded RNA for its genome. Which of the following best describes the immediate fate of this (–)-RNA when the virus enters the host cell?

A. It is used directly to encode viral proteins.
B. It is used as a template to synthesize a (+)-strand viral messenger RNA (mRNA).
C. It is used as a template to synthesize viral DNA.
D. It is converted to a provirus.
E. It is integrated into the host cell genome.

[2.2] If a double-stranded DNA molecule undergoes two rounds of replication in an in vitro system that contains all of the necessary enzymes and nucleoside triphosphates that have been labeled with ^{32}P, which of the following best describes the distribution of radioactivity in the four resulting DNA molecules?

A. Exactly one of the molecules contains no radioactivity.
B. Exactly one of the molecules contains radioactivity in only one strand.
C. Two of the molecules contain radioactivity in both strands.
D. Three of the molecules contain radioactivity in both strands.
E. All four molecules contain radioactivity in only one strand.

[2.3] A 48-year-old man has had a lengthy history of skin cancer. In the past 6 years he has had over 30 neoplasms removed from sun-exposed areas and has been diagnosed with xeroderma pigmentosum. Which of the following best describes the enzymatic defect in patients with xeroderma pigmentosum?

A. DNA polymerase α
B. DNA polymerase γ
C. DNA ligase
D. Excision repair enzymes
E. RNA polymerase III

Answers

[2.1] **B.** Class Vb viruses, since they have a minus single-stranded RNA genome, cannot use their genomic RNA directly to encode viral proteins. It is used to synthesize a (+)-stranded viral mRNA that is then used to encode the viral proteins. The influenza virus does this by the following mechanism: a virus-specific polymerase first cleaves an oligonucleotide from a host cell mRNA. It uses this as a primer that is elongated by the polymerase to synthesize the viral (+)-mRNA using the (–)-viral RNA genome as the template.

[2.2] **C.** After two rounds of replication using ^{32}P-labeled nucleoside triphosphates (NTPs), all four DNA molecules will be radioactive; two will be radioactive in both strands, two will be radioactive in only one strand.

[2.3] **D.** Xeroderma pigmentosum is a genetic disease in which the ability to remove pyrimidine dimers caused by exposure to UV light is impaired. The mechanism used to remove these pyrimidine dimers (also used to repair DNA that has formed adducts with carcinogenic compounds) is excision repair. The enzymes used in this repair mechanism cleave the affected strand on either side of damaged nucleotides. The oligonucleotide containing the damaged nucleotides is removed and the gap is filled in by DNA polymerase and DNA ligase.

BIOCHEMISTRY PEARLS

- ❖ The most stable DNA structure is formed when two polynucleotide chains are joined by hydrogen bonding between the side chain bases. The base pairing is specific in that adenine pairs with thymine and guanine pairs with cytosine (A-T; G-C)
- ❖ Experimentally, the double helix can be separated, or denatured, by increasing the temperature to well above 50°C (122°F); if the temperature is carefully decreased, renaturation occurs when the base pairs reform.
- ❖ Ribavirin is a purine nucleoside analog exhibiting in vitro antiviral activity against a broad spectrum of DNA and RNA viruses, including clinical efficacy against both influenza A and B virus.
- ❖ Ribavirin seems to cause mutagenic changes to RNA viruses, with ribavirin being inserted into newly synthesized copies of their RNA genome.

REFERENCES

Crotty S, Maag D, Arnold JJ, et al. The broad-spectrum antiviral ribonucleoside ribavirin is an RNA virus mutagen. Nat Med 2001;6:1375–9.

Gilbert BE, Knight V. Biochemistry and clinical applications of ribavirin. Antimicrob Agents Chemother 1986;30:201–5.

Maag D, Castro C, Hong Z, et al. Hepatitis C virus RNA-dependent RNA polymerase (NS5B) as a mediator of the antiviral activity of ribavirin. J Biol Chem 2001;276:46094–8.

CASE 3

A 32-year-old female is being treated with methotrexate for a recently diagnosed choriocarcinoma of the ovary, and presents with complaints of oral mucosal ulcers. The patient recalls being advised not to take folate-containing vitamins during therapy. An uncomplicated surgical exploration was performed 5 weeks ago with removal of the affected ovary. The patient has been taking methotrexate for 2 weeks and has never had any of the above symptoms before. On examination, patient was afebrile and appeared ill. Several mucosal ulcers were seen in her mouth. The patient also had some upper abdominal tenderness. Her platelet count is decreased at 60,000/mm^3 (normal 150,000 to 450,000/mm^3).

- **What is the most likely etiology of her symptoms?**
- **What is the biochemical explanation of her symptoms?**
- **What part of the cell cycle does methotrexate act on?**

ANSWERS TO CASE 3: METHOTREXATE AND FOLATE METABOLISM

Summary: A 32-year-old female has oral ulcerations and thrombocytopenia (low platelet count) after beginning methotrexate initiated for recently diagnosed ovarian cancer. She recalls being instructed to avoid folate during therapy.

- **Likely cause of her symptoms:** Side effects of methotrexate (antimetabolite chemotherapy) affecting rapidly dividing cells such as oral mucosa.
- **Biochemical explanation of her symptoms:** Related to effects of methotrexate on cell cycle of all cells (particularly rapidly dividing cells). Folate antagonists inhibit dihydrofolate reductase (tetrahydrofolate needed for purine synthesis).
- **Cell cycle affected by methotrexate:** Deoxyribonucleic acid (DNA) synthesis (S) phase.

CLINICAL CORRELATION

Chemotherapeutic agents are used to treat various types of cancers. Although some are specific for cancer cells, most chemotherapeutic agents are toxic for both normal and cancer cells. Methotrexate acts as a folate antagonist, affecting DNA synthesis. Because cancer cells divide faster than normal cells, a higher proportion of these neoplastic cells will die. Nevertheless, normal cells that also are rapidly dividing, such as the gastrointestinal mucosa, the oral mucosa, and the bone marrow cells, may be affected. The patient was advised to avoid folate during therapy, since folate would be an "antidote," and would allow the cancer cells to escape cell kill.

APPROACH TO THE CELL CYCLE

Objectives

1. Understand the components of the cell cycle.
2. Know how folate is involved in DNA synthesis.
3. Be familiar with the terminology of nucleoside and nucleotide.

Definitions

Cell cycle: The time interval between cell divisions in proliferating cells. The cell cycle is divided into four phases: M phase, in which mitosis takes place; G_1 phase, prior to synthesis of DNA; S phase, in which

DNA and histones are synthesized to duplicate the chromosomes; and G_2 phase, during which there is cell growth and synthesis of macromolecules. Under certain conditions, the cell can enter a quiescent, or G_0 phase, which is not part of the regular cell cycle.

Dihydrofolate reductase (DHFR): The enzyme required to convert folic acid to its active form, tetrahydrofolate. It requires the cofactor NADPH as a source of reducing equivalents to reduce folate first to DHFR and then to tetrahydrofolate.

Methotrexate: A drug that has a similar structure to DHFR. It binds to DHFR reductase and competitively inhibits it, thus decreasing the levels of tetrahydrofolate in the cells. It effectively stops DNA synthesis in rapidly dividing cells such as cancer cells.

Tetrahydrofolate (THF): The active form of the vitamin folic acid. THF is one of the major carriers of one-carbon units at various oxidation states for biosynthetic reactions. It is required for the synthesis of the nucleotide thymidylate (dTMP). Although bacteria can synthesize folic acid, eukaryotes must obtain folate from the diet. Dietary sources of folate include leafy green vegetables (e.g., spinach and turnip greens), citrus fruits, and legumes. Many breakfast cereals, breads, and other grain products are fortified with folate.

DNA: Two large molecules composed of deoxynucleotides attached by hydrogen bonds in a helical, antiparallel relationship.

Nucleosides: Nitrogenous base plus a sugar (bases in DNA are adenine [A], thymine [T], cytosine [C], and guanine [G]; bases in RNA are A, C, G, but uracil [U] instead of T; sugar moiety in DNA is deoxyribose, and in RNA is ribose).

Nucleotide: Nucleoside plus phosphate group (see Table 3-1).

Deoxythymidylate (dTMP): DNA nucleotide consisting of the deoxyribose sugar, thymidine as nitrogenous base, and one phosphate.

Table 3-1
TERMINOLOGY OF NUCLEIC ACIDS

Bases	Purine (Adenine [A], Guanine [C]) Pyrimidine (Uracil [U], Cytosine [C], Thymine [T])
Sugar	Deoxyribose (DNA), Ribose (RNA)
Base + Sugar	Nucleoside
Base + Sugar + phosphate	Nucleotide

DISCUSSION

The **cell cycle** is defined as the time interval between cell divisions in proliferating cells. It is important to note that the cell cycle is not a simple "clock." Movement through the cell cycle is controlled by a variety of proteins that allow the cell to respond to various stimuli. The eukaryote cell cycle is composed of the following **four phases** (Figure 3-1):

1. M phase—mitosis
2. G_1 phase (gap 1)—between mitosis and initiation of DNA synthesis
3. S phase—when DNA synthesis occurs
4. G_2 phase (gap 2)—cell growth and macromolecule synthesis

Although there is great variation in the length of the mammalian cell cycle (hours to days), as a generalization, mammalian cells divide once every 24 hours. The M and S phases of the cell cycle are relatively constant. Therefore, the length of the mammalian cell cycle is determined by length of the G_1 and G_2 phases. **Cell division** occurs during **M phase, and the mitotic interphase** is composed of (G_1 + S + G_2). **M phase, or mitosis,** is divided into four subphases, **prophase, metaphase, anaphase, and telophase.** In **prophase,** the nuclear membrane breaks down while the replicated chromosomes condense and are

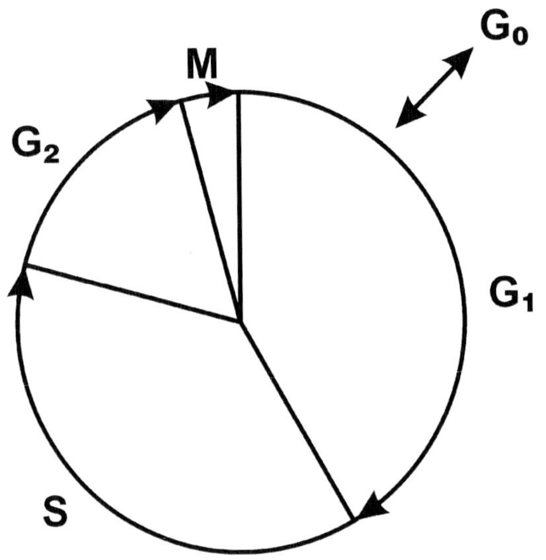

Figure 3-1. Cell cycle. (G_1 = cell growth; S = DNA synthesis and replication; G_2 = proteins made in preparation for cell division; M = mitosis; G_0 = rest phase.)

released into the cytoplasm. The chromosomes become aligned on the equatorial plate of the cell during **metaphase,** and they move from the equatorial plate to the poles during **anaphase.** The final step in M phase, **telophase,** is reformation of the nuclear membrane around the chromosomes followed by cytokinesis, or formation of two daughter cells. During the G_1 phase, the cell "monitors" itself and its environment. The cell is metabolically active and undergoes continuous growth, but no DNA synthesis occurs during this phase. During G_1 the cell makes a "decision" either to continue in the cell cycle and divide or to "withdraw" from the cell cycle and differentiate. The synthesis of DNA and histones to form two sets of chromosomes occurs during S phase. The key event of the G_2 phase is for the cell to "make sure that its entire DNA has replicated." Continued cell growth and the synthesis of cellular macromolecules also occur during the G_2 phase, preparing for cell division.

Under certain conditions cells can leave G_1 and enter into the G_0 **phase** of the cell cycle. This phase is not part of the regular cell cycle and represents a specialized state, which can be temporary or permanent. Entry into the G_0 phase can be triggered by growth factor withdrawal, negative growth factors, or limited protein synthesis. The cells in G_0 are in a **nonproliferative or quiescent state,** which can vary tremendously in length. Some cells differentiate and never divide again. Others can resume proliferation to replace lost cells as a result of injury or cell death. An important point is that cancer cells do not generally have a G_0 phase.

Key regulatory proteins that govern the cell cycle are the **cyclin-dependent kinases (CDKs),** which are **serine/threonine protein kinases,** and the **cyclins,** which are regulatory proteins that bind to the CDKs. Proteins that inhibit the kinase activity (cyclin-dependent kinase inhibitors, or CDIs) are also present in the cell. The CDKs are regulated by the levels of cyclins and the CDK/cyclin complex phosphorylates proteins on serine or threonine residues. The cyclins bind to and activate CDKs via protein—protein interactions. Cyclins act as regulatory subunits controlling activity and specificity of the CDK activity. These cyclins can themselves be phosphorylated and dephosphorylated. **Cyclin accumulation and degradation controls normal cell cycle** activity. For example, the degradation of specific cyclins by proteolysis at the metaphase-anaphase transition ends mitosis.

Tetrahydrofolate (THF) is the major source of 1-carbon units used in the biosynthesis of many important biological molecules. This **cofactor** is derived from the **vitamin folic acid** and is a carrier of activated 1-carbon units at various **oxidation levels (methyl, methylene, formyl, formimino, and methenyl).** These compounds can be interconverted as required by the cellular process. The major donor of the 1-carbon unit is **serine** in the following reaction:

$$\text{Serine} + \text{THF} \rightarrow \text{Glycine} + N^5,N^{10}\text{-methylene-THF}$$

All cells, especially rapidly growing cells, must synthesize thymidylate (dTMP) for DNA synthesis. The difference between (T) and (U) is one methyl group at the carbon-5 position. **Thymidylate is synthesized by the methylation of uridylate (dUMP)** in a reaction catalyzed by the enzyme **thymidylate synthase.** This reaction requires a methyl donor and a source of reducing equivalents, which are both provided by N^5,N^{10}-methylene THF (Figure 3-2). For this reaction to continue, the regeneration of THF from dihydrofolate (DHF) must occur.

$$DHF + NADPH + H^+ \rightarrow THF + NADPH^+$$

The enzyme **dihydrofolate reductase (DHFR)** catalyzes this reaction, which is a target of the **anticancer drugs aminopterin and methotrexate.** These drugs are analogs of DHF and act as **competitive inhibitors of DHFR.**

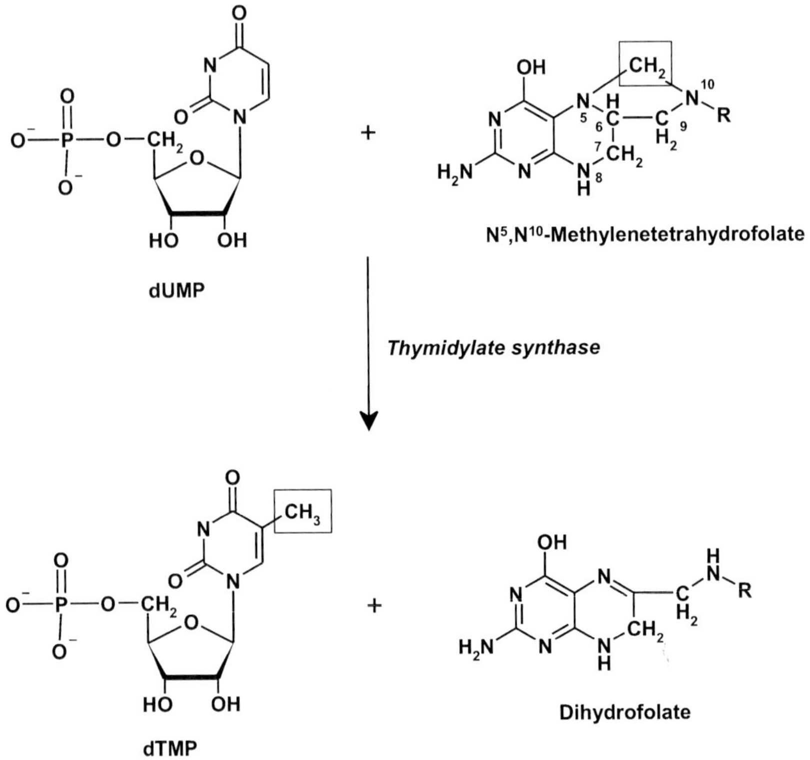

Figure 3-2. Thymidylate synthesized by the methylation of uridylate (dUMP) in a reaction catalyzed by the enzyme thymidylate synthase.

Inhibition of this enzyme prevents the regeneration of THF and **blocks dTMP synthesis** because of the lack of the methyl donor required for the reaction of thymidylate synthase.

THF is also required as a donor of two carbon atoms in the synthesis of the purine ring structure required for adenine and guanine. The carbon atoms donated by THF are indicated in Figure 3-3. Therefore, a lack of THF blocks the synthesis of the purine ring structure because of the lack of the ability of the cell to synthesize N^{10}-formyl-THF.

In summary, **DNA synthesis requires synthesis of dTMP and the purines adenine and guanine.** THF, derived from the **vitamin folic acid,** is required for the biosynthesis of these nucleotides. Treatment with methotrexate blocks the cell's ability to regenerate THF, leading to inhibition of these biosynthetic pathways. The lack of nucleotides prevents DNA synthesis, and these cancer cells cannot divide without DNA synthesis. Unfortunately, the effects of methotrexate are nonspecific and other rapidly dividing cells such as epithelial cells in the oral cavity, intestine, skin, and blood cells are also inhibited. This leads to the side effects associated with methotrexate (and other cancer chemotherapy drugs) such as mouth sores, low white blood cell counts, stomach upset, hair loss, skin rashes, and itching.

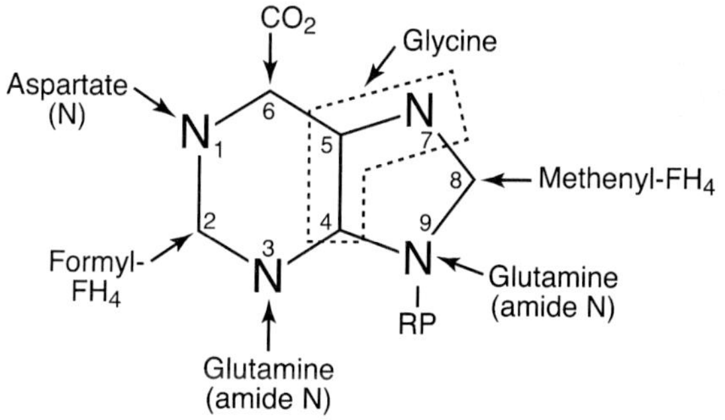

Figure 3-3. Origin of the atoms of the purine base ring system (RP = pentose phosphate).

COMPREHENSION QUESTIONS

[3.1] A 44-year-old woman who recently lost her job because of absenteeism, presents to her physician complaining of loss of appetite, fatigue, muscle weakness, and emotional depression. The physical examination reveals a somewhat enlarged liver that feels firm and nodular, and there is a hint of jaundice in the sclerae and a hint of alcohol on her breath. The initial laboratory profile included a hematological analysis that showed that she had an anemia with enlarged red blood cells (macrocytic). A bone marrow aspirate confirmed the suspicion that she has a megaloblastic anemia because it showed a greater than normal number of red and white blood cell precursors, most of which were larger than normal. Further analyses revealed that her serum folic acid level was 2.9 ng/mL (normal = 6 to 15), her serum B_{12} level was 153 pg/mL (normal = 150 to 750), and her serum iron level was normal.

The patient's megaloblastic anemia is most likely caused by which of the following?

A. A decreased synthesis of methionine
B. A decreased conversion of dUMP to dTMP
C. A decrease in the synthesis of phosphatidyl choline
D. A decrease in the levels of succinyl CoA
E. A decreased synthesis of dUTP

[3.2] A patient presents with a urinary tract infection and is prescribed a combination drug containing trimethoprim and sulfamethoxazole. These drugs are effective because they do which of the following?

A. Bind to operons to prevent synthesis of bacterial mRNA
B. Block transport across bacterial cell walls
C. Inhibit bacterial synthesis of cobalamin (B_{12})
D. Inhibit bacterial synthesis of THF
E. Inhibit synthesis of phospholipids in bacteria

[3.3] Methotrexate is often used as a chemotherapeutic agent to treat patients with leukemia. This drug is effective because it inhibits cells in which part of the cell cycle?

A. G_1 phase
B. S phase
C. M phase
D. G_2 phase
E. G_0 phase

[3.4] Leukemia patients are often given the compound Leucovorin (N^5-formyl THF) following treatment with the drug methotrexate. Why is Leucovorin useful as part of this treatment protocol?

A. It facilitates the uptake of methotrexate by cells
B. It can be converted to THF by bypassing DHFR
C. It acts as an activator of thymidylate synthase
D. It prevents the uptake of methotrexate by normal cells
E. It stimulates cells of the immune system

Answers

[3.1] **B.** Megaloblastic anemia is caused by a decrease in the synthesis of deoxythymidylate and the purine bases usually caused by a deficiency in either THF or cobalamin or both. This results in decreased DNA synthesis, which results in abnormally large hematopoietic cells created by perturbed cell division and DNA replication and repair. This patient exhibits signs of chronic alcoholism, which often leads to a folate deficiency. This can occur due to poor dietary intake, decreased absorption of folate due to damage of the intestinal brush border cells and resulting conjugase deficiency, and poor renal resorption of folate.

[3.2] **D.** Bacteria must synthesize the folate that is required for their biosynthetic processes; they do not have a transporter to bring folate into the cell. Trimethoprim inhibits prokaryotic DHFR (eukaryotic is not affected) and sulfamethoxazole is an analog of *p*-aminobenzoic acid (PABA), a precursor to folic acid. Bacteria will use this analog instead of PABA and produce a nonfunctional folate.

[3.3] **B.** Methotrexate inhibits the synthesis of deoxythymidine by preventing the regeneration of THF by inhibiting the enzyme DHFR. Inhibiting the synthesis of deoxythymidylate prevents the cell from synthesizing its DNA. DNA synthesis occurs exclusively during S phase of the cell cycle.

[3.4] **B.** Leucovorin (N^5-formyl THF, folinic acid) is used as an antidote for cells that have decreased levels of folic acid. Treatment of leukemia patients with methotrexate kills the tumor cells but also other normal rapidly dividing cells. N^5-formyl THF is normally administered 24 hours following treatment with methotrexate; it can be converted to THF by these normal cells by bypassing the block caused by methotrexate. Therefore, these normal cells can synthesize deoxythymidine and carry out DNA synthesis.

BIOCHEMISTRY PEARLS

- The eukaryote cell cycle is composed of the following four phases: M phase, mitosis; G_1 phase (gap 1), between mitosis and initiation of DNA synthesis; S phase, DNA synthesis; and G_2 phase (gap 2), cell growth and macromolecule synthesis.
- The cells in G_0 are in a nonproliferative or quiescent state that can vary tremendously in length.
- DNA synthesis requires the formation of dTMP and the purines adenine and guanine. THF, derived from the vitamin folic acid, is required for the biosynthesis of these nucleotides.
- The cell cycle is regulated by specific proteins: CDKs, which are serine/threonine protein kinases, and the cyclins, which are regulatory proteins that bind to the CDKs.
- Treatment with methotrexate blocks the cell's ability to regenerate THF, leading to inhibition of these biosynthetic pathways. Thus methotrexate affects DNA synthesis.

REFERENCES

Alberts B, Johnson A, Lew J, et al. The cell cycle and programmed cell death. In: Molecular Biology of the Cell, 4th ed. New York: Garland, 2002.

Cory JG. Purine and pyrimidine nucleotide metabolism. In: Devlin TM, ed. Textbook of Biochemistry with Clinical Correlations, 5th ed. New York: Wiley-Liss, 2002.

CASE 4

A 47-year-old female is brought to the emergency department with complaints of malaise, nausea and vomiting, and fatigue. The patient reveals a long history of alcohol abuse for the last 10 years requiring drinks daily especially in the morning as an "eye opener." She has been to rehab on several occasions for alcoholism but has not been able to stop drinking. She is currently homeless and jobless. She denies cough, fever, chills, upper respiratory symptoms, sick contacts, recent travel, hematemesis, or abdominal pain. She reports feeling hungry and has not eaten very well in a long time. On physical exam she appears malnourished but in no distress. Her physical exam is normal. Her blood count reveals a normal white blood cell count but does show an anemia with large red blood cells. Her amylase, lipase, and liver function tests were normal.

- **What is the most likely cause of her anemia?**

- **What is the molecular basis for the large erythrocytes?**

ANSWERS TO CASE 4: FOLIC ACID DEFICIENCY

Summary: 47-year-old alcoholic white female has fatigue, malaise, nausea, vomiting, and poor nutritional intake with macrocytic anemia and no evidence of pancreatitis, liver disease, or peptic ulcer disease.

- **Cause of anemia:** Folic acid deficiency.
- **Molecular basis of macrocytosis:** Abnormal proliferation of erythroid precursors in the bone marrow, since folate deficiency encumbers the maturation of these cells by inhibition of deoxyribonucleic acid (DNA) synthesis.

CLINICAL CORRELATION

Folate is an essential vitamin, found in green leafy vegetables. It is essential for many biochemical processes in the body, including DNA synthesis and red blood cell synthesis. Recently, folate supplementation has been found to be important in the prevention of fetal neural tube defects such as anencephaly (absence of brain cerebral cortex and no skull or skin covering the brain), and spina bifida (spinal cord malformation whereby the meninges are exposed leading to neurologic deficits). Alcoholics in particular are at risk for folate deficiency because of impaired gastrointestinal absorption and poor nutrition. Macrocytic anemia (large red blood cells) may be seen with folate deficiency. Treatment consists for folic acid replacement (usually 1 mg/day) by mouth with correction of anemia over the following 1 to 2 months. The diet usually requires adjustment, and correctable causes addressed (malnutrition in this case). Notably, folate deficiency in pregnancy has been associated with neural tube defects (NTDs) in fetuses. It is recommended that mothers take at least 400 μg of folic acid 3 months prior to conception to reduce the risk of NTD. At times, more than 400 μg of folic acid per day is recommended prior to conception. Some specific examples include a history of previous NTD, sickle cell disease, multiple gestations, and Crohn disease.

APPROACH TO FOLATE AND DNA SYNTHESIS

Objectives

1. Understand the important metabolic roles of folic acid with production of thymine, purine synthesis, and methionine.
2. Be aware of how folate deficiency causes megaloblastic anemia.

Definitions

***S*-Adenosyl methionine:** An important carrier of activated methyl groups. It is formed by the condensation of ATP with the amino acid methionine catalyzed by the enzyme methionine adenosyltransferase in a reaction that releases triphosphate.

Dihydrofolate reductase: The enzyme that reduces folic acid (folate) first to dihydrofolate and then to the active tetrahydrofolate. Dihydrofolate reductase uses NADPH as the source of the reducing equivalents for the reaction.

Folic acid: An essential vitamin composed of a pteridine ring bound to *p*-aminobenzoate, which is in an amide linkage to one or more glutamate residues. The active form of the enzyme is tetrahydrofolate (THF, FH_4), which is an important carrier of 1-carbon units in a variety of oxidation states.

Megaloblastic anemia: An anemia characterized by macrocytic erythrocytes produced by abnormal proliferation of erythroid precursors in the bone marrow due to a limitation in normal DNA synthesis.

Methotrexate: One of a number of antifolate drugs. Methotrexate is an analog of folate which competitively inhibits dihydrofolate reductase. Since a plentiful supply of THF is required for ongoing synthesis of the pyrimidine nucleotide thymidylate, synthesis of this nucleotide is inhibited resulting in decreased DNA synthesis.

Methyl trap: The sequestering of tetrahydrofolate as N^5-methyl THF because of decreased conversion of homocysteine to methionine as a result of a deficiency of methionine synthase or its cofactor, cobalamin (vitamin B_{12}).

DISCUSSION

Folate (folic acid) is an essential vitamin which, in its **active form of tetrahydrofolate** (**THF,** Figure 4-1), **transfers 1-carbon groups to intermediates in metabolism.** Folate plays an important role in **DNA synthesis.** It is required for the de novo synthesis of **purines** and for the **conversion of deoxyuridine 5-monophosphate (dUMP) to deoxythymidine 5′-monophosphate (dTMP).** Additionally, folate derivatives participate in the biosynthesis of choline, serine, glycine, and methionine. However, in situations of folate deficiency, symptoms are not observed from the lack of these products as adequate levels of choline and amino acids are obtained from the diet. (See also Case 3.)

Folate deficiency results in **megaloblastic anemia.** Megaloblastic anemia is characterized by **macrocytic erythrocytes** produced by **abnormal proliferation of erythroid precursors in the bone marrow.** Folate deficiency encumbers the maturation of these cells by **inhibition of DNA synthesis.** Without an adequate supply of folate, DNA synthesis is limited by **decreased purine and dTMP levels.**

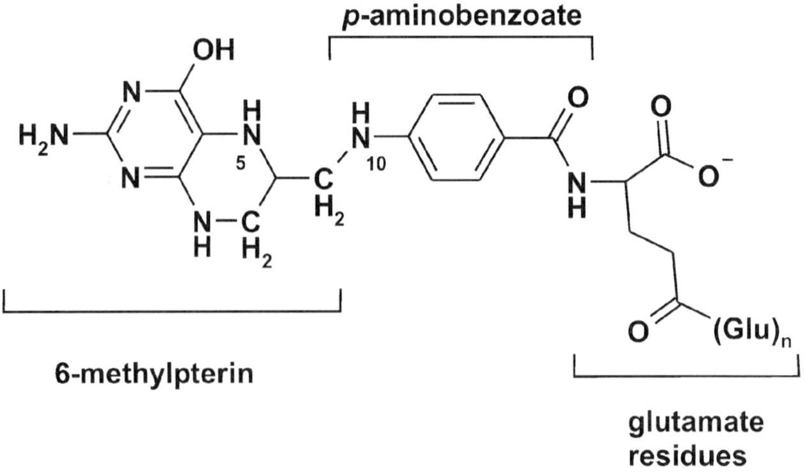

Figure 4-1. The structure of tetrahydrofolate, the active form of folic acid.

Folate exists in a pool of interconvertible intermediates each carrying a 1-carbon fragment in several different oxidation states (Figure 4-2). The total body stores of folate is approximately 110 mg/70 kg and approximately 420 μg/70 kg is lost each day via the urine and feces. **Two different forms of folate are required for different aspects of nucleotide biosynthesis. N^{10}-formyl THF** provides the **C-2 and C-8 carbons** for the de novo synthesis of **purine rings,** and thus is critical for DNA metabolism.

The **methylene form, N^5,N^{10}-methylene THF,** is required for the production of **dTMP from dUMP.** This reaction involves the transfer of a CH_2 group and a hydrogen from N^5,N^{10}-methylene THF. In this process, THF is oxidized to dihydrofolate (DHF). For subsequent dTMP production, THF must be regenerated. THF is produced from DHF by the enzyme DHF reductase (DHFR) in a reaction requiring NADPH. DHFR is the target of methotrexate, an antifolate cancer chemotherapeutic, which by limiting the available pool of N^5, N^{10}-methylene THF, inhibits DNA synthesis in rapidly dividing cancer cells. Methotrexate therapy can produce side effects resembling folate deficiency. Additionally, bacterial DHFR is a target for antimicrobials.

Outside of DNA synthesis, **folate plays a role in methylation metabolism.** The major **methyl donor is *S*-adenosyl methionine (SAM),** which is required for many reactions. For example, SAM is needed for the **production of norepinephrine from epinephrine and for DNA methylation,** which can influence **gene transcription.** After methyl group transfer, SAM is converted to ***S*-adenosyl homocysteine (SAH),** which is hydrolyzed to **homocysteine and**

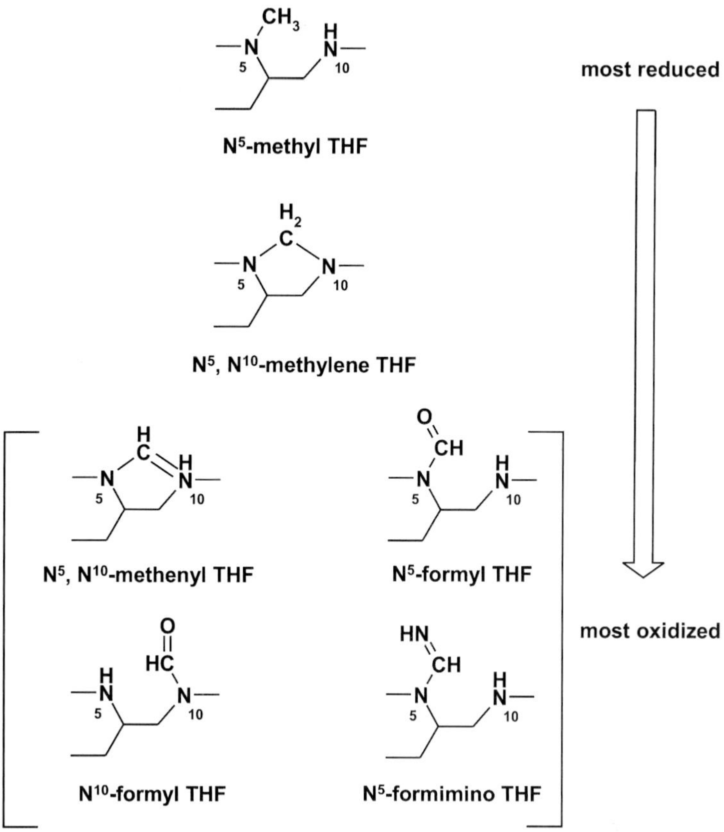

Figure 4-2. Structures of the various 1-carbon carriers of tetrahydrofolate (THF). THF can carry one-carbon units in the oxidation states of methanol (N^5-methyl THF), formaldehyde (N^5,N^{10}-methylene THF) or formic acid (remaining structures).

adenosine. To restore the levels of methionine (an essential amino acid), homocysteine must be methylated (Figure 4-3). This reaction is dependent on N^5-methyl THF and vitamin B_{12}. **Methionine levels can be limiting,** making the availability of N^5-methyl THF for conversion of homocysteine to methionine critical. Further, in the absence of B_{12}, THF can be trapped in the N^5-methyl THF form and thus be removed from the THF pool. This is referred to as **the "methyl trap,"** which can impact other areas of 1-carbon metabolism, such as dTMP production.

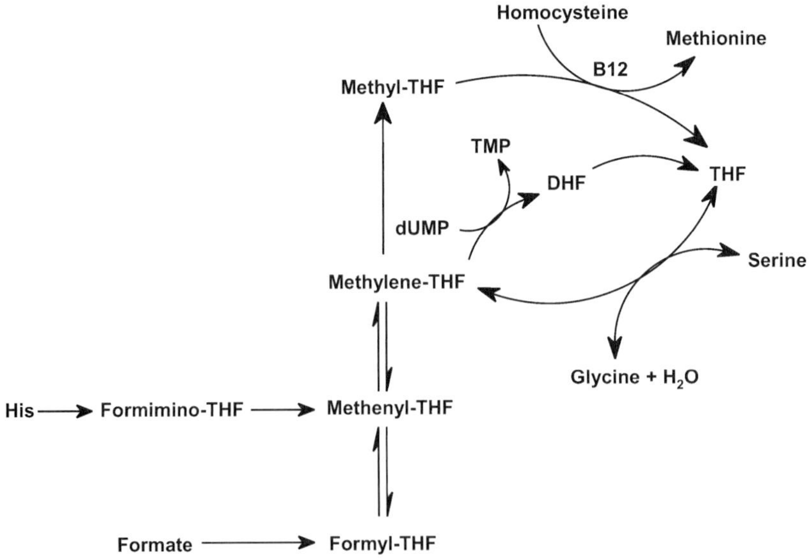

Figure 4-3. Reaction pathways showing the interconversion of 1-carbon carriers of tetrahydrofolate (THF). Note that all interconversions are reversible except for the conversion of N^5,N^{10}-methylene THF to N^5-methyl THF.

Folate deficiency perturbs DNA metabolism and methylation reactions. **Leafy green vegetables are good sources of folate;** however, folate is labile and may be damaged during food preparation. Another dietary source of folate is cereal products, especially breads and breakfast cereals that have been fortified with folic acid. Folate is also produced within the gut lumen by certain intestinal bacteria; however, the amount of folate absorbed from this source is minor in humans. In its simplest form, folate consists of three connected chemical moieties: a pteridine ring (6-methylpterin), *p*-aminobenzoic acid (PABA), and glutamate. In nature, folate is generally polyglutamated (decorated by two to seven additional glutamic acid residues). Conjugases (γ-glutamyl carboxypeptidases) in the intestinal lumen cleave off extra glutamic acid residues, and folate is absorbed by the mucosa of the small intestine (Figure 4-4). In cases of chronic alcoholism, folate deficiency may result from poor nutrition or from poor absorption of folate secondary to a conjugase deficiency. Once folate deficiency occurs, abnormal megaloblastic replication of epithelial mucosa can occur, which further impairs folate uptake.

After absorption, folate is reduced to THF by dihydrofolate reductase. The majority of circulating folate is in the form of N^5-methyl THF. Cells use specific transporters for THF uptake, and cellular machinery polyglutamates the folate to aid in cellular retention.

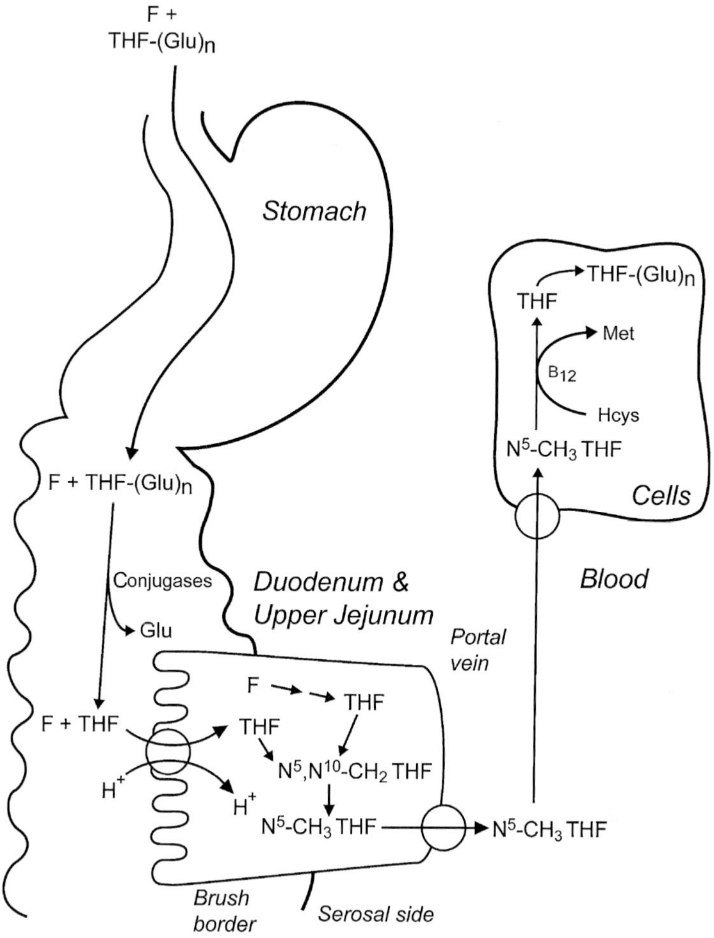

Figure 4-4. Intestinal absorption of dietary folates (THF-[Glu]$_n$) and folic acid (F) from fortified cereal products and vitamin supplements. In the duodenum and upper jejunum, extra glutamate residues are cleaved by conjugases (γ-glutamyl carboxypeptidases). Folate (F) and reduced folate (THF) are absorbed by a proton-coupled, high affinity folate transporter into the mucosal cell, converted to N^5-methyl THF and exported into the portal circulation. N^5-methyl THF is taken up into cells by facilitative diffusion, converted to THF by the B_{12}-requiring methionine synthase and then converted to a polyglutamate.

In summary, **folate** is a vitamin acquired from the diet that is **essential for 1-carbon metabolism.** Inadequate folate levels inhibit **DNA synthesis** by limiting **purine nucleotide and dTMP levels,** which results in the abnormal cellular proliferation observed in **megaloblastic anemia.** Folate is also required to replenish the **methionine pool for SAM-dependent methylation reactions.**

COMPREHENSION QUESTIONS

[4.1] Because of the close interrelationship between the vitamins, patients with deficiencies of either folate or vitamin B_{12} exhibit similar symptoms. Which of the following tests would best help distinguish between a folate and vitamin B_{12} deficiency?

A. Activity of methionine synthase
B. Blood level of cystathionine
C. Blood level of homocysteine
D. Blood level of methionine
E. Blood level of methylmalonate

For Questions [4.2] to [4.4] refer to the following list of vitamins:

A. Ascorbic acid (vitamin C)
B. Biotin
C. Cobalamin (vitamin B_{12})
D. Folic acid
E. Niacin (vitamin B_3)
F. Pantothenic acid
G. Riboflavin (vitamin B_2)
H. Thiamine (vitamin B_1)

For each patient, described in the Questions [4.2] to [4.4], select the vitamin that is most likely to be deficient.

[4.2] A muscular 25-year-old male presents with dermatitis and an inflamed tongue. A history reveals that he has been consuming raw eggs as part of his training regimen for the past 6 months.

[4.3] A 30-year-old male goes to his dentist complaining of loosening teeth. Examination also reveals his gums are swollen, purple, and spongy. The dentist also notes that the patient's fingers have multiple splinter hemorrhages near the distal ends of the nail and that a wound on the patient's forearm has failed to heal properly.

[4.4] A female neonate is found to have a small spina bifida in her lower spinal column that could affect bladder and lower limb function.

Answers

[4.1] **E.** Vitamin B_{12} is a cofactor in two biochemical reactions, the conversion of homocysteine to methionine by the enzyme methionine synthase and the conversion of L-methylmalonyl-CoA to succinyl-CoA by methylmalonyl-CoA mutase. N^5-methyl THF is a methyl donor in the methionine synthase reaction. A folate deficiency would result in decreased methionine synthase activity and decreases in methionine and cystathionine concentrations, while homocysteine levels would be increased. A vitamin B_{12} deficiency would also yield these same results, but in addition methylmalonate levels would increase as a consequence of a decrease in the activity of methylmalonyl-CoA mutase activity.

[4.2] **B.** Raw eggs contain a protein, avidin, which binds biotin strongly. Because native avidin is resistant to hydrolysis by digestive proteases, when it binds biotin it prevents its absorption. Avidin that has been denatured by cooking will be broken down during the digestive process. A biotin deficiency manifests itself as an erythematous, scaly skin eruption and can also cause hair loss and conjunctivitis. A biotin deficiency can also occur following prolonged total parenteral nutrition if biotin is not supplemented.

[4.3] **A.** The patient exhibits the classic symptoms of scurvy, a deficiency in vitamin C. In addition to being an important biological antioxidant, ascorbic acid is required for the hydroxylation of proline and lysine residues of procollagen in the synthesis of collagen. A deficiency leads to defects in collagen synthesis, which adversely affects the intercellular cement substances in connective tissue, bones, and dentin.

[4.4] **D.** The fetus needs a constant supply of cofactors for normal development. Folic acid supplements of 400 μg/day prior to conception have been shown to decrease the incidence of neural tube defects such as spina bifida.

BIOCHEMISTRY PEARLS

- Folate (folic acid) is an essential vitamin that, in its active form of tetrahydrofolate, transfers 1-carbon groups to intermediates in metabolism and plays an important role in DNA synthesis.
- THF is necessary for the de novo synthesis of purines and the conversion of deoxyuridine 5′-monophosphate (dUMP) to deoxythymidine 5′-monophosphate (dTMP).
- The major metabolic perturbation in folate deficiency occurs in megaloblastic anemia.

REFERENCES

Devlin TM, ed. Textbook of Biochemistry with Clinical Correlations, 5th ed. New York: Wiley-Liss, 2002.

Frenkel EP, Yardley DA. Clinical and laboratory features and sequelae of deficiency of folic acid (folate) and vitamin B_{12} (cobalamin) in pregnancy and gynecology. Hematol Oncol Clin North Am 2000;14(5):1079–100.

Lin Y, Dueker SR, Jones AD, et al. Quantitation of in vivo human folate metabolism, Am J Clin Nutr 2004;80:680–91.

Que A., et al. Identification of an intestinal folate transporter and the molecular basis for hereditary folate malabsorption, Cell 2006;127:917–28.

Murray RK, Granner DK, Mayes PA, et al. Harper's Illustrated Biochemistry, 26th ed. New York: Lange Medical Books/McGraw-Hill, 2003.

CASE 5

A 32-year-old male presents to your clinic with complaints of a sore throat. He reports numerous upper-respiratory infections over the last 3 months. Patient states that he required antibiotics for some of the infections. The patient's sore throat has been present for 4 days and is progressively worsening. He is no longer able to eat solid foods because of the pain. Nobody else in contact with him has been ill. Patient gives a history of intravenous (IV) drug use in the past, but no other significant medical history is given. On exam, patient is found to have a temperature of 37.8°C (100.0°F) and is in minimal distress from the sore throat. His pharynx is erythematous and has numerous white plaques coating the throat. There is also prominent cervical lymph node enlargement. His chest is clear to auscultation and heart is regular rate and rhythm. A CD4 T lymphocyte cell count is performed and is less than 200 cells/mm^3 (normal >500 cells/mm^3). The responsible organism is composed of ribonucleic acid (RNA) genome.

- **What is the most likely diagnosis?**

- **What is the biochemical mechanism that the pathogen uses to affect the patient's cells?**

- **What enzyme is required for this pathogen to affect the host genome?**

ANSWERS TO CASE 5: HUMAN IMMUNODEFICIENCY VIRUS

Summary: A 32-year-old male with history of IV drug use has numerous upper-respiratory symptoms and adenopathy, and he presents now with sore throat with numerous white plaques. He also has a low CD4 count. The pathogen has a ribonucleic acid (RNA) genome.

- **Most likely diagnosis:** Candida esophagitis secondary to human immunodeficiency virus (HIV).
- **Biochemical mechanism of disease:** The HIV genome is composed of single-stranded RNA. Reverse transcription of the viral RNA is required to infect the host cells. HIV causes immunosuppression because it has a propensity for attaching to helper T cells (CD4 cells), the cells that support cell mediated immunity.
- **Enzyme necessary:** Reverse transcriptase.

CLINICAL CORRELATION

This 32-year-old male likely acquired HIV infection by sharing needles with an infected individual. Initially, an HIV infection may cause systemic flu-like symptoms, adenopathy, and fatigue. During this phase, the patient experiences a viremia. The next phase is largely asymptomatic as the virus slowly causes attrition of the CD4 helper T cells. When the levels of these important cells drop to low levels, the patient will not be able to fend off organisms that are commonly colonizing the skin, gastrointestinal tract, or in the air. Treatment for HIV infection includes agents that attack the unique aspects of the virus such as nucleoside analogue reverse transcriptase inhibitors and protease inhibitors.

APPROACH TO HIV

Objectives

1. Understand normal transcription/translation.
2. Be familiar with reverse transcriptase and mechanism by which HIV works.
3. Know the mechanism of action of HIV medications.

Definitions

Deoxyribonucleic acid (DNA) replication: The process by which DNA is duplicated in the cell. This process takes place during the S-phase of the cell cycle. DNA is duplicated in a semiconservative manner; that is, each new DNA double-strand contains one of the original strands (parent strands) and one of the newly synthesized strands (daughter strands).

HIV: Human immunodeficiency virus; a retrovirus that causes acquired immunodeficiency syndrome (AIDS).

Retrovirus: A virus in which the genetic material located in the virus is RNA. The genetic information in the retrovirus must be converted to DNA in the host by the process of reverse transcription.

Reverse transcription: The synthesis of DNA from an RNA template.

Transcription: The process by which the information contained in the nuclear DNA is converted to cytosolic mRNA. This is accomplished by RNA polymerases that synthesize an mRNA that has a complementary sequence to the DNA strand that was used as the template.

Translation: The synthesis of proteins from mRNA. This is a process that takes place on ribosomes and requires the participation of aminoacyl-charged tRNAs, mRNA, and various initiation and elongation factors.

Exons: Sequence of nucleotides that appear in mature RNA.

Introns: Sequences of nucleotides that do not appear in mature RNA but are excised and do not appear in messenger RNA.

Spliceosome: A complex of small nuclear ribonucleoprotein particles (snRNPs) that catalyze the splicing, or removal of introns, of mRNA precursors.

DISCUSSION

DNA is the biological blueprint material, which authentically carries all the necessary cellular information and passes it from generation to generation. Thus, it is essential for the cells to preserve the integrity of the DNA and keep it as error free as possible. One is amazed with the versatile DNA molecule, which dictates both the unique and the similar features of the offspring from its parents. What are the processes that take place in a cell to duplicate and interpret this genetic code into functional signals? DNA is duplicated in a **semiconservative fashion** in a process called **DNA replication.**

DNA interpretation follows the central dogma (Figure 5-1): first, the DNA is **decoded to form** messenger RNA **(mRNA)** in the nucleus by a process called **transcription. Transcription** is a complex process involving an **enzyme RNA**

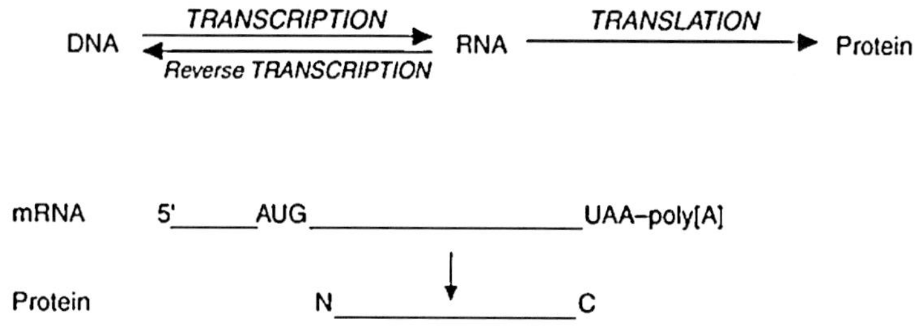

Figure 5-1. The central dogma.

polymerase and several transcription factors. The DNA strand that directs RNA synthesis via **complementary base pairing** is called the **template. RNA synthesis is always unidirectional from 5′ (phosphate) to 3′ (hydroxyl).** One can organize transcription in a stepwise fashion: (1) **binding of RNA polymerase** to the DNA; (2) formation of the **transcription bubble** (separation of the DNA strands); (3) **addition of the first ribonucleic acid residue;** and (4) addition of the **second residue,** formation of a **phosphodiester bond** between the ribonucleic acid residues, and **release of pyrophosphate.** The addition process continues until the **termination signal** is encountered. Once the **nascent RNA chain emerges, it is capped with methylguanylate at the 5′ end and polyadenylated in the 3′ end.** This primary RNA transcript is processed further by splicing machinery to yield the mature functional mRNA. **Heterogenous nuclear RNA (hnRNA) contains exons and introns.** The introns are excised and **mature RNA enters the cytoplasm.**

The **mRNA** is then processed in the cytoplasm by a process called **translation.** Translation involves the **ribosomes** along with a battery of **initiating (IF), elongation (EF), and release (RF) factors.** An important rule to remember in translation is that each **triplet nucleotide codon codes for a specific amino acid (the three-letter genetic code).** Maintaining the order of amino acids is important to obtain a functional protein. mRNA is translated into protein with the help of **transfer RNA (tRNA),** which **brings an amino acid along with it (aminoacyl tRNA) and ribosomes.** Ribosomes are in turn composed of numerous proteins and several rRNA molecules. The stepwise process of **translation** is as follows: (1) The ribosomal unit along with the **initiation factors and formyl-methionine (fMet)-tRNA bind close to the initiating codon AUG in the 5′ region** of the mRNA. (2) The **ribosomal complex consists of three sites, namely, A, P, and E.** The **aminoacyl tRNA binds the triplet codon at the A site.** The **formyl-methionyl group** forms a peptide bond with the amino group of the aminoacyl tRNA in **the P site,** releasing the now **uncharged $tRNA_{fMet}$,** which exits the ribosome assembly through the **E site.** (3) **Elongation** continues until the ribosome encounters one of the **stop codons UAA, UAG, and UGA.** (4) **Termination of protein synthesis** occurs with the help of release factors which cleave the peptide chain from the tRNA and dissolve the ribosomal complex.

In **retroviruses such as HIV,** where the genetic material is **RNA** rather than DNA, **reverse transcription** occurs. The life cycle of HIV begins when the virion binds to cell surface receptors (CD4 receptors) on helper T lymphocytes. This results in a conformational change that enables the viral coat to fuse with the lymphocyte membrane, thus releasing the viral RNA and viral proteins into the cytosol, including reverse transcriptase and integrase. The **RNA genome (Figure 5-2) is reverse transcribed into a double-stranded DNA** molecule by utilizing the **reverse transcriptase enzyme.** This enzyme is an unusual **DNA polymerase, because it uses both DNA and RNA as template.** First, it makes DNA from the RNA and uses this in turn to make the

LTR	**gag**	**pol**	**env**	**LTR**

Figure 5-2. Simplified HIV genome.

second strand of DNA. This double-stranded DNA is transported into the nucleus and is recognized by the **viral enzyme integrase,** which catalyzes the insertion of the viral DNA into the host genome thus establishing a permanent infection. The final step is the transcription of the integrated viral DNA producing a large amount of **viral RNA, which is packaged in a capsid** along with the other essential proteins and can bud from the plasma membrane. Based on the pathogenesis of HIV, it is very important to understand the transcription mechanism of this virus. The HIV life cycle can be simplified and drawn out as shown in Figure 5-3.

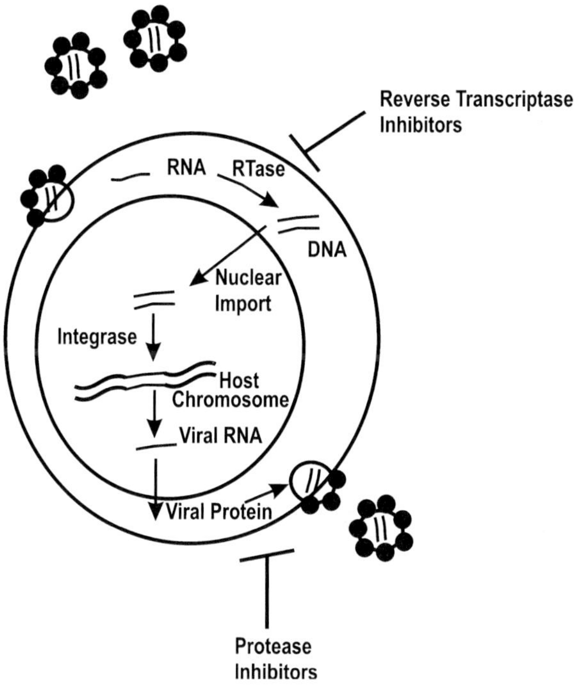

Figure 5-3. Life cycle of HIV.

The **long terminal repeat (LTR) contains the enhancer and the promoter** regions of HIV. The ***gag*** encodes the structural proteins, which help in packaging the RNA of the virus to generate new virus particles. The ***pol*** **gene** encodes both the critical enzymes **reverse transcriptase and integrase.** The ***env*** **codes for envelope proteins** of the virus that along with the host plasma membrane help the complete virus particle to bud off from the cell. Apart from these proteins, the **three regulatory proteins,** namely, the negative factor **Nef,** trans-activator of transcription **Tat,** and the regulator of viral gene expression, **Rev,** can affect viral transcription output. Tat specifically promotes transcriptional activity, while Rev is responsible for switching from early to late HIV gene expression and is located in the 3′ end of the *env* gene. With the latest research identifying new proteins in addition to the preexisting knowledge, one wonders what would be the most effective antiretroviral therapy. The available drugs in the market include **nucleoside reverse transcription inhibitors and protease inhibitors.**

Nucleoside reverse transcriptase inhibitors act as substrates for the reverse transcriptase enzyme and require a phosphorylation event in order to be activated. The **nucleoside drugs lack the 3′-hydroxyl group;** therefore, their incorporation into viral DNA will effectively **terminate the elongation process.** This class includes zidovudine, didanosine, stavudine, lamivudine, and the latest, abacavir. As these agents affect very early events in pathogenesis of HIV they prevent acute infection of susceptible cells but have little effect on already infected cells. The common side effects of these drugs are lactic acidosis, hepatomegaly, anemia, anorexia, fatigue, nausea, and insomnia. A subcategory to this class of drugs is the nonnucleoside inhibitors, which act by binding close to the active site, inducing a conformation change and thereby inhibiting enzyme function. Some of the members of the class are delavirdine, nevirapine, and efavirenz. A major drawback with these nonnucleoside drugs is that they are effectively metabolized by the cytochrome P450 system and are prone to drug interaction. The most common adverse effects are rash, elevated liver function, and impaired concentration.

Protease inhibitors are targeted towards the **HIV proteases, which are essential to activate precursors of *gag-pol.* HIV protease is essential for infectivity** and cleaves the viral polyprotein (*gag-pol*) into active viral enzymes and structural proteins. The mechanism of action of these drugs is by binding reversibly to the active site of HIV protease and blocking subsequent viral maturation. This major class includes saquinavir, ritonavir, indinavir, and nelfinavir. Toxicities of protease inhibitors include nausea, vomiting, and diarrhea.

HIV inhibitor treatment eventually leads to the selection of resistant mutations, one key reason being that the reverse transcriptase is error prone since it lacks the 3′ exonuclease activity. Currently, combinatorial inhibitors with different protein targets, for example, two reverse transcriptase (RT) nucleoside inhibitors and one protease inhibitor (PI), are used in highly active antiretroviral therapy (HAART). To overcome drug resistance and find a definitive cure for HIV infection several efforts are being made to develop vaccines and also use ribozymes to target HIV mRNA. Because of the molecular understanding of HIV, the future of anti-HIV therapy seems bright and promises an effective cure for the AIDS patient.

COMPREHENSION QUESTIONS

[5.1] The current therapeutic strategy for patients who have been infected with HIV is a multidrug regimen known as highly active antiretroviral therapy (HAART). One type of drug used in this therapy is a nucleoside/nucleotide analog, such as didanosine. Which of the following best describes the mechanism of action of these drugs?

A. They inhibit the synthesis of viral proteins.
B. They directly bind to and inhibit reverse transcriptase.
C. They prevent the hydrolysis of the viral polyprotein.
D. They prematurely terminate the DNA synthesized by reverse transcriptase.
E. They inhibit the viral enzyme integrase.

[5.2] Enzyme-linked immunosorbent assay (ELISA) is used for routine screening of HIV infection and the Western blot test has been successfully used as a confirmatory test. Which of the following best describes the strategy used to confirm the presence of an HIV infection by Western blot analysis?

	GEL ELECTROPHORESIS AND BLOT PERFORMED ON	BLOT INCUBATION WITH	BLOT DETECTS
A.	HIV protein standards	Patient serum	Presence of anti-HIV antibodies in serum
B.	Patient serum	Anti-HIV antibodies	Presence of HIV proteins in serum
C.	Anti-HIV antibody standards	Patient serum	Presence of HIV proteins in serum
D.	Patient serum	Labeled oligonucleotides	Presence of HIV RNA in serum

[5.3] The following diagram schematically represents the life cycle of the human immunodeficiency virus (HIV). Which of the labeled steps best indicates the site at which an HIV protease inhibitor disrupts the cycle?

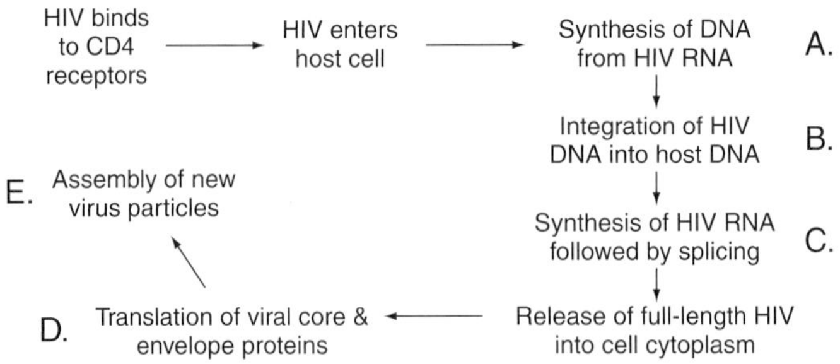

[5.4] Given the mRNA nucleotide sequence, choose the best protein sequence that will likely result. (Use the amino acid table in Questions [13.2] and [13.3] in Case 13, keeping in mind that [T] and [U] are analogous). mRNA 5′ AUCGGAUGUCUCGGGUUCUGUAAAGGUAAUC 3′

A. Met-Ser-Arg-Val-Leu
B. Ser-Arg-Val-Leu
C. Met-Leu-Ser-Val
D. Ser-Arg-Val-Phe-Phe
E. Pro-Ser-Val-Gly

Answers

[5.1] **D.** The nucleoside/nucleotide analogs like azidothymidine (AZT) and didanosine are incorporated into the DNA synthesized by HIV reverse transcriptase. Because they do not have a 3′-hydroxyl group, they cannot form a bond with the next nucleotide and the chain is terminated. Host cell DNA synthesis is not affected because of the nuclear DNA repair mechanisms.

[5.2] **A.** In the Western blot confirmatory test, standard HIV proteins (*gag, pol,* and *env*) are separated by gel electrophoresis and then blotted onto a nitrocellulose membrane. The membrane is then incubated with the patient's serum. Any anti-HIV antibodies will bind to the respective HIV protein on the membrane. Finally, a labeled antihuman antibody is added to indicate the presence of any anti-HIV antibody binding. The test is highly specific; that is, a positive result is highly indicative of HIV infection.

[5.3] **E.** The HIV proteins synthesized by the host cell are produced as a long polyprotein that must be cleaved to the active HIV enzymes and structural proteins. HIV protease inhibitors bind to and inhibit the aspartic protease that hydrolyzes the polyprotein, thus preventing the assembly of infective viral particles.

[5.4] **A.** The initiator codon is AUG and codes for Met; UCU = Ser; CGG = Arg; GUU = Val; CUG = Leu; UAA = stop codon. The mRNA is read in the 5′ → 3′ direction.

BIOCHEMISTRY PEARLS

- ❖ DNA is decoded to form messenger RNA (mRNA) in the nucleus by a process called transcription. RNA synthesis is always unidirectional from 5′ (phosphate) to 3′ (hydroxyl), and begins at the 3′ end of the DNA chain toward the 5′ end.
- ❖ The mRNA is processed in the cytoplasm (translation) involving the ribosomes. Each triplet nucleotide codon codes for a specific amino acid (the three-letter genetic code).
- ❖ Messenger RNA is translated into protein with the help of tRNA, which brings an amino acid along with it (aminoacyl tRNA) and ribosomes.
- ❖ Some viruses such as HIV have RNA genomes, and they usually are reverse transcribed into a double-stranded DNA molecule by utilizing the reverse-transcriptase enzyme.

REFERENCES

Levy JA. HIV and the Pathogenesis of AIDS, 2nd ed. Washington DC: ASM Press, 1998.

Lodish H, Berk A, Zipursky SL, et al. Molecular Cell Biology, 4th ed. New York: Freeman, 2000.

Raffanti S, Haas DW. Anti-bacterial agents. In: Goodman AG, Gilman LS, eds. The Pharmacological Basis of Therapeutics, 10th ed. New York: McGraw-Hill, 2001.

CASE 6

A 21-year-old female presents to the physician's office with complaints of very painful vulvar ulcers. The patient states she has had the symptoms for approximately 3 days, and they are worsening. She says that prior to the ulcerations, there was a burning and tingling sensation of the skin in the same area. She has noted similar symptoms like this before and has been told it is sexually transmitted. On examination you see multiple extremely tender, vesicular (blister-like) lesions on an erythematous (red) base on both labia majora of the vulva. She has a moderate amount of tender inguinal lymph nodes bilaterally. The physician uses a swab to sample the ulcer and send it off for diagnostic analysis.

- **What is the most likely diagnosis?**

- **If the test is for deoxyribonucleic acid (DNA), what method would be used to amplify the fragments of DNA sampled?**

ANSWERS TO CASE 6: HERPES SIMPLEX VIRUS/ POLYMERASE CHAIN REACTION

Summary: A 21-year-old female presents with recurrent episodes of painful ulcerations on vulva. The physician uses a swab to sample the ulcer for diagnosis. She also has some neurologic symptoms at the vulva prior to appearance of lesions.

- **Most likely diagnosis:** Herpes simplex virus 2 (HSV-2) outbreak.
- **Biochemical Technique:** Polymerase chain reaction to amplify the small amount of HSV DNA. This is a very sensitive diagnostic technique that is able to detect small amounts of HSV DNA, and through in vitro techniques, it is able to rapidly produce large quantities of DNA.

CLINICAL CORRELATION

This 21-year-old female has a recurrent episode of vulvar ulcers accompanied by burning and tingling in the same region. These symptoms are caused by the herpes virus affecting the afferent nerve. After a primary infection, the herpes virus lays dormant in the dorsal ganglia of the nerve root. Then in times of stress or for unknown reasons, the virus becomes active and travels down the nerve to the skin. Thus, the patient often has neurologic symptoms even before the outbreak on the skin. Viral culture is an accurate method of diagnosis. Perhaps even better is polymerase chain reaction (PCR), which can also be used for detection of numerous infectious organisms, genetic mutations, and forensic testing.

APPROACH TO POLYMERASE CHAIN REACTION

Objectives

1. Describe the life cycle of herpes simplex viruses.
2. Describe the process of PCR.
3. Know the definitions and purpose of restriction endonucleases and oligonucleotides.
4. Be familiar with how PCR may be used for identifying infections and mutations.
5. Cite the advantages of PCR over other biotechnology involving recombinant DNA.

Definitions

Annealing: The process of allowing single-stranded lengths of DNA to base-pair to form double-stranded DNA. It is used most frequently when referring to the process of binding an oligonucleotide primer or probe to a longer DNA fragment.

Denaturation: The process by which the secondary and tertiary structure of proteins and nucleic acids is broken down to form random chains. In the case of DNA, it specifically refers to the separation of double-stranded DNA into single strands by the breaking of the hydrogen bonds that form the complementary base pairing, usually by increasing the temperature.

PCR (polymerase chain reaction): PCR is a method by which DNA or DNA fragments can be amplified through several steps of denaturation, annealing, and elongation. Through this process, the amount of target DNA can be increased by a factor of 10^6 to 10^9.

Restriction enzymes: Endonucleases isolated from bacteria that will selectively cleave DNA having specific nucleotide sequences. The enzyme recognizes particular palindromic nucleotide sequences and will hydrolyze both strands within that sequence. Molecular biologists use these enzymes in various recombinant DNA techniques.

Taq DNA polymerase: A DNA polymerase isolated from the thermophilic bacteria *Thermus aquaticus.* It has the advantage that it remains active during the denaturation steps of PCR in which the temperature is increased to separate the DNA strands.

Palindrome: having the same sequence in the complimentary strand as the original strand of DNA or RNA when read from the 5′ to the 3′ direction.

Acyclovir (Acycloguanosine): Acycloguanosine, a prodrug that is activated to the active acyclo-GTP, which inhibits viral DNA polymerase by acting as a DNA chain terminator.

DISCUSSION

Herpes simplex viruses (HSV) are members of Herpesviridae which also includes Epstein-Barr virus (infectious mononucleosis), varicella zoster virus (chicken pox), and cytomegalovirus. Herpesviruses are large (150 nm diameter) enveloped linear double-stranded DNA viruses. Because of the phospholipid envelope, herpesviruses are sensitive to acids, detergents, solvents, and drying. In general, herpesvirus replication begins with the attachment of the viral particle to the host cell surface, followed by the fusion of the viral envelope and cell membrane and release of the viral nucleocapsid into the cell cytoplasm. The nucleocapsid attaches to the nuclear envelope and then the viral genome is released into the cellular nucleus. The herpesvirus genome is transcribed and translated by host factors but it is replicated by a viral encoded DNA polymerase. This HSV DNA polymerase is an important target for drug therapy. The viral nucleocapsids are formed in the nucleus and the envelope is acquired from the nuclear or Golgi membrane. Mature viral particles are released from the host cell by exocytosis or cell lysis. Depending on the herpesvirus and type of host cell that is infected, a lytic, persistent (macrophages and lymphocytes), or latent (nerve cells) infection can be established.

HSV can infect many cell types (including macrophages, lymphocytes, and neurons) causing lytic, persistent, and latent infections. Because of nerve cell involvement in latent infections, a recurrence of the disease is often preceded by a prodrome (sensations such as burning or tingling). Initial HSV-1 and HSV-2 infections are usually established on mucous membranes. While HSV-1 and HSV-2 are respectively referred to as oral-herpes and genital-herpes based on their typical site of infection, HSV-1 and HSV-2 can be found on both oral and genital tissue.

Diagnosis can be made by examining infected tissue or cells for characteristic cytopathologic effects, virus isolation and culture, or serology. Molecular methods (such as, DNA *in situ* hybridization, and PCR of vesicle fluid or scrapings) are also used for diagnosis. The molecular methods are gaining favor because they yield rapid results and identify the viral type or strain. Because of these diagnostic advantages, PCR is the preferred method for diagnosis of HSV encephalitis.

The **polymerase chain reaction (PCR)** is an *in vitro* method for the **exponential amplification of a specific DNA region.** Today, PCR is one of the most important biochemical techniques and is applied in virtually all fields of modern molecular biology. The ingredients for PCR reactions are very simple: the **target DNA to be amplified,** a **heat-stable DNA polymerase, the four deoxyribonucleotides, two oligonucleotide primers, and a reaction buffer.** The entire PCR reaction is carried out in a single tube containing all these necessary components. The **target DNA segment is amplified** with a high degree of specificity by using **two DNA oligonucleotide primers** that are complementary to sequences on the 3′-flanking regions of opposite strands of the target segment. The PCR amplification occurs by repeated cycles of **three temperature-dependent steps** called **denaturation, annealing, and elongation** as schematically presented in Figure 6-1. Native DNA exists as a double helix; therefore the first step, denaturation, of the process separates the two DNA chains by **heating** the reaction mixture to 90°C to 95°C (194°F to 203°F). In the second step, **annealing,** the reaction mixture is **cooled** to 45°C to 60°C (113°F to 140°F) so that the **oligonucleotide primers can bind or anneal** to the separated strands of the target DNA. In the final step of the process, **elongation,** the **DNA polymerase adds nucleotides to the 3′ ends of the primers to complete a copy of the target DNA template.** A heat-stable DNA polymerase obtained from the thermophilic bacterium ***T. aquaticus*** is used to synthesize the new strands in most PCR reactions. Since this Taq DNA polymerase works the best at around 72°C (161.6°F), the temperature of the PCR mixture is raised to that temperature for elongation to proceed efficiently. At the end of a three-step cycle, each target region of DNA in the vial has been duplicated. This three-step cycle is then repeated multiple times. Each new DNA strand can then act as a new template in the next cycle, yielding approximately **one million copies of a target DNA after 20 cycles.** Therefore PCR is a method able to amplify any DNA sequence virtually without limit and allows the separation of the nucleic acid of interest from its context.

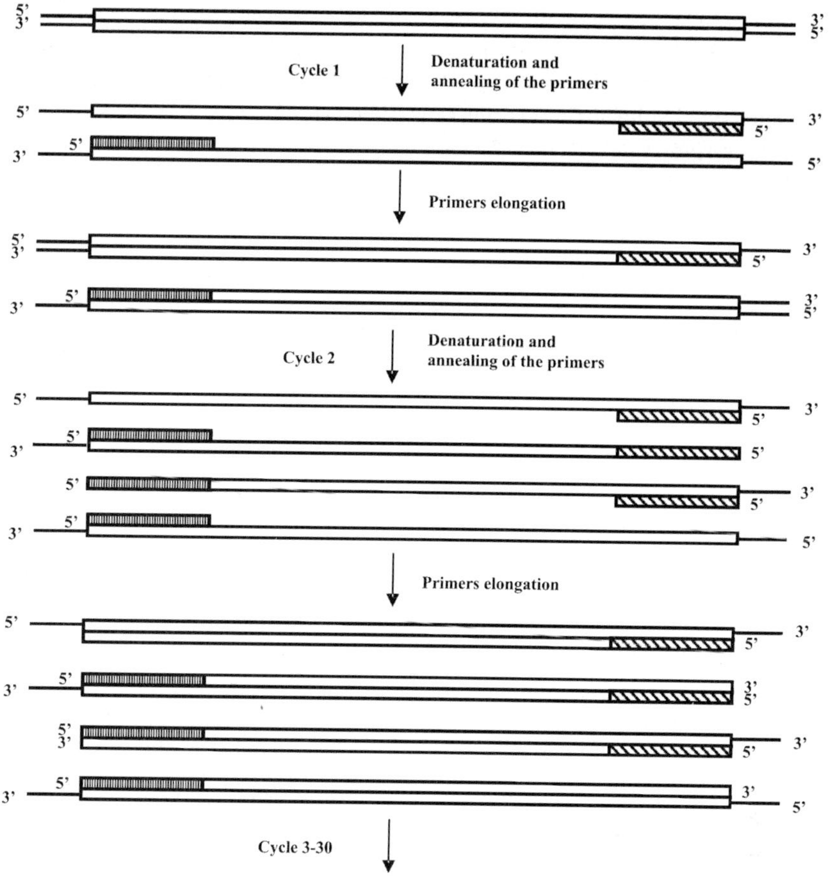

Figure 6-1. PCR reaction diagram.

PCR is also becoming the leading method for detection of the continuously increasing number of human pathogens. Examples include herpes simplex virus (HSV), human papillomavirus, human immunodeficiency virus (HIV), human T lymphotropic virus type I and type II, cytomegalovirus (CMV), Epstein-Barr virus, human herpesvirus-6 (HHV-6), hepatitis B virus, B16 parvovirus, JC and BK viruses, rubella virus, mycobacteria, *Toxoplasmosis gondii, Trypanosoma cruzi,* and malaria, with many more to follow. PCR can be applied as a detection method for virtually any pathogen for which even limited nucleotide sequence information is known and for which a specimen of infected tissue can be readily obtained. In most cases PCR assays are more discriminative than conventional serology. For example, it is difficult to distinguish HSV-I from HSV-II or HIV-1 from HIV-2 by serology, yet such distinctions can be readily made on the basis of PCR amplification of type-specific genetic sequences. PCR yields rapid results, typically in 1 to 2 days in a clinical setting.

It is applicable to a wide variety of clinical, pathologic, or forensic specimens, as well as to formalin-fixed tissue, inactivated bacterial cultures, and archaeological specimens. However, since **PCR detects nucleic acids from both living and dead microbes,** this must be taken into account if PCR is used to monitor response to therapy.

Prior to the PCR method for detection of HSV-1 and HSV-2, clinical specimens were analyzed using immunologic methods to confirm the virus identity following a viral culture. Such methods are time-consuming (3 to 12 days) and expensive. In addition, cultivation of the virus is not always successful. Turnaround time for viral culture averaged 108 hours for positive results and 154 hours for negative results. In comparison, the present PCR method of HSV detection is specific, relatively fast and accurate. The method allows detection of one to ten virus particles in the presence of microgram quantities of cellular or heterogeneous DNA. The PCR assay offers increased sensitivity, specificity, and improved turnaround time (24 to 48 hours) when compared with traditional viral culture techniques. These methods have allowed a better understanding of the physiopathogeny of the disease. In particular, PCR has revealed the importance of asymptomatic viral shedding in infected patients. PCR also helps diagnosis in many situations where viral isolation by culture proves difficult or impossible, for example, in treated or atypical lesions, or in newborn central nervous system infections. PCR has demonstrated the existence of prolonged viremia in infected newborns. PCR helps sequencing of the viral genome for further epidemiologic studies or analysis of resistance to antiviral drugs. Recently, PCR-derived techniques have been developed to quantify viral load in real time, thus allowing a diagnosis in a few hours.

After **amplification of the samples** using PCR the identification of the virus species can be achieved through **restriction enzyme digestion,** which yields a unique pattern of **different fragment sizes** characteristic for each herpesvirus (or other pathogen). **Restriction endonucleases are enzymes able to cut double-stranded DNA at specific palindromic recognition sequences** that are mostly four to six nucleotides long. Furthermore, a well-designed restriction enzyme panel allows the discrimination between human herpes virus 6 variant A and variant B. In this way, this method can readily detect human herpesviruses, including occasional multiple infections, in a variety of clinical samples. When PCR assay was compared to isolation and electron microscopy for the detection of HSV in clinical samples, all specimens positive by conventional methods were also positive by PCR. However, in a number of clinical specimens in which HSV could not be detected by conventional methods, PCR was able to demonstrate the presence of the virus.

Treatment of HSV typically involves the use of nucleoside analogs such as acyclovir (acycloguanosine) that inhibit the viral DNA polymerase. Acyclovir is a prodrug that is activated by viral thymidine kinase. The activated drug, acyclo-GTP, inhibits the viral DNA polymerase by acting as a chain terminator when it is incorporated into viral DNA.

COMPREHENSION QUESTIONS

[6.1] One of the steps in the PCR amplification of DNA fragments is the denaturation step in which the temperature is raised to break the hydrogen bonds that make up the base pairing. Which of the following DNA fragments would most likely require the greatest increase in temperature to cause complete denaturation?

A. 5′-C-A-A-T-G-T-A-A-T-T-G-C-A-T-3′
3′-G-T-T-A-C-A-T-T-A-A-C-G-T-A-5′
B. 5′-A-T-A-T-A-T-A-T-A-T-A-T-A-T-3′
3′-T-A-T-A-T-A-T-A-T-A-T-A-T-A-5′
C. 5′-A-A-C-C-G-G-A-C-C-G-C-G-A-T-3′
3′-T-T-G-G-C-C-T-G-G-C-G-C-T-A-5′
D. 5′-A-G-A-G-A-G-A-G-A-G-A-G-A-G-3′
3′-T-C-T-C-T-C-T-C-T-C-T-C-T-C-5′
E. 5′-G-A-C-T-G-T-A-A-T-A-C-G-A-T-3′
3′-C-T-G-A-C-A-T-T-A-T-G-C-T-A-5′

[6.2] Restriction enzymes are used to cleave genomic DNA into smaller fragments. Which of the following single-strand DNA sequences has the best potential to be a site of action for a restriction endonuclease?

A. T–A–G–C–T–T
B. C–T–G–C–A–G
C. A–A–C–C–A–A
D. G–T–G–T–G–T
E. A–A–A–C–C–C

[6.3] A 21-year-old woman was abducted when she went to the local convenience store. Her body was found the next morning in a wooded area behind the store. The autopsy revealed that she had been sexually assaulted and strangled. Crime scene investigators were able to collect a semen sample from vaginal fluid as well as tissue samples from underneath the victim's fingernails. DNA samples were obtained from three suspects besides the victim. A variable number of tandem repeats (VNTR) analysis was performed on the DNA samples from the evidence collected, the victim, and the suspect, and the results were compared.

Which of the following techniques is the most appropriately applied for this analysis?

A. Allele-specific oligonucleotide probes
B. DNA sequencing
C. Northern blot
D. Southern blot
E. Western blot

Answers

[6.1] **C.** The temperature at which the DNA strands separate is dependent on the number of hydrogen bonds that make up the base pairing. Since G-C pairs have three hydrogen bonds while A-T pairs only have two, then the strand that has the greatest number of G-C pairs will have the higher melting temperature, Tm, which is the temperature at which half of the base pairs are broken. Interestingly, the TATA box, the starting point of transcription in eukaryotes, has weaker bonding.

[6.2] **B.** Most of the sequences recognized by restriction endonucleases are palindromes; that is, they have the same nucleotide sequence on both strands when read in the 5′ to 3′ direction. Since choice B has the only sequence that will be a palindrome when paired with its complementary strand, it is the most likely to be a site recognized by a restriction endonuclease (it is the site recognized by the restriction endonuclease *Pst*1).

[6.3] **D.** A VNTR analysis examines the hypervariable regions of the human genome. These contain sequences that are repeated in tandem a variable number of times and the length is unique for each individual. Because it is DNA fragments that are being analyzed, the Southern blot is the most appropriate technique to use to separate and detect these regions. DNA sequencing is too time-consuming to be practical for forensic analyses. The Northern blot is used to separate and detect RNA, whereas the Western blot is used for proteins. Allele-specific oligonucleotide probes are used when testing for the presence of a genetic mutation that either introduces or removes a restriction site.

BIOCHEMISTRY PEARLS

- Polymerase chain reaction (PCR) is an in vitro method for the exponential amplification of a specific DNA region.
- PCR involves (1) denaturation, which separates the two DNA chains by heating the reaction mixture to 90°C to 95°C (194°F to 203°F); (2) annealing or cooling to 45°C to 60°C (113°F to 140°F) so that the oligonucleotide primers can bind or anneal to the separated strands of the target DNA; and (3) elongation with DNA polymerase (at 72°C [162°F] when *Taq* DNA polymerase used) adding nucleotides to the 3′ ends of the primers to a complete copy of the target DNA template.
- GC pairs have three hydrogen bonds, whereas AT pairs have only two hydrogen bonds and are therefore weaker.
- A heat-stable DNA polymerase obtained from the thermophilic bacterium *T. aquaticus* is used to synthesize the new strands in most PCR reactions.
- Endonucleases usually act at sites of palindromes.

REFERENCES

Granner DK, Weil PA. Molecular genetics, recombinant DNA, & genomic technology. In: Murray RK, Granner DK, Mayes PA, et al., eds. Harper's Illustrated Biochemistry, 26th ed. New York: Lange Medical Books/McGraw-Hill, 2003.

Johnson G, Nelson S, Petric M, et al. Comprehensive PCR-based assay for detection and species identification of human herpesviruses. J Clin Microbiol 2000; 38(9):3274–9.

CASE 7

A 39-year-old female presents to the clinic for a routine health maintenance exam. The patient reports she is feeling nervous and anxious all the time with frequent palpitations. On further questioning she reports having diarrhea and has been losing weight. She has also noticed a change in hair and fingernail growth and frequently feels warm while others are cold or comfortable. She denies any history of depression or anxiety disorder and is not taking any medications. On examination, her heart rate is 110 beats per minute. She has a slight tremor and has increased reflexes in all extremities. A nontender thyroid enlargement is appreciated in the thyroid region. Her TSH level is low at 0.1 mIU/mL. The patient is told that she has an autoimmune antibody process.

- **What is the most likely diagnosis?**
- **What is the biochemical mechanism for this disorder?**

ANSWERS TO CASE 7: HYPERTHYROIDISM/STEROID MESSENGER REGULATION OF TRANSLATION

Summary: A 39-year-old female has symptoms of nervousness, weight loss, gastrointestinal and skin alterations, heart palpitations, heat intolerance, and physical signs of hyperreflexia and a goiter.

- **Likely diagnosis:** Hyperthyroidism, likely Graves disease.
- **Biochemical mechanism:** The most frequent cause of hyperthyroidism, Graves disease, is an autoimmune process in which thyroid hypersecretion is caused by circulating immunoglobulins that bind to the TSH receptor on the thyroid follicular cells and stimulate thyroid hormone production. The diagnosis is confirmed by increased thyroid stimulating IgG antibodies and is frequently seen in other family members.

CLINICAL CORRELATION

This 39-year-old woman has symptoms of hyperthyroidism, thyroid excess. This causes a tachycardia, tremor, nervousness, thin skin, weight loss through the hypermetabolic state, and hyperreflexia. If unchecked, the high levels of thyroid hormone can sometimes even cause adrenergic crisis (so called thyroid storm), which has a high rate of mortality. Normally, the thyroid hormone (thyroxine) is under tight control. The pituitary release of thyroid stimulating hormone is stimulated by insufficient thyroxine, and suppressed by excess thyroxine. In Graves disease, the most common cause of hyperthyroidism in the United States, an autoimmune immunoglobulin is produced that stimulates the TSH receptor of the pituitary. This is confirmed by either assaying for the thyroid-stimulating immunoglobulin, or a radionuclide scan revealing diffuse increased uptake throughout the thyroid gland. Treatment acutely includes β-adrenergic antagonists and agents that inhibit the catabolism of thyroid hormone such as propylthiouracil (PTU).

APPROACH TO STEROID MESSENGER REGULATION OF TRANSLATION

Objectives

1. Understand the general mechanisms of hormone action.
2. Know about some of the specific mechanisms by which hormones activate receptors.
   a. Know about hormones that bind to cell membrane receptors.
   b. Know about hormones that act through cyclic nucleotides.
   c. Know about hormones that act through calcium and the PIP_2 system.
   d. Know about hormones that bind intracellular receptors and activate genes.

3. Be familiar with how immunoglobulins of Graves disease cause hyperthyroidism.
4. Be aware of the mechanism of action of PTU and methimazole.
5. Understand the regulation of hormone levels.

Definitions

Effector: The protein in a signal transduction pathway (e.g., a hormonal response) that produces the cellular response.

G protein: A guanosine triphosphate (GTP)–binding protein that serves as a transducer in a signal transduction pathway. On binding GTP and releasing guanosine diphosphate (GDP), a G protein is able to activate the effector enzyme (e.g., adenylate cyclase).

Graves disease: An autoimmune disease in which the B lymphocytes synthesize an immunoglobulin (thyroid-stimulating immunoglobulin, TSIg) that binds to and activates the TSH receptor in such a way that the thyroid hormones do not feedback inhibit the receptor–effector interaction, leading to a hyperthyroid condition.

Hormone: A chemical signal that is produced in one set of cells and directs the activity in another set of cells that can be endocrine, paracrine, or autocrine.

Receptor: A protein that will perceive the signal of a hormone or other chemical signal (e.g., a neurotransmitter) by binding it and transmitting that signal further down the signal transduction pathway.

Second messenger: A molecule that is synthesized within a cell in response to a receptor binding a hormone.

Transducer protein: A protein (such as the α-subunit of a G protein) that transmits the signal from a hormone-bound receptor to the effector protein.

DISCUSSION

Cellular biochemistry is a complex system of reactions and processes that must be efficiently regulated and integrated with processes underway in other cells and tissues. One method by which regulation is achieved is through the interaction of **hormones and their associated receptors** located either within or on the surface of the cell. A hormone is any substance in an organism that carries a signal to change metabolic processes within a cell.

The **hormonal-signaling process** is summarized in Figure 7-1. Hormones are released from secretory tissues in response to metabolic signals as well as electrical or chemical signals from the nervous system. The **released hormone binds to a receptor,** which can either be on the **cell surface** or, as in the case of **steroid and similar hormones, within the cell.** The hormone–receptor complex starts a series of events in which the signal is converted to other chemical forms that bring about **changes in the biochemical reactions within the cell.**

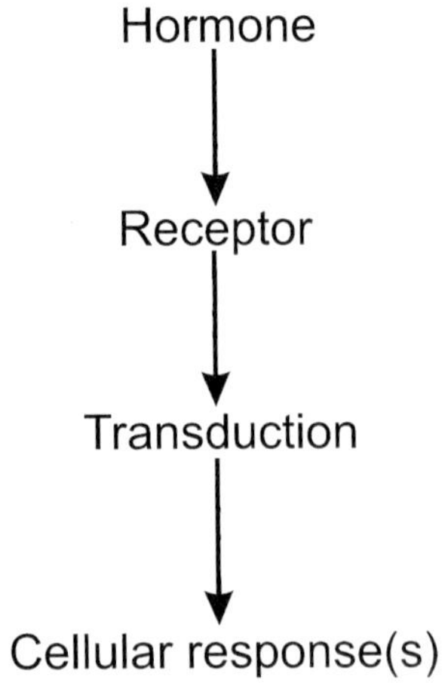

Figure 7-1. The hormonal-signaling process.

Hormones can be grouped into three main classifications based on their chemical structure and how they are synthesized. **Peptide hormones** include the **polypeptide hormones insulin and glucagon,** as well as the smaller peptides such as thyrotropin-releasing hormone (TRH) and the enkephalins. Some peptide hormones, like **thyrotropin** (TSH) are **glycoproteins. Amino acid-derived hormones** are synthesized from amino acid precursors. The catecholamines and serotonin are included in this grouping, as are the thyroid hormones, T_3 and T_4. The third classification is the steroid and steroid-like hormones, which includes the **progestogens, corticosteroids, androgens, estrogens, and calcitriol (the active form of vitamin D).**

The **cell-associated receptors** are the molecular entities primarily responsible for **recognizing the hormonal signal.** When the hormone binds to the receptor, the hormone–receptor complex initiates the events that culminate in the cellular response. These cell-associated hormone receptors can be classified into two main categories, nuclear receptors and cell-surface receptors. **Nuclear receptors** are intracellular **proteins present in either the cytosol or the nucleus** that bind hormones which cross the cell membrane by simple diffusion. When **nuclear receptors** bind their cognate hormone, they undergo a **conformational change that enables them to bind to DNA at specific sites** called ***cis*-response elements.** When the hormone-nuclear receptor binds specifically to these response elements, they **influence transcription of the DNA genome into messenger RNA, either by activating or repressing it.**

Cell-surface receptors, as their name implies, **bind hormones on the extracellular side of the cell membrane.** When the hormone binds to its receptor on the cell surface, the **hormone-bound receptor either activates or forms a complex with a transducer protein in the membrane** that will cause the activation of some **enzymatic activity** in the cytoplasm of the cell. The enzymatic activity, sometimes called the **effector,** catalyzes the production of a second messenger that **mediates the intracellular response.**

Most transducer proteins that interact with cell-surface receptors are **GTP-binding proteins, commonly called G proteins.** G proteins typically are composed **of three subunits: α, β, and γ.** When the receptor is in the unbound state, all three subunits are bound together to form a heterotrimer that is in close association with the membrane-bound receptor. In this state, GDP is bound to the α-subunit (Figure 7-2). Binding of the hormone by the receptor causes a conformational change such that the α-subunit is able to exchange the bound GDP for a molecule of GTP. The GTP-bound α-subunit separates from the βγ-dimer and is able to interact with the effector enzyme in either a stimulatory or inhibitory manner, depending on the nature of the α-subunit. Some

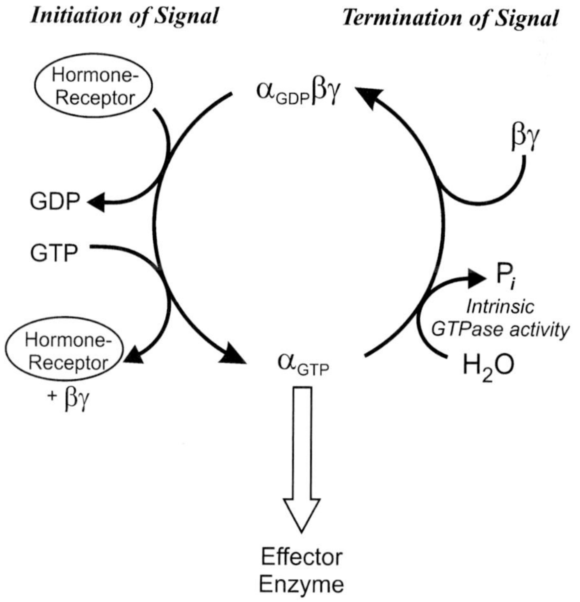

Figure 7-2. Transduction by heterotrimeric G proteins.

different G protein families are presented in Table 7-1. **Common effector enzymes are adenylate cyclase, which converts ATP into 3′,5′-cyclic AMP (cAMP) and phospholipase C (PLC),** which hydrolyzes the membrane lipid phosphatidylinositol-4,5-bisphosphate (PIP_2) to diacylglycerol and inositol-1,4,5-triphosphate (IP_3). While GTP is bound to the α-subunit, the interaction with the effector enzyme continues. However, the α-subunit also contains an intrinsic **GTPase activity** that will hydrolyze GTP to GDP and inorganic phosphate (P_i). This provides a mechanism by which the hormonal signal can be switched off, because GTP hydrolysis to GDP causes the α-subunit to release the effector enzyme and reassociate with the βγ subunits.

The hormonal cascade known as the **hypothalamic-pituitary-thyroid axis** employs **G-protein transducers that activate both adenylate cyclase and phospholipase C.** After receiving an electrochemical signal from the central nervous system (CNS), thyrotropin-releasing hormone (TRH) is released from the arcuate nucleus and median eminence of the **hypothalamus.** TRH is carried to the anterior pituitary via long portal vessels. In the anterior pituitary, **TRH binds to G protein–coupled TRH receptors on the cell surface of thyrotrophs.** This activates phospholipase C, releasing both DAG and IP_3. **DAG activates protein kinase C (PKC), which phosphorylates target proteins.** The release of IP_3 opens Ca^{2+} channels in the endoplasmic reticulum, thus releasing Ca^{2+} from storage and increasing cytoplasmic [Ca^{2+}]. Both of these events result in the increase in synthesis and release of thyrotropin, also known as thyroid-stimulating hormone or TSH, which is stored in secretory granules within the thyrotroph. Circulating TSH binds to TSH receptors on the basolateral membranes of the thyroid follicular cell. The **TSH receptor is a G protein–coupled receptor that activates adenylate cyclase** to produce **cAMP** when hormone is bound. The **increase in cAMP triggers** a series of events

Table 7-1

G-PROTEIN FAMILIES AND THEIR FUNCTIONS

G_α CLASS	INITIATING SIGNAL	DOWNSTREAM SIGNAL
G_s	β-Adrenergic amines, glucagon, parathyroid hormone, thyrotropin (TSH), corticotrophin, many others	Stimulates adenylate cyclase
G_i	Acetylcholine, α-adrenergic amines, many neurotransmitters, chemokines	Inhibits adenylate cyclase
G_q	Acetylcholine, α-adrenergic amines, TRH, many neurotransmitters	Increases IP_3 and intracellular calcium ion
G_t (Transducin)	Photons	Stimulates cGMP phosphodiesterase

within the follicular cell that result in the **secretion of the thyroid hormones thyroxine (T_4; 3,5,3′,5′-tetraiodo-L-thyronine) and 3,5,3′α-triiodo-L-thyronine (T_3).** These events include increases in

1. The transport of iodide ions across the basolateral membrane into the thyroid follicular cell via the Na/I cotransporter (NIS)
2. Iodination of tyrosine residues on colloidal thyroglobulin in the lumen of the follicle
3. Conjugation of iodinated tyrosines to form T_3 and T_4 on thyroglobulin
4. The endocytosis of thyroglobulin into the follicular cell from the lumen
5. The hydrolysis of thyroglobulin in the lysosome to release T_3 and T_4
6. The secretion of T_3 and T_4 into the bloodstream

Although the **thyroid hormones** are derived from amino acids, they **act in a fashion similar to that of the steroid hormones.** In the circulation, almost all of the thyroid hormones (99.98 percent of T_4 and 99.5 percent of T_3) are tightly bound to the proteins thyroxine binding globulin (TBG), albumin, and transthyretin. However, the free, or "unbound," T_3 and T_4 **are responsible for the biological effects** of the thyroid hormones. The **thyroid hormones enter the cell either by simple or carrier-mediated diffusion across the cell membrane.** Within the cytoplasm, approximately 50 percent of T_4 is deiodinated to form T_3. The **hormones enter the nucleus where they bind to the thyroid hormone receptor (THR).** The affinity of the THR is approximately 10 times greater for T_3 than it is for T_4, one reason why T_3 is more biologically active than T_4. When T_3 or T_4 bind to THR, the complex is able to bind to the thyroid response element of specific genes. The net effect of increased thyroid hormones is an increase in basal metabolic rate. This is accomplished by increasing the expression of genes that code for enzymes in both the **catabolic and anabolic pathways of fats, carbohydrates and proteins, as well as the expression of the Na-K pump.** The careful regulation of thyroid hormone secretion is therefore important in the overall regulation of metabolism. Thus, T_3 and T_4 exert a negative feedback inhibition on the synthesis of TRH by the hypothalamus and the secretion of TSH by the anterior pituitary, so that the levels of T_3 and T_4 are tightly controlled.

Graves disease is an autoimmune disease in which the B lymphocytes synthesize an immunoglobulin that binds to and activates the TSH receptor on the thyroid follicular cell membrane. This **thyroid-stimulating immunoglobulin (TSIg)** binds to and activates the TSH receptor in such a way that T_3 and T_4 **do not feedback inhibit** the receptor–effector interaction. Thus, all of the processes enumerated above that are sequelae of the TSH receptor binding hormone are increased resulting in a hyperthyroid condition.

Nonsurgical treatments for Graves disease include administration of **thionamide antithyroid drugs,** which include propylthiouracil (PTU) and methimazole, and/or treatment with radioactive iodine. The **thionamides act by inhibiting the enzyme thyroid peroxidase,** an enzyme on the apical membrane of the thyroid follicular cell that faces the lumen. This enzyme **catalyzes**

the oxidation of iodide ion (I^-) to atomic iodine (I^0), which is required for the iodination of tyrosyl residues on thyroglobulin; these drugs decrease the amounts of T_3 and T_4 synthesized. They also decrease the amount of TSIg that is synthesized by mechanisms that are not understood. Radioactive iodine treatments cause the progressive destruction of thyroid cells, thus reducing thyroid function over a period of several weeks to months.

COMPREHENSION QUESTIONS

[7.1] The thyroid hormones T_3 and T_4 are synthesized in the follicular cells of the thyroid gland. From which of the following essential amino acids are the thyroid hormones synthesized?

A. Isoleucine
B. Lysine
C. Methionine
D. Phenylalanine
E. Valine

[7.2] The thyroid hormones T_3 and T_4 bind to the thyroid hormone receptor (THR) in the target cells. Which of the following mechanisms best describes the role of the THR?

A. It activates adenylyl cyclase to produce cAMP.
B. It activates the phosphoinositide cascade.
C. It is a soluble guanylyl cyclase.
D. It is a tyrosine kinase.
E. It is a transcription factor.

[7.3] A 26-year-old male presents complaining of heat intolerance, heavy sweating, tremulousness, and feeling "jittery inside." Physical examination reveals reddened conjunctiva and warm and moist palms, but the thyroid gland was not visibly enlarged. Which of the following tests would be most helpful to obtain an accurate diagnosis?

A. Electrocardiogram
B. Free thyroxine level
C. Serum cortisol levels
D. Serum electrolytes
E. Serum glucose level

Answers

[7.1] **D.** Thyroid hormones are synthesized from tyrosyl residues on thyroglobulin in the colloidal space of the thyroid follicular cells. Although tyrosine can be obtained from the diet, it can also be synthesized from the essential amino acid phenylalanine by the action of phenylalanine hydroxylase.

[7.2] **E.** Thyroid hormones are similar to steroid hormones in that they bind to receptors in the nucleus of the cell. On binding to the receptor, the receptor–hormone complex binds to DNA affecting the transcription of messenger RNA.

[7.3] **B.** This patient appears to have a hyperthyroid condition even though the thyroid does not appear to be enlarged. Thyroid function tests would be most helpful to determine if this is the case. The free thyroxine level is a direct measure of the amount of free T_4, the biologically active T_4, in the serum. Elevation of the free T_4 indicates a hyperthyroid condition.

BIOCHEMISTRY PEARLS

- Hormones bind to receptors, which can either be on the cell surface or, as in the case of steroid and similar hormones, within the cell.
- The hormone–receptor complex starts a series of events in which the signal is converted to other chemical forms that bring about changes in the biochemical reactions within the cell.
- Hormones that bind to cell-surface receptors activate or form a complex with a transducer protein in the membrane that will cause the activation of some enzymatic activity in the cytoplasm of the cell, producing a second messenger.
- Nuclear receptors are intracellular proteins present in either the cytosol or the nucleus that bind hormones (steroids) which cross the cell membrane by simple diffusion. The nuclear receptors undergo conformational changes that enable them to bind to DNA at specific sites.

REFERENCES

Barrett EJ. The thyroid gland. In: Boron WF, Boulpaep EL, eds. Medical Physiology: A Cellular and Molecular Approach. Philadelphia, PA: W.B. Saunders, 2003.

Farfel Z, Bourne HR, Iiri T. Mechanisms of disease: the expanding spectrum of G protein diseases. N Engl J Med 1999;340:1012–20.

Litwack G, Schmidt TJ. Biochemistry of hormones I: polypeptide hormones. In: Devlin TM, ed. Textbook of Biochemistry with Clinical Correlations, 5th ed. New York: Wiley-Liss, 2002.

❖ CASE 8

A 3-year-old Caucasian boy is brought to the clinic for a chronic productive cough not responding to antibiotics given recently. He has no fever or sick contacts. His medical history is significant for abdominal distention, failure to pass stool, and emesis as an infant. He continues to have bulky, foul-smelling stools. No diarrhea is present. He has several relatives with chronic lung and "stomach" problems, and some have even died at a young age. The examination reveals an ill appearing, slender male in moderate distress. The lung exam reveals poor air movement in the base of lungs bilateral and coarse rhonchi throughout both lung fields. A chloride sweat test was performed and was positive, indicating cystic fibrosis (CF).

◆ **What is the mechanism of the disease?**

◆ **How might gel electrophoresis assist in making the diagnosis?**

ANSWERS TO CASE 8: CYSTIC FIBROSIS

Summary: A 3-year-old Caucasian male has a history of chronic pulmonary and gastrointestinal problems and has a positive chloride sweat test. A family history of similar symptoms is also present.

- **Likely diagnosis:** Cystic fibrosis, which is an autosomal recessive disorder resulting in defective chloride ion channels of exocrine glands and epithelial tissues of the pancreas, sweat glands, and mucous glands in the respiratory, digestive, and reproductive tracts.
- **Gel electrophoresis:** Separates deoxyribonucleic acid (DNA) chains of varying length and can allow identification of a specific gene sequences.

CLINICAL CORRELATION

Cystic fibrosis is an inherited condition affecting approximately 1 in 2500 white individuals. Affected patients usually have abnormal mucus secretion and eccrine sweat glands leading to respiratory infections, gastrointestinal obstruction, pancreatic enzyme dysfunction leading to malabsorption of nutrients, and excessive electrolyte secretion. The protein cystic fibrosis transmembrane conductance regulator (CFTR) is defective, leading to abnormal chloride transport. Approximately 70 percent of mutations are accounted for by deletion of three specific base pairs at the F 508 position of the CFTR. Oligonucleotide probes can be used to assay for this mutation, but other tests are required for the less common mutations.

APPROACH TO GEL ELECTROPHORESIS AND CLONING

Objectives

1. Understand the process of gel electrophoresis.
2. Know the difference between various types of blots.
3. Be familiar with DNA sequencing using the Sanger dideoxynucleotide method.
4. Understand the process and uses of cloning DNA.

Definitions

Blotting: The transfer of proteins or nucleic acids from an electrophoresis gel to a membrane support (such as nitrocellulose or nylon). The membrane blot is then incubated with probes that bind the molecules of interest.

Complementary DNA (cDNA): A sequence of DNA copied from messenger ribonucleic acid (mRNA), which does not contain introns that were present in native DNA.

CFTR: Cystic fibrosis transmembrane conductance regulator; a ligand-gated chloride channel that is regulated by phosphorylation. It is a member of the adenine nucleotide binding cassette (ABC) family of transport proteins.

Electrophoresis: The technique by which charged molecules in solution are separated in an electric field according to their different mobilities in the supporting medium.

Gel electrophoresis: Electrophoresis using a gel as the supporting medium. Common gels are polyacrylamide and agarose.

Northern blot: The technique by which molecules of ribonucleic acid **(RNA)** are separated by gel electrophoresis, transferred to a membrane support, and incubated with labeled oligonucleotide probes. Specificity is obtained by using oligonucleotide probes that have sequences complementary to the target RNA.

Southern blot: The technique by which molecules of **DNA** are separated by gel electrophoresis, transferred to a membrane support, and incubated with labeled oligonucleotide probes. Specificity is obtained by using oligonucleotide probes that have sequences complementary to the target DNA.

Western blot: The technique by which protein molecules are separated by gel electrophoresis, transferred to a membrane support, and incubated with labeled antibodies. Specificity is obtained using antibodies that will bind the protein molecule of interest.

DISCUSSION

Cystic fibrosis is the most common lethal autosomal recessive disease affecting the Caucasian population. It has a frequency of approximately 1 in 2500 and a carrier frequency of approximately 1 in 25. The protein affected is the cystic fibrosis transmembrane conductance regulator (CFTR), which is a chloride ion channel. There are over 1000 mutations that have been discovered in the CFTR gene and over 80 percent of these mutations lead to disease. The mutations lead to (1) defective or decreased protein production, (2) defective processing of the protein, (3) protein that is defective in the regulation of the chloride channel, or (4) defect in the transport of chloride ions. The most common mutation, a deletion of a phenylalanine residue at amino acid position 508 (ΔF_{508}), results in misfolding of the protein; it consequently does not traffic to the membrane.

Defects in the CFTR decrease the ability of cells to transport Cl^- in a number of tissues, particularly the pancreas, airway epithelia, and sweat glands. When Cl^- transport is defective in the pancreas, it leads to decreased HCO_3^- secretion and decreased hydration that leads to thick secretions that block the pancreatic ducts and destruction of the organ. In the lungs, the decreased absorption of Cl^- ions is thought to increase the absorption of the airway surface liquid thus increasing the viscosity of mucous, decreasing mucociliary clearance and increasing the incidence of airway infections. In sweat glands, defects in CFTR prevent the reabsorption of Cl^- in the sweat gland duct, thus increasing the concentration of NaCl in sweat. The presence of salty sweat in children with CF gives rise to the Northern Europe adage: "Woe to that child which when kissed on the forehead tastes salty. He is bewitched and soon must die."

Gel electrophoresis is a method routinely used in biochemistry and molecular biology to **separate, identify, and purify peptides, proteins, and DNA fragments based on their size. Electrophoresis** is a method by which **charged molecules in a solution are separated in an electric field because of their different mobilities.** The **mobility of an ion** in the electric field is dependent on the **charge of the ionized molecule, the voltage gradient of the electric field, and the frictional resistance** of the supporting **medium.** The supporting medium of choice for most protein and peptide analyses, as well as small DNA fragments, is **polyacrylamide gel.** For large DNA fragments, **agarose gel** gives the best results.

In general, **DNA molecules have a high negative charge-to-mass ratio** owing to their **sugar-phosphate backbone.** Therefore, when an electric field is applied, **DNA** will **migrate** through the agarose (or polyacrylamide) gel **toward the positive pole** in the electrophoresis apparatus. Because of the high charge-to-mass ratio, the relative electrophoretic mobility of DNA depends primarily on the size of the DNA molecule and porosity of the gel matrix, which can be varied by the concentration of agarose or polyacrylamide in the gel. **Small DNA strands will move fast, whereas large fragments will lag behind smaller fragments** as the DNA migrates through the gel. To estimate the size of DNA in the samples, molecular weight markers (a mixture of DNA fragments of known molecular size) are applied to one of the lanes in the gel. Gel electrophoresis separates DNA molecules based on their frictional coefficient, which is dependent on the length and conformation of the fragment. For most linear double-stranded DNA molecules, separation will be according to their molecular weight. However, **circular DNA molecules** (e.g., from bacteria or plasmids) can adopt a super **coiled structure and therefore will pass through the gel with less resistance than the rod shape of linear DNA.**

Gel electrophoresis can be applied as a method for detection of genetic diseases such as Cystic fibrosis (CF). Larger deletions in the **cystic fibrosis transmembrane conductance regulator (CFTR) gene** will cause it to move differently than the full-length wild type gene. In most cases, the DNA sample of the patient is amplified by a polymerase chain reaction (PCR) using specific primers. After the electrophoresis of the PCR samples the bands are **visualized under ultraviolet (UV) light using ethidium bromide staining.**

Bands of molecules separated by electrophoresis can be detected in the gel by various staining and destaining methods, such as the above mentioned ethidium bromide. However, more specific information can be obtained by application of **blotting** techniques. There are three different blotting techniques based on whether DNA, RNA, or protein is being analyzed. However, **all blotting techniques have major steps in common.** These include (1) the

isolation of the biomolecules of interest (DNA, RNA, or protein); (2) **separation of the mixture using gel electrophoresis;** and (3) subsequent transfer **of the sample molecules from the gel to a nitrocellulose or nylon membrane.** The final identification step involves **incubation of the membrane blot with probes** that bind specifically to the molecule of interest. DNA is analyzed using **Southern blotting,** named after E. M. Southern, who first described this procedure in 1975. Based on this initial terminology, **Northern** and **Western** blotting were given as names referring to RNA and protein-blotting transfer, respectively. The major difference between the procedures involves the probes used for detection and identification of the molecules of interest. **Southern and Northern blots take advantage of the ability of complementary nucleic acid strands to hybridize.**

In a Southern blot, after the DNA fragments have been transferred from the electrophoretic gel to the membrane, they are allowed to interact with a single-stranded DNA oligonucleotide that contains the sequence of nucleic acids in the gene of interest. The **oligonucleotide probe is usually radioactively labeled,** which allows the DNA band to be identified following exposure to x-ray film. **Northern blots are used to monitor the expression of a gene of interest by determining the amount of mRNA present.** Then the mRNA is isolated, separated by electrophoresis, and, after transfer to a membrane, is allowed to interact with a **radioactive DNA polynucleotide** that has a sequence of the gene of interest. **Visualization is by exposure to x-ray film.** In the case of a **Western blot** of a protein mixture, after the proteins have been transferred to the membrane, the membrane is **probed with an antibody that binds specifically to the target protein/antigen.** The antibody may be radioactive or coupled to a fluorescent chromophore to allow easy detection.

The ability to sequence DNA and analyze genes at the nucleotide level has become a powerful tool in many areas of analysis and research. The **Sanger dideoxynucleotide method of DNA sequencing** was first developed in 1975 by Frederick Sanger, after whom the technique was named. The strategy used in this method is to create **four sets of labeled fragments, corresponding to the four deoxyribonucleotides.** It involves the enzymatic duplication of a DNA strand that is complementary to the strand of interest. In the Sanger method, **2′,3′-dideoxyribonucleotide triphosphates (ddNTPs) are used** in addition to deoxyribonucleotide triphosphates (dNTPs). Since the **ddNTPs lack the 3′-OH group, they will terminate DNA chain elongation** because they cannot form a phosphodiester bond with the next dNTP. Each of the four sequencing reaction tubes (A, C, G, and T) would then contain a single-stranded DNA template; a primer sequence; DNA polymerase to initiate synthesis where primer is hybridized to the template; mixture of the four deoxyribonucleotide triphosphates (dATP, dTTP, dCTP, and dGTP) to extend the DNA strand; one labeled dNTP (using a radioactive element or dye); and one ddNTP, which terminates the growing chain wherever it is incorporated. **Tube A would have ddATP, tube C ddCTP, and so forth.** Since the concentration of the ddNTP is low (approximately 1 percent of dNTP), the **chain will terminate**

randomly at various positions throughout the chain, thus yielding an array of different length DNA fragments each terminated with the particular dideoxyribonucleotide. The **fragments of varying length are then separated by electrophoresis in parallel lanes** (each corresponding to the four deoxyribonucleotides) and the positions of the fragments analyzed to determine the sequence. The fragments are separated on the basis of size, with the shorter fragments moving faster and appearing at the bottom of the gel. The **sequence is then read from bottom to the top, yielding the 5′ to 3′ sequence.**

The ability to rapidly sequence DNA has become an important tool for molecular biology. For example, the polymerase chain reaction (PCR) method needs the information of the sequence flanking the region of interest to be able to design **specific oligonucleotides (primers) to amplify the specific DNA region.** Another important use is identifying restriction sites in plasmids, which is useful in cloning a foreign gene into the plasmid. Before DNA sequencing, molecular biologists had to sequence proteins directly, which was a challenging and laborious process. Now amino acid sequences can be determined more easily by **sequencing a piece of cDNA and finding an open reading frame.** In eukaryotic gene expression, sequencing made it possible to identify conserved sequence motifs and determine important sites in the promoter regions. Furthermore, sequencing can be used to identify the site of a point mutation or other changes in the genome.

Recombinant DNA technology allows for a **transfer of a DNA fragment of interest into a self-replicating genetic element such as a bacterial plasmid or virus.** This technology has allowed many human genes to be cloned in *Escherichia coli,* yeast, or even mammalian cells, leading to easy production of human recombinant proteins in vitro. These include insulin for diabetics, factor VIII for males suffering from hemophilia A, human growth hormone (GH), erythropoietin (EPO) for treating anemia, three types of interferons, several interleukins, adenosine deaminase (ADA) for treating some forms of severe combined immunodeficiency (SCID), angiostatin and endostatin for trials as anticancer drugs, and parathyroid hormone with many more in development.

Efforts toward gene therapy make use of recombinant DNA technology. In the case of CF, which is the result of a defect in a single gene, direct insertion of the normal gene should theoretically restore the function of CFTR in the treated cells of the CF patient. Since **respiratory failure is the major cause of deaths (95 percent) in CF,** lung cells have become the primary target for efforts at treating CF with gene therapy. Another advantage is the ability of delivering vectors containing the functional CFTR gene directly into the patient's airways by aerosol, without causing trauma to any other part of the body. To date, **CF trials rely on adenovirus, adeno-associated virus (AAV), and cationic liposomes to mediate gene transfer** to nasal epithelial cells and respiratory epithelial cells. Adenoviruses do not integrate into the host chromosome; therefore, the vectors derived from these viruses have the advantage of negligible oncogenic potential. The disadvantages are the development of inflammation and the transient expression of the recombinant genes. Adeno-associated viruses are unique among animal viruses, because they require coinfection with an unrelated helper virus (either adenovirus or herpesvirus) for productive infection in cell culture. However, AAV has the ability to integrate into the host genome in the absence of a helper virus, but because it integrates at a specific location, the risk of virus-induced mutagenesis and oncogenesis is reduced. **Liposomes,** however, are synthetic lipid bilayers forming spherical vesicles, with size ranging from 25 nm to 1 mm in diameter depending on how they are made. Liposomes can transfect a variety of cell types, are biodegradable and hypoimmunogenic, and can be manufactured to a drug standard. Theoretically, there is no restriction on the size of the DNA to be delivered by liposome. The disadvantages are short-term expression of the transferred gene, no specific targeting ability, and a low transfection rate in vivo.

COMPREHENSION QUESTIONS

[8.1] A 30-month-old female child whose growth rate has been in the lower 10th percentile over the last year presents with chronic, nonproductive cough and diarrhea with foul-smelling stools. She is diagnosed as having cystic fibrosis. For which of the following vitamins is this child most likely to be at risk of deficiency?

A. Ascorbic acid (vitamin C)
B. Biotin
C. Folic acid
D. Retinol (vitamin A)
E. Riboflavin (vitamin B_2)

[8.2] Some forms of genetic diseases such as cystic fibrosis and sickle cell anemia can be diagnosed by detecting restriction fragment length polymorphisms (RFLPs). Which of the following is most likely to be used in an RFLP analysis?

A. Dideoxynucleotides
B. Mass spectrometry
C. Northern blot
D. Southern blot
E. Western blot

[8.3] Your patient has been diagnosed with cystic fibrosis and has been determined to have the most common mutation, the ΔF_{508} gene. Which of the following is the most cost- and time-effective method for testing family members to see who are carriers of the mutation?

A. Allele-specific oligonucleotide probe analysis
B. DNA fingerprinting analysis
C. DNA sequencing
D. Restriction length fragment polymorphism analysis

Answers

[8.1] **D.** Because cystic fibrosis leads to pancreatic damage and diminution of the ability to secrete HCO_3^- and pancreatic digestive enzymes with the result that fat and protein are absorbed poorly. Retinol is a fat-soluble vitamin that must be absorbed along with lipid micelles; other fat-soluble vitamins are E, D, and K. The other vitamins listed are water-soluble and their absorption is not significantly affected.

[8.2] **D.** Restriction length fragment polymorphism analysis detects mutations in the DNA that either introduce or eliminate a recognition site for a restriction enzyme. This is detected by performing a Southern blot analysis on patient DNA after incubation with a specific restriction enzyme and comparing it to one performed on DNA obtained from a normal gene.

[8.3] **A.** If the mutation is known, radioactive or fluorescent probes can be produced for alleles that contain the mutation and for those that have a normal DNA sequence. The samples of DNA are spotted onto nitrocellulose paper in narrow bands. The paper is then incubated with the probe either for the normal or the mutant sequence. Since it does not involve an electrophoresis step, this type of analysis is much less time-consuming (and less expensive).

BIOCHEMISTRY PEARLS

❖ Northern blot uses oligonucleotide probes to identify RNA, Southern blot identifies DNA, and Western blot identifies protein molecules.

❖ Recombinant DNA technology allows for a transfer of a DNA fragment of interest into a self-replicating genetic element such as a bacterial plasmid or virus.

❖ Restriction length fragment polymorphism analysis detects mutations in the DNA that either introduce or eliminate a recognition site for a restriction enzyme.

❖ Oligonucleotide probes can be directed against precise DNA sequences and yields a quick and accurate result.

REFERENCES

Schwiebert LM. Cystic fibrosis, gene therapy, and lung inflammation: for better or worse? Am J Physiol Lung Cell Mol Physiol 2004;286: L715–L716.

Welsh MJ, Ramsey BW, Accurso F, et al. Cystic fibrosis. In: Scriver CR, Beaudet AL, Sly WS, et al., eds. The Metabolic & Molecular Bases of Inherited Disease, 8th ed. New York: McGraw-Hill, 2001.

❖ CASE 9

A 40-year-old male returned from a deer-hunting trip approximately 6 weeks ago, and presents to clinic with multiple complaints. He states that recently he has had worsening joint pain and "arthritis" in multiple joints that seems to move to different spots. Patient also complains of some numbness in his feet bilaterally. The patient denies any medical problems and he had a normal annual physical prior to hunting trip. On further questioning, he remembered having a rash on his body and the lesions were circular and appeared to be resolving in the center. He noted that he felt really bad once he got home with muscle ache (myalgias), joint ache (arthralgias), stiff neck, and severe headache. He also remembered that many of his hunting friends had experienced flea and tick bites and is quite sure he was bitten as well. The physical exam is essentially normal except some joint tenderness of left knee and right shoulder. After making your diagnosis you gave him a prescription for erythromycin.

- **What is the most likely diagnosis?**

- **What is the biochemical mechanism of action of erythromycin?**

ANSWERS TO CASE 9: ERYTHROMYCIN AND LYME DISEASE

Summary: A 40-year-old male who presents with migrating arthralgias and neurologic changes which were preceded by a centrally clearing circular rash, myalgias, and headache after a hunting trip where he was exposed to fleas and ticks.

- **Diagnosis:** Lyme disease
- **Biochemical site of action of erythromycin:** Inhibits bacterial protein biosynthesis at the translocation step of translation

CLINICAL CORRELATION

Lyme disease is a multisystem disease caused by the spirochete (spiral-shaped bacteria), *Borrelia burgdorferi,* nearly always transmitted by tick bite. Lyme disease is much more common in the New England area, usually during the late spring and early summer months. The deer tick is the main vector of transmission. The first stage is an acute infection with a red papule developing at the site of the bite, sometimes with lymphadenopathy and fever. The second stage includes disseminated infection and can lead to involvement of the heart, brain, joints, and skin. The typical target-shaped skin lesions may be seen. The third stage is a chronic infection and may last for years. Antibiotics are the treatment of choice, usually tetracyclines or penicillins.

APPROACH TO PROTEIN BIOSYNTHESIS

Objectives

1. Understand protein biosynthesis (initiation, elongation, translocation, termination).
2. Know key differences between prokaryotic and eukaryotic protein synthesis.
3. Know the role of ribosomal ribonucleic acid (rRNA).
4. Know the mechanism of action of different antibiotics with protein biosynthesis.

Definitions

A-site: The acceptor site on the ribosome into which an aminoacyl-charged transfer RNA (tRNA) is brought that has an anticodon that is complementary to the messenger RNA (mRNA) codon. The ribosome will then catalyze the formation of a peptide bond with the aminoacyl group on this tRNA and the growing peptide chain.

Anticodon: The three-base sequence on a tRNA that will base pair with the three-base sequence on the mRNA. The anticodon is specific for an amino acid; it translates the DNA sequence into an amino acid sequence in the protein produced.

Codon: The three-base sequence of an mRNA that code for a particular amino acid.

Elongation factors: Proteins required for bringing the aminoacyl-tRNA to the A-site, codon recognition, and translocation of the newly elongated peptidyl-tRNA from the A-site to the P-site.

Initiation factors: Proteins required for assembly of the ribosomal complex with mRNA and Met-tRNA so that protein synthesis can proceed.

Nucleolar organizing region (NOR): Area of the nucleolus where a great deal of rRNA transcription and synthesis occurs.

P-site: The peptidyl site on the ribosome to which Met-tRNA is brought to base pair with the mRNA sequence AUG. It is also the site to which the peptidyl RNA is moved in a process known as translocation following the formation of a new peptide bond.

Posttranscription modification of tRNA: The synthesis of tRNA involves modification of some uridine nucleotides to unusual nucleotides, such as pseudouridine, ribothymidine, and dihydrouridine.

Ribosomes: Complexes of proteins and rRNA on which protein synthesis occurs. There are two major subunits to all ribosomes, a larger subunit (50S for prokaryotes, 60S for eukaryotes) and a smaller subunit (30S for prokaryotes, 40S for eukaryotes).

DISCUSSION

The **synthesis of proteins** involves **converting the nucleotide sequence of specific regions of DNA into mRNA (*transcription*)** followed by the **formation of peptide bonds** in a complex set of reactions that occur on **ribosomes (*translation*).** The amino acids incorporated into the protein are first **activated by being attached to a family of tRNA molecules,** each of which recognizes, by **complementary base-pairing interactions,** particular sets of **three nucleotides (codons) in the mRNA.** Proteins are synthesized on ribosomes by linking amino acids together in the specific linear order stipulated by the sequence of codons in an mRNA. **Ribosomes are compact ribonucleoprotein particles found in the cytosol of all cells. All ribosomes are composed of a small and a large subunit.** The two subunits contain rRNAs of different lengths, as well as a different set of proteins. **All ribosomes consist of two major rRNA molecules** (23S and 16S rRNA in bacteria, 28S and 18S rRNA in eukaryotes) and **one or two small RNAs.** Small and large subunits are brought together by an mRNA molecule and protein synthesis starts immediately. After the protein is synthesized, the ribosomal subunits separate and are reused.

Protein synthesis is divided into three stages: initiation, elongation, and termination.

Chain initiation: **Initiation requires the smaller ribosomal subunit, an initiation tRNA (with a 5′-CAU-3′ anticodon), mRNA with its initiator codon (5′-AUG-3′),** and several initiation factors, all of which form the ribosomal initiation complex. The initiation complex is near completion when a **tRNA carrying the amino acid methionine hydrogen bonds to the AUG codon on the mRNA.** When these components are in place, the larger ribosomal subunit joins the complex in such a way that the initiator met-$tRNA_{met}$ is localized in the *P*- or peptidyl site (Figure 9-1). The chain initiation phase ends and a second amino acid can be inserted.

Chain elongation: **Elongation** occurs by bringing an **aminoacyl-charged tRNA to the *A*- or acceptor site,** followed by **formation of the peptide bond and the translocation of a tRNA from the *A*- to the *P*-site.** Proteins called elongation factors are the workhorses in the process of elongating the nascent polypeptide chain by one amino acid at a time. Translation elongation factors are involved in the following three-step cycle:

1. **Codon recognition:** A hydrogen bond forms between the mRNA codon and the anticodon of the next aminoacyl tRNA at the empty *A*-site of the ribosome.
2. **Peptide bond formation:** An enzyme called peptidyl transferase embedded in the large ribosomal subunit catalyzes the formation of a peptide bond between the polypeptide of the peptidyl-tRNA in the *P*-site and the newly arrived aminoacyl tRNA in the *A*-site.
3. **Translocation:** The tRNA in the *P*-site is ejected so that the newly formed peptidyl-tRNA located in the *A*-site can shift over to the *P*-site freeing up the *A*-site for the next aminoacyl-charged tRNA.

Chain termination: **Elongation comes to an end** when one (or more) of three **stop codons (UAA, UAG, UGA) is encountered.** A protein called a release factor binds directly to the termination codon in the *A*-site. The newly synthesized protein is hydrolyzed from the tRNA and both the tRNA and the protein are released from the ribosome.

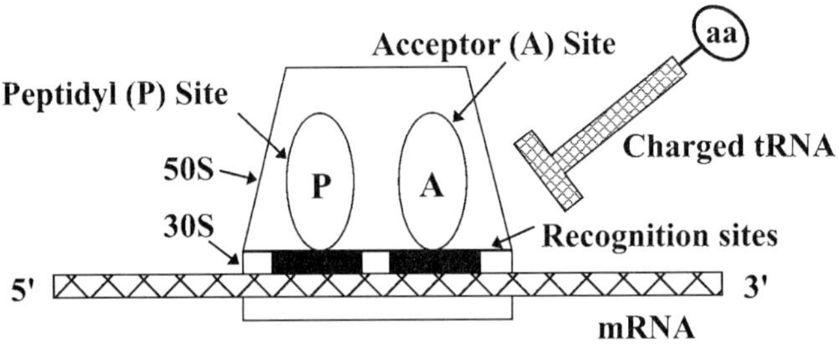

Figure 9-1. Various sites of antibiotic action in protein synthesis.

A list of key differences between prokaryotes and eukaryotes with respect to protein synthesis is shown in Table 9-1. These include the existence of multiple eukaryotic initiation factors that facilitate the assembly of the ribosomal protein synthetic machinery, whereas there are **only three for prokaryotes.** An **initiation site on bacterial mRNA consists of the AUG initiation codon** preceded with a gap of approximately 10 bases by the Shine-Dalgarno polypurine hexamer, whereas the 5′ Cap (a 7-methylguanylate residue in a 5′→5′ triphosphate linkage) acts as an initiation signal in eukaryotes. In **prokaryotes, the first or *N*-terminal amino acid is a formyl-methionine (fMet),** but in eukaryotes it is usually a **simple methionine.** Additionally, the size and nature of the prokaryotic ribosomes are quite different from the eukaryotic ribosomes.

Ribosomal RNA (rRNA) is a component of the ribosomes, the protein synthesis factories in the cell. **rRNA molecules are extremely abundant, making up at least 80 percent of the RNA** molecules found in a typical eukaryotic cell. Virtually all ribosomal proteins are in contact with rRNA. Most of the contacts between ribosomal subunits are made between the 16S and 23S rRNAs such that the interactions involving rRNA are a key part of ribosome function. The environment of the tRNA-binding sites is largely determined by rRNA. The rRNA molecules have several roles in protein synthesis. 16S rRNA plays an active role in the functions of the 30S subunit. It interacts directly with mRNA, with the 50S subunit, and with the anticodons of tRNAs in the *P*- and *A*-sites. Peptidyl transferase activity resides exclusively in the 23S rRNA. Finally, the rRNA molecules have a structural role. They fold into three-dimensional shapes that form the scaffold on which the ribosomal proteins assemble.

Many antibiotics act to inhibit protein synthesis in prokaryotes without affecting eukaryotic cells much. Several important targets in protein synthesis have been identified that are blocked by these agents to curb microorganism

Table 9-1
DIFFERENCES IN PROKARYOTIC AND EUKARYOTIC PROTEIN SYNTHESIS

ORGANISM	PROKARYOTE	EUKARYOTE
Start	fMet-$tRNA_f^{Met}$	Met-$tRNA_i^{Met}$
Recognition sequence	Shine-Dalgarno sequence	5′ caps direct e-IFs
Initiation factors	IF-1, IF-2, IF-3	multiple e-Ifs (>10)
Elongation factors	EF-Tu, EF-G, EF-Ts	Multi-subunit eEF-1, eEF-2, eEF-3

growth (see Figure 9-1). *Aminoglycosides* interfere with binding of fMet-tRNA to the ribosome, thereby preventing correct initiation of protein synthesis and partially freezing the complex. *Puromycin,* on the other hand, causes premature termination of protein synthesis because it resembles tyrosyl-tRNA. It can bind to the *A*-site on the ribosome, and the peptidyl transferase will form a peptide bond between the growing peptide and puromycin. However, since there is no anticodon to bind to the mRNA, the peptidyl-puromycin is released from the ribosome following formation of the peptide bond, thus stopping the synthesis. *Tetracyclines* inhibit protein synthesis in **bacteria by blocking the *A*-site on the ribosome and inhibiting binding of aminoacyl tRNAs. *Macrolides* (see below) bind to the 50S subunit near the *P*-site to cause conformational changes and inhibit translocation of the peptidyl tRNA from the *A*-site to the *P*-site.** *Lincosamides* bind near the *P*-site and interfere with binding of the aminoacyl end of the AA-tRNA. They occupy the site or change the ribosomal conformation such that it destabilizes the ribosomes and the growing chains fall off the mRNA. *Chloramphenicol* inhibits protein synthesis by bacterial ribosomes by blocking peptidyl transfer. It inhibits peptide bond formation between AA-tRNA and the growing chain on the *P*-site by inhibiting peptidyl transferase. *Neomycin, kanamycin,* and *gentamicin* interfere with the decoding site in the vicinity of nucleotide 1400 in 16S rRNA of 30S subunit. This region interacts with the wobble base in the anticodon of tRNA and blocks self-splicing of group I introns. *Streptomycin,* a basic trisaccharide, causes misreading of the genetic code in bacteria at relatively low concentrations but can inhibit initiation at higher concentrations.

Macrolide antibiotics constitute a group of 12- to 16-membered lactone rings substituted with one or more sugar residues, some of which may be amino sugars. Macrolides such as erythromycin (Figure 9-2) are generally bacteriostatic, although some of these drugs are bactericidal only at very high concentrations. Gram-positive bacteria accumulate approximately 100 times more erythromycin than do gram-negative microorganisms. Cells are considerably more permeable to the nonionized form of the drug, which explains increased antimicrobial activity observed at alkaline pH. The newer macrolides have structural modifications, such as methylation of the nitrogen atom in the lactone ring, that improve acid stability and tissue penetration of these agents. Macrolides act by binding reversibly to the ribosomal subunits of sensitive microorganisms and thereby inhibiting protein synthesis. Resistance to macrolides in clinical isolates is most frequently a result of posttranscriptional methylation of an adenine residue of 23S ribosomal RNA, which leads to coresistance to macrolides. Other mechanisms of resistance involving cell impermeability or drug inactivation have also been detected. It is believed that erythromycin does not inhibit peptide bond formation directly but rather inhibits the translocation step wherein a newly synthesized peptidyl tRNA molecule moves from the acceptor site on the ribosome

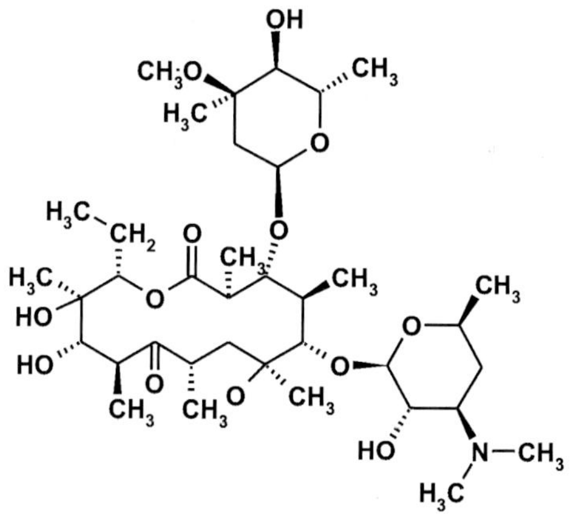

Figure 9-2. Chemical structure of erythromycin.

to the peptidyl or donor site. Erythromycin, which does not reach the peptidyl transferase center, induces dissociation of peptidyl tRNAs containing six, seven, or eight amino acid residues.

COMPREHENSION QUESTIONS

For each of the following steps in prokaryotic protein synthesis (Questions [9.1] to [9.3]), indicate the most appropriate antibiotic (A–J) to inhibit the process.

A. Aminoglycosides
B. Chloramphenicol
C. Erythromycin
D. Gentamicin
E. Kanamycin
F. Lincosamides
G. Neomycin
H. Puromycin
I. Streptomycin
J. Tetracycline

[9.1] Transfer of the peptide from the peptidyl tRNA to the aminoacyl-tRNA and formation of a peptide bond.

[9.2] Binding of aminoacyl-tRNA in the *A*-site of the ribosomal complex.

[9.3] Translocation of the peptidyl tRNA from the *A*-site to the *P*-site.

[9.4] The 6-year-old son of a migrant worker is brought to a clinic with chills, headache, nausea, vomiting, and sore throat. The examining physician notes a persistent grayish colored membrane near the tonsils. History reveals that the patient has not been immunized against diphtheria. Diphtheria toxin is potentially lethal in this unimmunized patient because it causes which of the following?

A. Inactivates an elongation factor required for translocation in protein synthesis
B. Binds to the ribosome and prevents peptide bond formation
C. Prevents binding of mRNA to the 60S ribosomal subunit
D. Inactivates an initiation factor
E. Inhibits the synthesis of aminoacyl-charged tRNA

[9.5] Replication of a particular DNA sequence is noted to be under inhibitory control usually. However, when substance "A" is added, it binds to a repressor, rendering the repressor inactive and allowing transcription to occur. Which of the following terms describes agent "A"?

A. Histone
B. Operon
C. Polymerase
D. Transcriber
E. Inducer

Answers

[9.1] **B.** Chloramphenicol inhibits protein synthesis by inhibiting peptidyl transferase. This peptidyl group cannot be transferred to the aminoacyl-tRNA in the *A*-site.

[9.2] **J.** Tetracyclines bind to the *A*-site of the prokaryotic ribosome and prevent aminoacyl-tRNAs from binding. Thus protein synthesis is halted because new amino acids cannot be added to the growing protein.

[9.3] **C.** Erythromycin and other macrolide antibiotics bind the 50S subunit near the *P*-site and cause conformational changes that inhibit the translocation of peptidyl tRNA from the *A*-site to the *P*-site.

[9.4] **A.** Diphtheria toxin has two subunits. The B subunit binds to a cell surface receptor and facilitates the entry of the A subunit into the cell. The A subunit then catalyzes the ADP-ribosylation of elongation factor 2 (EF2). EF2 is thus inhibited from participating in the translocation process of protein synthesis; hence, protein synthesis stops.

[9.5] **E.** An inducer is a small molecule that binds to and inactivates a repressor, which allows the sequence of DNA to be transcribed. An operon is a set of prokaryotic genes in close proximity that are coordinated as "all off" or "all on." An inducer may act to "turn on" the operon. One classic example is the *lac operon*. When allolactose is present, it serves as an inducer, and the operon is turned on, allowing proteins to be formed that metabolize lactose.

BIOCHEMISTRY PEARLS

- ❖ The synthesis of proteins involves converting the nucleotide sequence of specific regions of DNA into mRNA ***(transcription),*** followed by the formation of peptide bonds in a complex set of reactions that occur on ribosomes ***(translation).***
- ❖ Protein synthesis is divided into three stages: initiation, elongation, and termination.
- ❖ Ribosomal RNA (rRNA) is a component of the ribosomes, the protein synthetic factories in the cell.
- ❖ Many antibiotics take advantage of the differences of the rRNA between eukaryotic and prokaryotic cells.

REFERENCES

Alberts B, Johnson A, Lewis J, et al. Molecular Biology of the Cell, 4th ed. New York and London: Garland, 2002.

Lodish H, Berk A, Zipursky SL, et al. Molecular Cell Biology, 4th ed. New York: Freeman, 2000.

Petri WA. Anti-microbial agents. In: Goodman AG, Gilman LS, eds. The Pharmacological Basis of Therapeutics, 10th ed. New York: McGraw-Hill, 2001.

Prescott LM, Harley JP, Klein DA. Microbiology, 3rd ed. Boston, MA: W.C. Brown, 1996.

Voet D, Voet JG, Pratt CW. Fundamentals of Biochemistry, upgrade ed. New York: John Wiley, 2002.

CASE 10

A 20-year-old female presents to the ER with complaints of fever, pelvic pain, and some nausea and vomiting increasing over the last 2 days. She denies diarrhea or sick contacts. She is currently sexually active with a new partner. On examination she has a temperature of 38.9°C (102°F) and appears ill. She has moderate bilateral lower abdominal tenderness and minimal guarding without rebound or distention. Bowel sounds are present and normal. Pelvic exam revealed a foul-smelling discharge through cervix with severe cervical motion tenderness and bilateral adnexal tenderness. Cervical cultures were obtained. Patient was begun on a quinolone antibiotic.

- **What is the most likely diagnosis?**
- **What is the biochemical mechanism of action of the quinolone?**
- **What is the role of deoxyribonucleic acid (DNA) topoisomerases?**

ANSWERS TO CASE 10: PELVIC INFLAMMATORY DISEASE

Summary: A 20-year-old female with history of new sexual partner, fever, abdominal and pelvic pain, foul-smelling discharge through cervical os, and severe cervical motion tenderness.

- **Most likely diagnosis:** Pelvic inflammatory disease
- **Biochemical mechanism of action of quinolone:** Inhibits DNA gyrase
- **Function of topoisomerases:** Enzymes that assist in formation of superhelices and regulate the breaking and rejoining of DNA chains

CLINICAL CORRELATION

Pelvic inflammatory disease (PID) is usually an acute infection affecting the fallopian tubes and possibly the uterus and ovaries. It is generally sexually transmitted, caused by organisms such as *Chlamydia* or *Neisseria gonorrhoeae* (gonorrhea). The diagnosis is made clinically based on the typical history and physical examination. A purulent cervical discharge is highly suggestive. Nearly all patients have cervical motion tenderness, that is, pain with motion and palpation of the cervix. The treatment is with antibiotics. Quinolone antibiotics have been popular in the past; however, increasing bacterial resistance, particularly in Southeast Asia and California, has rendered these agents less desirable. Complications of PID include infertility or ectopic pregnancy (pregnancy in the tube) as consequence of tubal damage.

APPROACH TO DNA TOPOLOGY

Objectives

1. Know about DNA superhelices.
2. Understand the role of topoisomerases and DNA gyrase.
3. Understand the importance of histones and nonhistone proteins.
4. Be familiar with nucleosomes and polynucleosomes.
5. Know about the chromosomal structure.

Definitions

DNA gyrase: A type II topoisomerase enzyme present in bacteria that introduces negative supercoils into the DNA double helix in advance of the replication fork.

Histones: Proteins containing a large number of positively charged amino acids (lysine, arginine) that associate with DNA to form nucleosomes.

Nucleosome: Disk-shaped particles that consist of a core of histone protein around which DNA is wrapped. They are a structural unit of chromatin.

Supercoiling: The act of DNA winding on itself as a result of unwinding caused by replication forks.

Topoisomerase: Enzymes that control the amount of supercoiling in DNA. Type I topoisomerases will cleave one strand of DNA to relieve supercoiling, whereas type II topoisomerases will cleave both strands of the DNA double helix.

DISCUSSION

The **quinolones are broad-spectrum synthetic antibiotic** drugs that contain the **4-quinolone ring** (Figure 10-1). The first quinolone, *nalidixic acid,* was synthesized in 1962. More recently, a family of **fluoroquinolones** has been produced that contain a fluorine substituent at position 6 and a carboxylic acid moiety in the 3 position of the basic ring structure. R_1, R_7, and X are substituted with different side chains for the purpose of increasing bioavailability of the compound. A typical example of a fluoroquinolone is ciprofloxacin. The quinolone antibiotics target bacterial DNA gyrase in many gram-negative bacteria such as *Escherichia coli, Klebsiella pneumoniae, Pseudomonas aeruginosa,* and the like.

The normal biological functioning of DNA-like ribonucleic acid (RNA) transcription and DNA replication occurs only if it is in the **proper topological state.** In duplex DNA, the two strands are wound about each other once every 10 bp, that is, once every turn of the helix. Double-stranded circular DNA can form either **negative supercoils** when the strands are underwound or **positive supercoils** when they are overwound (Figure 10-2). Negative supercoiling introduces a torsional stress that promotes unwinding or separation of the right-handed B-DNA double helix, while positive supercoiling overwinds such a helix.

Supercoiling is controlled by a remarkable group of enzymes known as **topoisomerases,** which alter the topology of the circular DNA but not its

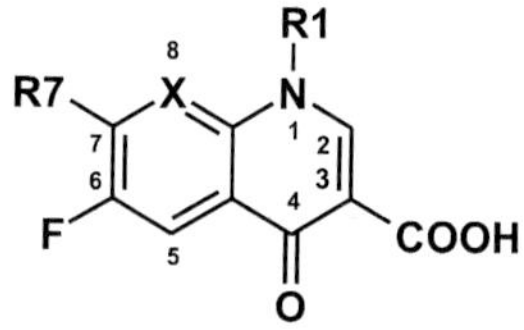

Figure 10-1. Structural formula for a fluoroquinolone. R_1, R_7, and X are side chains.

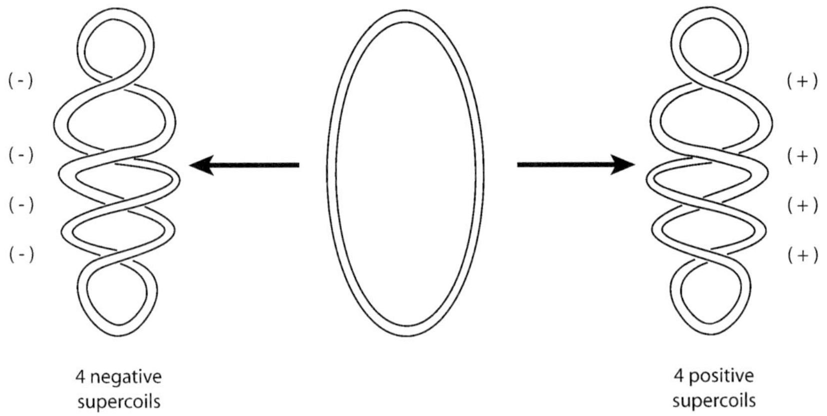

Figure 10-2. Positive and negative supercoiling in circular DNA.

covalent structure. There are two classes of topoisomerases. **Type I topoisomerases** relax DNA from negative supercoils formed by the action of **type II topoisomerase** by creating transient single-strand breaks in DNA without any expense of ATP. **Type II topoisomerases** (also called DNA gyrases) change DNA topology by **making transient double-strand breaks in DNA and require ATP consumption** (Figure 10-3).

During DNA replication, **type II topoisomerase, or TOPO II,** plays an important role in the fork progression by continuous removal of the excessive positive supercoils that stem from the unwinding of the DNA strands. TOPO II has the ability to **cut both strands of a double-stranded DNA molecule,** pass another portion of the duplex through the cut, and reseal the cut in a process that uses ATP. Hydrolysis of ATP by TOPO IIs inherent ATPase activity powers the conformational changes that are critical for the enzyme's operation. Based on the DNA substrate, TOPO II can change a positive supercoil into a negative supercoil or increase the number of negative supercoils by two.

The **DNA gyrase** of *E. coli* is composed of two 105,000-dalton A subunits and two 95,000-dalton B subunits encoded by ***gyr*A** and ***gyr*B genes,** respectively. The A subunits, which carry out the strand-cutting function of the gyrase, are the site of action of the quinolones. **DNA gyrase inhibition disrupts DNA replication and repair, transcription, bacterial chromosome separation during division,** and other cell processes involving DNA. The drugs inhibit gyrase-mediated DNA supercoiling at similar concentrations that are required to inhibit bacterial growth (0.1 to 10 μg/mL). Mutations in *gyr*A gene that encodes the A subunit of the polypeptide can confer resistance to these drugs. Eukaryotic cells lack DNA gyrase but have a similar type of

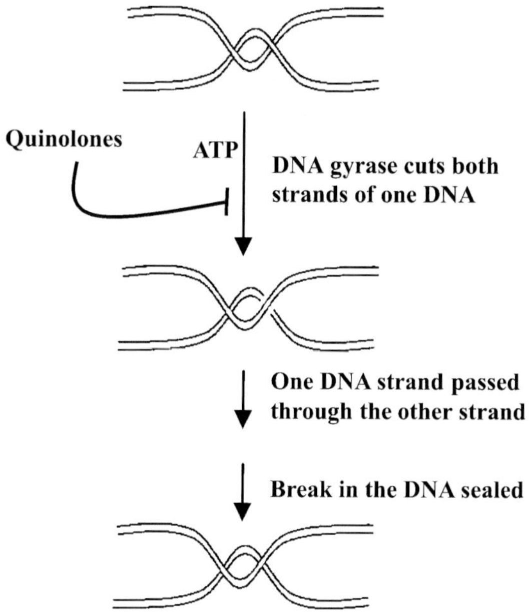

Figure 10-3. DNA gyrase action and quinolone inhibition.

topoisomerase that can remove positive supercoils from eukaryotic DNA to prevent its tangling during replication. Quinolones inhibit eukaryotic topoisomerase at much higher concentrations (100 to 1000 μg/mL).

Although **bacterial DNA is compacted as large circular chromosomes** with a single replication origin, most **eukaryotic DNA is much more highly organized** and is associated with many proteins to form chromatin that contains multiple replication origins. The general structure of chromatin has been found to be remarkably similar in the cells of all eukaryotes. The most abundant proteins associated with eukaryotic DNA (somewhat more than half its mass) are ***histones,* a family of basic proteins rich in the positively charged amino acids lysine and/or arginine,** which interact with the negatively charged phosphate groups in DNA. There are **five types of histones** that are evolutionarily conserved: **H1, H2A, H2B, H3, and H4.** Eight histone molecules (two each of H2A, H2B, H3, and H4) form an ellipsoid approximately 11 nm long and 6.5 nm in diameter. DNA coiled around the surface of this ellipsoid is termed the **nucleosome core** particle and has approximately $1^{3}/_{4}$ turns or 166 bp before it proceeds on to the next (Figure 10-4). The complex of histones plus DNA resembles a bead like structure and is called a **nucleosome.**

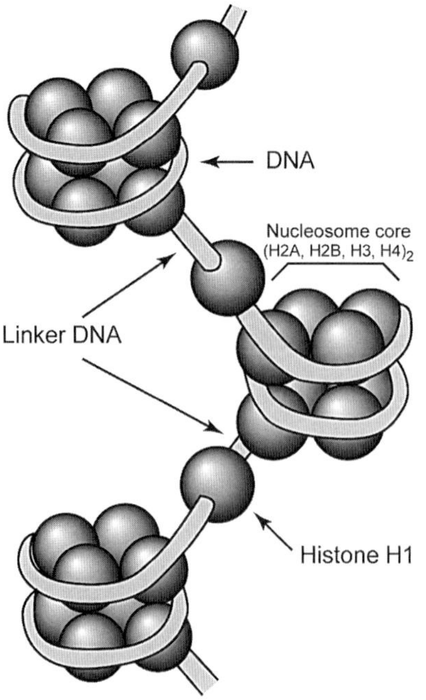

Figure 10-4. Schematic representation of nucleosomes showing DNA wrapped around a core of histones H2A, H2B, H3, and H4. Histone H1 associates with the linker region of DNA.

Composed of DNA and histones, nucleosomes are approximately 10 nm in diameter and are primary structural units of chromatin. **Between every two nucleosomes there is a stretch of DNA, the *linker* region,** which varies in length from 15 to over 55 bp. Histone H1 can associate with the linker region to aid in folding of DNA into more complex chromatin structures. The question of how the highly ordered nucleosome is formed can be explained by the fact that nucleosome assembly is facilitated by molecular chaperones. In the presence of **nucleoplasmin** (an acidic protein) and DNA topoisomerase I (nicking-closing enzyme) the nucleosome assembly proceeds rapidly without histone precipitation. Nucleoplasmin binds to histones but not to DNA or nucleosomes. It functions as a molecular chaperone to bring histones and DNA together in a controlled fashion and prevents their nonspecific aggregation through their otherwise strong electrostatic interactions. The nicking-closing enzyme acts to provide the nucleosome with its preferred level of supercoiling. Six nucleosomes are packed per turn further into a spiral or solenoid to form a regular array of *polynucleosomes* of 30 nm length.

As a 30-nm fiber, **the typical human chromosome would be 0.1 cm in length** and could span the nucleus more than 100 times. Clearly, there must be a still higher level of folding. This higher order packaging is one of the most fascinating but also most poorly understood aspects of chromatin. Chromosomes are generally decondensed during interphase. Several studies of interphase chromosomes have suggested that each long DNA molecule in a chromosome is divided into a large number of discrete domains that are folded differently. The regions that are least condensed correlate well with the regions that are actively synthesizing RNA. All chromosomes adopt a highly condensed conformation during mitosis and reflect a coarse heterogeneity of chromosome structure. Other complex chromosomal structures found in eukaryotes are in histone-depleted metaphase chromosomes. These structures form radial loops around a central fibrous protein called *scaffold.* Most of these loops are organized into condensed 300 Å filaments but the exact nature of these formations is not known. Nonhistone proteins, whose hundreds of varieties constitute approximately 10 percent of chromosomal proteins, are thought to be involved in these processes.

COMPREHENSION QUESTIONS

[10.1] A 38-year-old woman, who works as an administrative assistant for a large company, opened a package and found a suspicious white powder. Analysis of the powder indicates that it contained traces of the bacterium *Bacillus anthracis.* The woman was treated with ciprofloxacin, an effective antibiotic. Ciprofloxacin's mechanism of action is best described as an inhibition of which of the following?

A. Bacterial dihydrofolate reductase
B. Bacterial peptidyl transferase activity
C. Bacterial RNA polymerase
D. DNA gyrase
E. DNA polymerase III

[10.2] The Rubenstein-Taybi syndrome (RTS) is a genetic disease that is characterized by distinctive facial features, broad thumbs, broad big toes, and mental retardation. The affected gene is *CBP* (CREB-binding protein gene), which codes for a transcriptional activator. The RTS phenotype is best expressed by a haploinsufficiency model, in which two functional copies of the gene are required to produce sufficient *CBP* for proper development. *CBP* has a histone acetyltransferase activity, which does which of the following ?

A. Inhibits RNA polymerase II
B. Helps expose the promoters of genes
C. Inhibits the splicing of heterogeneous nuclear RNA (hnRNA) to messenger RNA (mRNA)
D. Prevents the addition of a poly-A tail to mRNA
E. Activates the formation of nucleosomes

[10.3] Acetylation and deacetylation of lysine residues on histone proteins provide one mechanism by which transcription can be activated or repressed. Which one of the histone proteins is least likely to participate in this process?

A. H1
B. H2A
C. H2B
D. H3
E. H4

Answers

[10.1] **D.** Ciprofloxacin is a fluoroquinolone antibiotic that will inhibit the strand-cutting function of A subunit of DNA gyrase, a bacterial topoisomerase II that introduces negative supercoils ahead of the replication fork. This disrupts DNA replication and repair, transcription, bacterial chromosome separation, and other bacterial processes involving DNA. At much higher concentrations, the type II topoisomerases of eukaryotic cells can be inhibited.

[10.2] **B.** When lysine residues in the *N*-terminal portion of histones are acetylated, it decreases the positive charge of the histone proteins and weakens the interaction between the histones and the DNA. As a result, the nucleosomes are "opened up" and lead to gene activation.

[10.3] **A.** Nucleosomes are disk shaped particles that consist of a core of histone protein around which DNA is wrapped. A short linker region of DNA joins nucleosomes. The core of the nucleosomes is made up of two copies each of histones H2A, H2B, H3, and H4. These histone proteins have *N*-terminal "tails" that contain lysine residues that can be reversibly acetylated, affecting the electrostatic interaction of the DNA with the histones. Histone H1 is not part of the nucleosomes core, therefore acetylation would likely not affect the protein–DNA interaction of the nucleosomes.

BIOCHEMISTRY PEARLS

- The quinolone antibiotics target bacterial **DNA gyrase** in many gram-negative bacteria.
- Supercoiling is controlled by topoisomerases which alter the topology of the circular DNA but not its covalent structure.
- **Type I topoisomerases relax** DNA from negative supercoils formed by the action of type II topoisomerase by creating transient single-strand breaks in DNA.
- **Type II topoisomerases** (also called **DNA gyrases**) change DNA topology by making transient double-strand breaks in DNA.
- Histones are abundant proteins associated with eukaryotic DNA and are a family of basic proteins rich in the positively charged amino acids lysine and/or arginine, which interact with the negative charges of DNA.

REFERENCES

Alberts B, Johnson A, Lewis J, et al. Molecular Biology of the Cell, 4th ed. New York and London: Garland, 2002.

Lodish H, Berk A, Zipursky SL, et al. Molecular Cell Biology, 4th ed. New York: Freeman, 2000.

Petri WA. Anti-microbial agents. In: Goodman AG, Gilman LS, eds. The Pharmacological Basis of Therapeutics, 10th ed. New York: McGraw-Hill, 2001.

Prescott LM, Harley JP, Klein DA. Microbiology, 3rd ed. Boston, MA: W.C. Brown, 1996.

Voet D, Voet JG, Pratt, CW. Fundamentals of Biochemistry, upgrade ed. New York: John Wiley, 2002.

CASE 11

A 32-year-old female presents to your clinic with concerns over a recently detected right breast lump. A mammogram performed revealed a right breast mass measuring 3 cm with numerous microcalcifications suggestive of breast cancer. During your discussion with the patient, she revealed that she had a sister who was diagnosed with breast cancer at the age of 39, a mother who passed away with ovarian cancer at age 40 years, and a maternal aunt who had both breast and colon cancer. Patient underwent an examination which revealed a fixed and nontender breast mass on right side measuring 3 cm with mild right axillary lymphadenopathy. No skin involvement is noted. A biopsy was performed and revealed intraductal carcinoma.

- **What cancer gene might be associated with this clinical scenario?**

- **What is the likely mechanism of the cancer gene in this case?**

ANSWERS TO CASE 11: ONCOGENES AND CANCER

Summary: A 32-year-old female with strong family history of breast, colon, and ovarian cancer, who now presents with a fixed breast lesion that is biopsy-proven carcinoma.

- **Most likely cancer gene:** Breast cancer (*BRCA*) gene
- **Likely mechanism:** Inhibition of tumor-suppressor gene

CLINICAL CORRELATION

This young woman has developed breast cancer at age 32 years. Moreover, she has two first-degree relatives with breast and/or ovarian cancer prior to menopause. This makes *BRCA* gene mutation likely. The *BRCA1* gene resides on chromosome 17. This gene encodes a protein which most likely is important in deoxyribonucleic acid (DNA) repair. Thus, a mutation of the *BRCA*1 gene likely leads to abnormal cells propagating unchecked. A woman with a *BRCA1* mutation has a 70 percent lifetime risk of developing breast cancer, and a 30 to 40 percent risk of ovarian cancer. The vast majority of breast cancer is not genetically based, but occurs sporadically. However, familial-based breast cancers are most common because of *BRCA1* mutation. *BRCA2* is another mutation that is more commonly associated with male breast cancer. Other genetic mechanisms of cancer include oncogenes, which are abnormal genes that cause cancer usually by mutations. Protooncogenes are normal genes that are present in normal cells and involved in normal growth and development, but if mutations occur, they may become oncogenes.

APPROACH TO ONCOGENES

Objectives

1. Know the definitions of oncogenes and protooncogenes.
2. Understand the role of promoter and repressor functions of DNA synthesis.
3. Know the normal DNA replication.
4. Be familiar with DNA mutations (point mutations, insertions, deletions).
5. Know the process of DNA repair.
6. Understand the recombination and transposition of genes.

Definitions

Okazaki fragment: Short segments of DNA (approximately 1000 nucleotides in prokaryotes, 100 to 200 nucleotides in eukaryotes) synthesized on the lagging strand during DNA replication. As the replication fork opens, the ribonucleic acid (RNA) polymerase primase synthesizes a short RNA primer, which is extended by DNA polymerase until it reaches the preceding Okazaki fragment. The RNA primer on the previous Okazaki fragment is removed, the gap filled in with DNA and the strands ligated.

Oncogene: Genes whose products are involved in the transformation of normal cells to tumor cells. Most oncogenes are mutant forms of normal genes (protooncogenes).

Primase: An RNA polymerase that synthesizes a short RNA primer complementary to a DNA template strand that is being replicated. This RNA primer serves as the starting point for addition of nucleotides in the replication of the strand serving as a template.

Recombination: Any process in which DNA strands are cleaved and rejoined resulting in an exchange of material between molecules of DNA.

Tumor suppressor genes: Genes that encode for proteins that normally inhibit the progress of a cell through the cell cycle. If these genes are mutated, a deficiency in the suppressor proteins creates unregulated cell growth, a condition permissive for tumorigenesis.

DISCUSSION

DNA replication normally occurs during the **S phase of the cell cycle.** DNA replication occurs in a semiconservative fashion and this is because of the **intrinsic antiparallel nature of the double helix. All known DNA polymerases synthesize DNA in a 5′ to 3′ direction;** this means that one strand will be continuously synthesized and the other must be made through a **discontinuous mechanism and the production of Okazaki fragments.** After this process is finished, the cell will divide the newly replicated material in mitosis. In some aberrant cases, the cell will replicate its DNA over and over again without performing an intervening mitosis. This generates cells with abnormally high content of DNA, an abnormal form of DNA replication control.

DNA replication begins by separating the parental DNA strands. This is accomplished by enzymes called **helicases.** They **separate the strands** and move along the strands in a fixed direction at the expense of ATP, **opening up the strands so that the DNA polymerases can bind to them.** To prevent reannealing of the two strands, single-stranded binding proteins bind the two complementary strands. The **next step in the replication process is the laying down of an RNA primer,** which is catalyzed by the enzyme **primase. DNA polymerases extend the chain by adding deoxyribonucleotides to the 5′ end of the RNA primer.** It does this in a continuous **fashion on the leading strand,** but for the **lagging strand, the one with its 5′ end toward the replication fork, a series of short segments termed Okazaki fragments must be synthesized.** The RNA primers of the Okazaki fragments are removed as the DNA polymerase reaches the previous Okazaki fragment and the DNA segments are then joined by DNA ligase.

DNA mutations arise from a variety of intrinsic and extrinsic factors. Our genome is constantly under the assault of various **genotoxic agents such as ionizing radiation, oxygen free radicals, and UV light.** These agents serve to introduce **DNA double-strand breaks (DSBs) as well as thymine–thymine dimers into the DNA.** DSBs may generate deletion or insertion mutations and could alter the reading frame of the genetic code, an event that can easily lead to the malfunction of a protein. UV-induced dimers may generate point mutations that also alter the reading frame.

To respond to the various forms of DNA damage, cells have evolved a host of DNA **repair mechanisms** that serve to restore the genetic material. Depending on the nature of the DNA damage (DSB versus ultraviolet [UV]-induced thymine dimers, etc.), the cell will invoke a different mechanism of repair. In addition, the stage of the cell cycle at which the lesion is detected and processed can activate independent DNA repair pathways. For example, if a daughter chromatid template is present in S or G_2 **phase of the cell cycle, the cell will use this unperturbed partner molecule to fix a DSB.** This process is referred to as **homologous recombination (HR)** and represents a major branch of the DNA repair process. However, if the damage occurs during the G_1 **phase of the cell cycle, a period devoid of an existing chromatid template** that can be used for repair, **a general end joining process** will be used. This process, referred to as **nonhomologous end joining (NHEJ),** ligates the broken ends together with little to no regard for the loss of intervening sequences. Therefore, **NHEJ is considered an error prone process,** but given the large size of the genome and the presence of many forms of "junk DNA," the NHEJ process may not necessarily disrupt DNA sequences that encode proteins. In contrast to NHEJ, **HR is a process that is error free** by virtue of the fact that the **daughter chromatid is used as a template for repair.**

Recombination and transposition of genes are two processes that **mutate the genetic material.** As discussed above, recombination is integral to the process of DNA repair. When DNA recombination is impaired through mutation of specific genes (like *BRCA1* or *Rad51*), aberrant recombination takes place, generating abnormal chromosomes that possess translocations from two or more chromosomes. **Translocations transpose genes** from one chromosomal environment to another and often this leads to a disruption in gene expression. Such events are known to cause **gene amplification,** a phenomenon often associated with cancer. They may also generate chromosomes that possess two centromeres (dicentrics), leading to a variety of cellular defects in mitosis.

DNA replication is tightly controlled by a variety of proteins that act to promote the process (i.e., DNA polymerases and *cis*-acting elements that bind to DNA and recruit factors involved in the process) as well as ones that inhibit the synthesis of DNA, either directly or indirectly. One factor that indirectly inhibits DNA synthesis is the **p53 tumor suppressor. Tumor suppressors refer to a general class of proteins that function to slow and alter cell growth** and development through a variety of mechanisms. In the absence of these factors, cells will have a reduced capacity to perform a variety of functions essential to maintain genomic stability. The **p53 functions to control the** G_1 **S boundary of the cell cycle,** and if the cell is not prepared to enter S phase, then DNA synthesis will be negatively regulated.

A major cause of familial breast cancer results from mutation in the breast cancer susceptibility gene, ***BRCA1.*** Originally identified in 1994, it was not until 1997 when David Livingston and colleagues demonstrated that ***BRCA1* is a nuclear protein** that its function began to be understood. They found that after treatment of cells with agents known to generate DNA damage (i.e., hydroxyurea, UV light, ionizing radiation), ***BRCA1* was found to localize to discrete nuclear structures typically called *foci.*** Such foci are known to correlate with sites of DNA damage or the sites of stalled replication forks in S phase of the cell cycle. In addition to the striking nuclear localization of *BRCA1* in response to DNA damaging agents, it was also observed that the **BRCA1 protein was phosphorylated in response to these agents.** Given this and the wealth of data obtained from studying the cell cycle in yeast model systems, it was concluded that *BRCA*1 functioned as a cell cycle regulator in response to DNA damage.

Further studies have implicated *BRCA1* in the cellular response to DNA double-stranded breaks (DSBs), a potentially lethal form of DNA damage. **Cells defective in *BRCA1*** possess numerous cytological and biological features that have been known for years to be correlated with perturbation in the maintenance of chromosome stability. This includes **aneuploidy, centrosome amplification, spontaneous chromosome breakage, aberrant recombination events, sensitivity to ionizing radiation, and impaired cell cycle checkpoints.** In addition, a variety of experiments have demonstrated roles for *BRCA1* in enforcing the G_2/M cell cycle transition, homologous recombination between sister chromatids, as well as the restart of stalled replication in S phase.

The ***BRCA1* tumor suppressor interacts with numerous cellular proteins in large complexes.** This includes a variety of proteins implicated in various DNA repair and cell cycle processes. In fact, *BRCA1* has been reported to interact with as **many as 50 proteins!** Moreover, *BRCA1* has been shown to function in various transcriptional mechanisms, suggesting that the function of this important protein may go well beyond its well-documented role in DSB repair. How the inactivation of *BRCA1,* a gene that appears to operate in DNA repair pathways that appear generic to various cell types and predisposes women to inherited forms of breast cancer, remains a mystery and the subject of much debate. Given that *BRCA1* appears to have key roles in transcriptional regulation, it has been suggested that *BRCA1* could influence mammary tissue through pathways that impinge on the biology of estrogen and estrogen-related metabolites. In addition, it has been suggested that breast tissue differs from other tissues in the types of DNA repair processes that are used. Perhaps there are redundant DNA repair pathways in non-mammary tissue. In any event, the complex behavior of *BRCA1* promises to challenge future researchers as they search for the underlying mechanisms that initiate familial breast cancer.

COMPREHENSION QUESTIONS

[11.1] Hereditary retinoblastoma is a genetic disease that is inherited as an autosomal dominant trait. Patients with hereditary retinoblastoma develop tumors of the retina early in life, usually in both eyes. The affected gene *(RB1)* was the first tumor suppressor gene to be identified. Which of the following best describes the function of the protein encoded by the *RB1* gene?

A. It binds transcription factors required for expression of DNA replication enzymes.
B. It allosterically inhibits DNA polymerase.
C. It binds to the promoter region of DNA and prevents transcription.
D. It phosphorylates signal-transduction proteins.

[11.2] Mutations in the tumor suppressor gene *BRCA1* are transmitted in an autosomal dominant fashion. When a cell is transformed to a tumor cell in individuals who have inherited one mutant allele of this tumor suppressor gene, which of the following most likely occurs?

A. A transcription factor is over expressed.
B. Deletion or mutation of the normal gene on the other chromosome.
C. Chromosomal translocation.
D. Gene duplication of the mutant gene.

[11.3] Women who inherit one mutant *BRCA1* gene have a 60 percent chance of developing breast cancer by the age of 50. The protein produced by the *BRCA1* gene has been found to be involved in the repair of DNA double-strand breaks. Which of the following processes is most likely to be adversely affected by a deficiency in the *BRCA1* protein?

A. Removal of thymine dimers
B. Removal of RNA primers
C. Removal of carcinogen adducts
D. Homologous recombination
E. Correction of mismatch errors

Answers

[11.1] **A.** The *RB1* protein binds to E2F transcription factors, preventing them from activating transcription of the genes encoding enzymes required for DNA replication, such as DNA polymerase. The *RB1*–E2F complex acts as a transcription repressor. Midway through G_1 phase, *RB1* is phosphorylated by cyclin-dependent kinases, releasing E2F to activate transcription. A deficiency of *RB1* leads to unregulated transcription of DNA replicatory enzymes and DNA synthesis.

[11.2] **B.** Since *BRCA1* is inherited in an autosomal dominant fashion, then one allele is sufficient to produce the *BRCA1* protein. However, if the normal allele is somatically mutated or deleted, then the affected cell cannot produce the *BRCA1* protein and can be transformed into a tumor cell.

[11.3] **D.** *BRCA1* plays a significant role in the repair of double-strand breaks, therefore, since homologous recombination requires a cleavage of both strands of a DNA molecule, this event is most likely to be affected by a deficiency in this protein. Thymine dimers, mismatches, and adducts of DNA with carcinogens are effectively removed by a process of excision repair, in which a section of one strand of DNA is removed.

BIOCHEMISTRY PEARLS

- ❖ DNA replication normally occurs during the S phase of the cell cycle.
- ❖ DNA replication occurs in a semiconservative fashion and this is because of the intrinsic antiparallel nature of the double helix.
- ❖ All known DNA polymerases synthesize DNA in a 5′ to 3′ direction.
- ❖ To respond to the various forms of DNA damage, cells have evolved a host of DNA repair mechanisms that serve to restore the genetic code.
- ❖ Tumor suppressors refer to a general class of proteins that function to slow and alter cell growth and development through a variety of mechanisms. *BRCA* tumor suppressor gene when mutated, increases the risk of breast cancer.
- ❖ Inactivation of *BRCA1,* a gene that appears to operate in DNA repair pathways that appear generic to various cell types, predisposes women to inherited forms of breast cancer, but the exact mechanism has been elusive.

REFERENCES

Couch FJ, Weber BL. Breast cancer. In: Scriver CR, Beaudet AL, Sly WS, et al., eds. The Metabolic and Molecular Basis of Inherited Disease, 8th ed. New York: McGraw-Hill, 2001:999–1031.

Edenberg HJ. DNA replication, recombination, and repair. In: Devlin TM, ed. Textbook of Biochemistry with Clinical Correlations, 5th ed. New York: Wiley-Liss, 2002.

Scully R Chen J, Ochs RL. Dynamic changes of BRCA1 subnuclear location and phosphorylation state are initiated by DNA damage. Cell 1997;90:425–35.

Scully R, Livingston DM. In search of the tumor-suppressor functions of BRCA1 and BRCA2. Nature 2000;408:429–32.

❖ CASE 12

A 25-year-old Mediterranean female presents to her obstetrician at 12-weeks gestation for her first prenatal visit. This is her first pregnancy, and she is concerned about her baby and the risk of inheriting a "blood" disease like others in her family. The patient reports a personal history of mild anemia but nothing as severe as her brother who required frequent transfusions and died at age 10. The patient was told by her physician that she did not need to take iron supplementation for her anemia. Patient denies having any anemic symptoms. Her physical exam is consistent with a 12-week pregnancy and ultrasound confirmed an intrauterine pregnancy at 12-weeks gestation. The patient's hemoglobin level shows a hypochromic, microcytic (small sized red cell) anemia (hemoglobin, 9g/dL) and hemoglobin electrophoresis demonstrated increased hemoglobin A2 level (4.0 percent) and increased fetal hemoglobin level, a pattern consistent with β-thalassemia minor. The patient underwent chorionic villus sampling to assess whether the fetus was affected, and the diagnosis returned in several hours.

◆ **What is the molecular genetics behind this disorder?**

◆ **What was the likely test and what is the biochemical basis?**

ANSWERS TO CASE 12: THALASSEMIA/ OLIGONUCLEOTIDE PROBE

Summary: A 25-year-old Mediterranean pregnant female has a history of asymptomatic mild hypochromic, microcytic anemia, elevated hemoglobin A2 and F on electrophoresis. Her brother had severe hemolytic disease that required transfusions and ultimately caused his premature death at age 10. She is diagnosed with β-thalassemia minor.

- **Molecular genetics:** Impaired production of β-globin peptide chain. Numerous mutations have been identified in the production of ribonucleic acid (RNA) including in the promoter region and splice junctions.
- **Likely test:** Oligonucleotide probe. After chorionic villus sampling is performed, a radioactive probe can be used and hybridized with specific genetic mutations in the fetus' deoxyribonucleic acid (DNA), allowing for prompt detection and prenatal diagnosis.

CLINICAL CORRELATION

Anemia is the abnormally low level of hemoglobin or red blood cell mass, which has the potential of limiting the delivery of oxygen to tissue. By far, the most common cause of anemia is iron deficiency, leading to small volume of red blood cells (microcytic). Another common cause of microcytic anemia is thalassemia. Certain ethnicities have higher incidences of thalassemia, for example, Mediterranean or East Asian descent.

This patient is of Mediterranean descent, making thalassemia more likely. Furthermore, the microcytic (small red blood cell size) anemia in the face of elevated hemoglobin A2 and F is consistent with β-thalassemia minor. Patients with β-thalassemia major (Cooley anemia) typically have severe anemia requiring frequent transfusions and shortened life expectancy. Infants will appear healthy after birth, but as the hemoglobin F levels fall, the infant becomes severely anemic. Females with β-thalassemia major who survive beyond childhood are usually sterile.

APPROACH TO OLIGONUCLEOTIDE PROBES

Objectives

1. Know about the use of oligonucleotide probes for detection of mutations.
2. Know how oligonucleotide segments are synthesized.
3. Be familiar with common mutations that cause thalassemias (substitutions in TATA box, mutations in splice junction, and changes in stop codon).

Definitions

Erythropoiesis: Erythropoiesis is the development of mature red blood cells containing hemoglobin (erythrocytes) from pluripotential stem cells via a linear cascade.

Frameshift mutation: Insertion or deletion of a number of nucleotides that are not divisible by three into a coding sequence, thereby causing an alteration in the reading frame of the entire sequence downstream of the mutation.

Locus control region: Regulatory region which is believed to regulate transcription by opening and remodeling chromatin structure. It may also have enhancer activity.

Promoter sequence: A regulatory region present at a short distance upstream from the 5′ end of a transcription start site that acts as the binding site for RNA polymerase to initiate transcription.

Thalassemia: A group of genetic disorders that is characterized by the absence of or reduced synthesis of one or more of the four globin chains in hemoglobin. The sequelae can range from benign to fatal, depending on the severity of the decrease in the globin chain.

Transposons: These are segments of DNA that can move around to different positions in the genome of a single cell. In the process, they may cause mutations and increase (or decrease) the amount of DNA in the genome. These mobile segments of DNA are sometimes called **"jumping genes."**

DISCUSSION

Thalassemia is a common condition in the Mediterranean region and its clinical features were described as early as 1925. The term *thalassemia* is coined from the Greek word, which means "the sea." Mendelian transmission of thalassemia was discovered in 1938. Later, it became apparent that thalassemia is not a single disease but a **group of genetic disorders,** all of which arise from **abnormalities in hemoglobin synthesis.** The thalassemias are characterized by the **absence or reduced synthesis of one or more of the four globin chains of hemoglobin.** There are four different genetic loci that control the formation of α-, β-, γ-, and δ-chains of hemoglobin (Hb). **Fetal Hb** (HbF) is formed from **two α- and two γ-chains.** In adults, the γ-chain is replaced by the β- and δ-chains, which combine with two α-chains to form **HbA** ($\alpha_2\beta_2$, approximately 97 percent) and HbA_2 ($\alpha_2\gamma_2$, approximately 3 percent) (Figure 12-1). In the late 1970s, the gene was found to be localized to chromosome 16, whereas the β-, γ-, and δ-genes were clustered on chromosome 11. Thus, although thalassemia does not occur in the fetus since the β-globin genes are activated only after birth, prenatal diagnosis is possible.

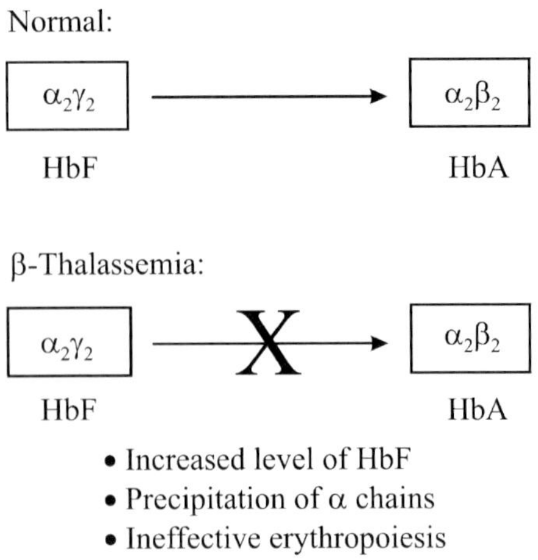

Figure 12-1. Maturation of hemoglobin from the fetal form to the adult form. In β-thalassemias, because of a defect in the β-globin gene, the γ-chain cannot be effectively replaced with the β-chain.

Thalassemias are named according to the affected globin chain: Thus, a disorder of the α-chain would be an **α-thalassemia,** and a disorder of the β-chain would be a **β-thalassemia.** There are two major forms of thalassemia, one in which the β-globin chain is not produced (β^0) and another in which there is significantly reduced level of β-globin chain (β^+). The positive category includes variants with unusually high levels of HbF or HbA_2, variants with normal levels of HbA_2, silent or benign variants, those that are inherited in a dominant fashion, and variants that are not linked to the β-gene cluster.

A severe decrease in β-globin levels leads to the precipitation of the α-chain, which in turn causes a defect in the maturation of the erythroid precursor, and erythropoiesis thus reducing red cell survival. The profound anemia in the affected individual stimulates the production of erythropoietin leading to the expansion of bone marrow and subsequent skeletal deformities. The hyperplasia of the bone marrow induces increased iron absorption leading to the deposition of iron in tissues. If the concentration of iron in the tissues becomes too high, it can lead to organ failure and death if appropriate therapeutic steps are not taken.

There are at least **200 β-thalassemia alleles** that have been characterized to date. **β-Thalassemias** are caused primarily by **point mutations** within the β-globin gene and the neighboring flanking region. There are two exceptions that are attributed to larger deletions ranging from 290 base pairs (bp) to greater than 60 kbp. One is a family of upstream deletions, which down regulates the locus control region, and the other affects the promoter region of β-globin gene. The **promoter region deletions** typically include the mRNA CAP site, TATA box, and CAAT elements. Apart from deletions, there are insertions of 6 to 7 kbp retrotransposons in the β-globin gene that decrease the transcript level to 15 percent in comparison to the control. In addition, there are also a few thalassemia mutations that segregate independently of the β-globin gene, such as those involving the trans-acting regulatory factors that affect the expression of the β-globin gene. One can thus broadly classify the different classes of mutations which result in β-thalassemia as follows:

1. **Transcription** (deletion, insertion, etc.)
2. **Messenger RNA (mRNA) processing** (cryptic splice sites, consensus sequence, poly A addition site, etc.)
3. **Translation** (initiation, nonsense, frameshift mutations)
4. **Posttranslation stability** (highly unstable β-chain)
5. Determinants unlinked to the β-globin gene cluster

The **most common point mutations** occur in the **CAAT or TATA box** in the **promoter** sequence. The single base substitution A to G at position −29 in individuals of African descent leads to a mild form of the disease, whereas the substitution A to C in those of Chinese descent leads to thalassemia major. Also, a single nucleotide substitution in the GT/AG splice junction can lead to a misspliced mRNA that does not allow the translation of a functional β-globin chain. The C to T mutation at position −101 causes an extremely mild deficit in β-globin gene. The effect is so mild that it is a "silent" or benign mutation in heterozygotes. All the point mutations that prevent the translation of the β-globin mRNA (for example, changing the start codon ATG to GTG, or TGT to the stop codon TGA) or a single nucleotide frame shift lead to β^0-phenotype.

There are several biochemical techniques that can be used to accurately determine the genetic defect that results in a particular thalassemia. Many of these techniques make use of the **polymerase chain reaction (PCR)**, which amplifies the DNA and allows detection of multiple mutations at once.

The ***amplification refractory mutation system*** (ARMS) is a technique in which the target DNA is amplified using a common primer and either mutation specific primer for β-thalassemia or the correct sequence primer to the target. This method provides quick screening of the DNA to detect if the patient carries the mutant gene or not. When both the primers are used in the same PCR reaction, they compete to amplify the target, and the technique is termed COP (competitive oligonucleotide priming), and the primers are labeled with different fluorescent dyes to allow detection.

A PCR restriction enzyme analysis takes advantage of the occurrence of approximately 40 thalassemia mutations that either introduce or remove a restriction endonuclease site. The PCR amplified target sequence is digested using restriction enzymes (enzymes that cleave DNA at particular nucleotide sequences) and the pattern of fragmentation on an agarose gel defines the presence or absence of a particular mutation.

Radio-labeled single stranded DNA oligonucleotide probes are allowed to anneal with the target strand in an ***oligonucleotide hybridization analyses.*** Typically, the target DNA is fixed onto nitrocellulose or nylon membrane. The probe forms stable duplexes with target sequences from the heterogenous mixture of many sequences in genomic DNA and is detected by autoradiography using an x-ray film. In populations that have predominantly one common mutation, a very successful and efficient hybridization analysis is the ***PCR allele-specific oligonucleotide (ASO) assay.*** This requires that the sequence of the common mutation is known. The amplified genomic DNA is transferred onto a nylon membrane and probed using an allele-specific oligonucleotide, that is, one that is complementary to the sequence of the DNA that contains the mutation. The membrane is probed both with oligonucleotides having the mutant and the correct DNA sequence of the β-globin gene. The genotype of the DNA is determined by the presence or the absence of the hybridization signal. Recently, **microchips,** which are an array of oligonucleotides immobilized on a glass plate, have been used to detect thalassemias. Fluorescent labeled DNA from an individual is hybridized onto this microchip and the reaction is monitored with a fluorescent microscope. This technique allows the study of a specific mutation using several hundred oligonucleotides simultaneously. It is highly sensitive with a low background; however, the resolving power is low.

Using one or a more of the DNA analysis tools described above, effective prenatal detection of various mutations in the β-globin gene cluster has been successfully achieved. Although gene therapy for genetic diseases is in its infancy, hopefully the future will bring effective treatment options for patients affected with thalassemia.

COMPREHENSION QUESTIONS

Micawley Talltwin is a 7-year-old child star who is brought to his pediatrician by his parents after they noticed that he felt very fatigued. They also noted that his abdomen seemed to be enlarged. Examination reveals an enlarged spleen. Further history reveals that he has been taking vitamins and iron supplements over the last few months. Laboratory tests show a microcytic anemia and elevated iron levels in tissues.

[12.1] Which of the following conditions is most consistent with the findings in this patient?

A. Aplastic anemia
B. Cooley anemia
C. Pernicious anemia
D. Thalassemia major
E. Thalassemia minor

[12.2] After diagnosing Micawley Talltwin and ascertaining he is not agranulocytic, his physician prescribes subcutaneous infusion of deferroxamine, an iron chelator, and monitors him for several weeks. What is his condition most likely to be on reexamination?

A. Iron levels decrease, but he remains anemic.
B. Iron levels decrease, and anemia recovers.
C. He develops irreversible and severe agranulocytosis.
D. Iron and vitamin C levels decrease.

[12.3] An electrophoretic analysis of Micawley Talltwin's hemoglobin indicates that although there is a decrease in the relative amount of the β-chain with respect to the α-chain, both the β- and the α-chains migrate at the same position as normal chains. Most likely his anemia is caused by which of the following?

A. A defect in an enzyme involved in heme synthesis
B. A point mutation in the coding region of the gene coding for the β-chain
C. A frameshift mutation in the coding region of the gene coding for the β-chain
D. A mutation in the promoter of the β-chain gene
E. A mutation in the structural gene of the β-chain

[12.4] As a medical geneticist, you analyze Talltwin's DNA and find that he is homozygous for thalassemia. Assuming the disease is autosomal recessive, what can you deduce about the genotype of Mr. and Mrs. Talltwin?

A. Dad Talltwin is a carrier of the disease, and Mom Talltwin is normal.
B. Mom Talltwin is a carrier of the disease, and Dad Talltwin is normal.
C. Dad Talltwin is homozygous, and Mom Talltwin is normal.
D. Mom Talltwin is homozygous, and Dad Talltwin is normal.
E. Both Mom and Dad Talltwin are carriers of the disease.

Answers

[12.1] **E.** The correct diagnosis is thalassemia minor because the patient had been asymptomatic until age 7 years. If he had thalassemia major or Cooley anemia, he would have exhibited symptoms as early as his first birthday. Pernicious anemia leads to a macrocytic or megaloblastic anemia, whereas aplastic anemia is characterized by normal sized erythrocytes.

[12.2] **A.** The iron chelator helps in excreting iron but has no role in increasing red blood cell (RBC) production to counteract anemia.

[12.3] **D.** Since the β-chain is decreased with respect to the α-chain, it is most likely that there is a mutation that decreases the expression of the β-chain gene, in which a mutation in the promoter region could result. A point mutation in the β-chain leading to an amino acid substitution could lead to changes in electrophoretic mobility but would not alter the levels of expression. A frameshift mutation in the β-chain would result in decreased β-chain on the electrophoregram.

[12.4] **E.** Thalassemia is autosomal recessive so Micawley Talltwin can get the disease only if both the parents were to be heterozygotes (or a carrier) of that particular trait.

BIOCHEMISTRY PEARLS

- The thalassemias are characterized by the absence or reduced synthesis of one or more of the four globin chains of hemoglobin.
- Thalassemias are named according to the affected globin chain; thus, a disorder of the α-chain would be an α-thalassemia and a disorder of the β-chain would be a β-thalassemia.
- The ***amplification refractory mutation system*** (ARMS) is a technique in which the target DNA is amplified using a common primer and either mutation specific primer for β-thalassemia or the correct sequence primer to the target.
- A ***PCR restriction enzyme analysis*** takes advantage of the occurrence of approximately 40 β-thalassemia mutations that either introduce or remove a restriction endonuclease site.
- Radio-labeled single stranded DNA oligonucleotide probes are allowed to anneal with the target strand in an ***oligonucleotide hybridization analyses.***
- With one common mutation, a very successful and efficient hybridization analysis is the ***PCR allele-specific oligonucleotide (ASO) assay.*** This requires that the sequence of the common mutation is known.

REFERENCES

Kanavakis E, Traeger-Synodino J, Vrettou C, et al. Prenatal diagnosis of the thalassaemia syndromes by rapid DNA analytical methods. Mol Human Reprod 1997;3:523–28.

Weatherall DJ, Clegg, JB. The Thalassaemia Syndromes, 4th ed. London: Blackwell Science, 2001.

CASE 13

An 8-year-old boy is brought to his pediatrician by his mother because she was concerned that he was having language-speech problems, was hyperactive, and was told by teachers that he may have mental retardation. The mother reports a strong family history of mental retardation in males. The boy on exam is found to have a large jaw, prominent ears, and enlarged testes (macroorchidism). The mother was told her family had a genetic problem causing the mental retardation. The patient underwent a series of blood tests and was scheduled to see a genetic counselor, who expressed that the etiology of the genetic defect was likely transmitted from his mother. The genetic counselor states that his mother likely has a silent mutation.

- **What is the most likely diagnosis?**
- **Which chromosome is likely to be affected?**
- **What are some types of biochemical mutations?**
- **What is the biochemical basis of the different types of mutations?**

ANSWERS TO CASE 13: FRAGILE X SYNDROME

Summary: An 8-year-old boy has mental retardation, speech-language problems, hyperactivity, physical findings of large jaw, prominent ears, and macroorchidism and has a strong family history of mental retardation. The genetic counselor informs the mother that she is the carrier and that she has a silent mutation.

- **Most likely diagnosis:** Fragile X syndrome (most common form of familial mental retardation).
- **Affected chromosome:** Chromosome X
- **Types of mutations:** *Silent* means protein product not affected; *missense* means single amino acid substitution leading to significant alteration (such as sickle cell); and *nonsense* means that a stop codon is formed.
- **Molecular basis of disease:** Mutation resulting in an increased number of CGG repeats on the X chromosome. When the number of repeats reaches a critical size, it can be methylated and inactivated resulting in the disorder. Individuals who carry 50 to 199 repeats are phenotypically normal and carry a premutation. If repeats exceed 200, the patient has a full mutation; and if methylation occurs, he or she will be affected.

CLINICAL CORRELATION

Fragile X is the most common inherited form of mental retardation, affecting primarily males. The clinical presentation can vary, but usually the affected male has moderate to severe mental retardation, hyperactivity, typical facies such as large jaw and large ears. Pigmented skin lesions (café au lait) can also be seen. Because females have two copies of the X chromosome, they are "resistant" to mutations on one gene. Fragile X affects the long arm of the X chromosome, with multiple copies of triplicate repeats, usually CGG, leading to methylation of the deoxyribonucleic acid (DNA). The fragile X mental retardation (FMR) gene product is affected and, through a little-understood mechanism, leads to mental retardation.

APPROACH TO MUTATIONS

Objectives

1. Know the definitions of point mutations (silent, missense, and nonsense), insertions, deletions, and frameshift mutations.
2. Be familiar with the defect in fragile X syndrome.

Definitions

Point mutation: The substitution of a single nucleotide in the genetic material of an organism. These include silent, missense, and nonsense mutations.

Silent mutation: A single nucleotide is exchanged by another, but this alteration does not change the amino acid for which the codon codes. The final protein product remains unchanged.

Missense mutation: A single nucleotide is exchanged by another, and this alteration does have an effect on the coding amino acid. The final protein product is also modified. The modification may or may not be deleterious to the final protein, depending on the function of the amino acid.

Nonsense mutation: A single nucleotide is exchanged by another, which produces a new stop codon at this position. This premature stop codon generally results in a truncated form of the protein and most often leaves it as an inactive form.

Deletion: One or more nucleotides are removed from the genetic sequence. If the deletions are multiples of one or two, a frameshift will be the result, which will likely damage the final protein product. A deletion of three or a multiple of three does not shift the reading frame, rather it would merely remove a codon(s). The final protein product would lose amino acid(s), which may or may not leave it inoperative.

Insertion: One or more nucleotides are added to the genetic sequence. These are the opposite of deletions.

Trinucleotide repeat expansion: Amplification from one generation to the next of three nucleotide repeats in the coding or noncoding regions of DNA. The mechanism may arise from DNA complimentary strand slippage. This is associated with fragile X syndrome and myotonic dystrophy.

DNA methylation: Process by which methyl groups are added to DNA bases (most often cytosine). Methylation functions to regulate gene expression because heavily methylated genes are not expressed. Also, bacteria use methylated DNA as a defense mechanism. Every organism has different patterns of methylated DNA, and bacteria take advantage of this by destroying foreign DNA via nucleases, enzymes that cut DNA at specific sites.

DISCUSSION

Replication often produces **changes in the chemical makeup of DNA.** Many of these changes are easily repaired; however, those **alterations in the DNA base sequence that do not get repaired are referred to as mutations.** There are various types of mutations including **point mutations, deletions, and insertions.**

Point mutations arise when **one base pair is substituted for another,** and they are the **most common types of mutations.** Point mutations can be further subdivided into three categories; **silent, missense, and nonsense. Silent mutations do not affect the translated mRNA or final protein product—thus the name *silent*.** The genetic code is degenerate, which means that most amino acids are encoded by several different codons or triplets of DNA/RNA bases. Therefore, it is possible to switch a single base but not alter the resulting protein product (Table 13-1a). The **sickle cell trait** is a result of a **missense mutation.** This type of point mutation actually **does alter the protein product.** In the case of sickle cell hemoglobin, an adenine is replaced by a thymine, causing a hydrophilic glutamic acid to be replaced with a hydrophobic valine in the resulting protein (Table 13-1b). In a **nonsense mutation, the mutation brings about a premature stop in the gene of interest** because the altered codon now represents a stop codon (Table 13-1c). These are commonly seen in muscular dystrophies and some cystic fibrosis cases. **Aminoglycosides are currently being used to treat premature stop codons** as they affect the translational accuracy of transfer ribonucleic acids (tRNAs), thereby allowing recognition of incorrect codons including the stop codon. Ultimately, the translation machinery reads through the premature stop codon because of altered amino acid insertion by the tRNAs and produces full-length protein instead of a truncated protein or degraded mRNA.

Table 13-1a

EXAMPLES OF MUTATIONAL EVENTS
SILENT MUTATION

GENETIC CODE		SILENT MUTATION ENCOUNTERED			
CODON	AMINO ACID	CODON	AMINO ACID	MUTANT CODON	AMINO ACID
GCC	Alanine	GCC	Alanine	GC**T**	Alanine
GCT	Alanine				
GCA	Alanine				
GCG	Alanine				

Table 13-1b

MISSENSE MUTATION

β-Chain of normal hemoglobin	GTG Gly	CAC His	CTG Leu	ACT Thr	CCT Pro	G**A**G **Glu**	GAG Glu
A missense mutation found in β-chain of sickle cell hemoglobin	GTG Gly	CAC His	CTG Leu	ACT Thr	CCT Pro	G**T**G **Val**	GAG Glu

Table 13-1c
NONSENSE MUTATION

Normal	GGC Gly	CTG Leu	ACA Thr	GGG Gly	CAA **Gln**	AAA Lys	CTG Leu
A nonsense mutation found in Duchenne muscular dystrophy	GGC Gly	CTG Leu	ACA Thr	GGG Gly	TAA **STOP**	AAA	CTG

Another type of mutation is a **deletion in which one or more DNA bases have been removed.** However, an insertion is a mutation wherein one or more DNA bases have been added. These two types of mutations can cause a shift in the open reading frame, called **frameshift mutations,** of a gene if the **insertion or deletion** is a **multiple of one or two bases.** Typically, a frameshift mutation will result in **disease either as result of altering the protein product, leaving it nonfunctional, or generation of a premature stop codon,** which confers a truncated form of the altered protein (Table 13-2). Both cases present major problems because the altered protein will not likely be able to carry out the normal duties required of it. Since codons are read as multiples of three, an insertion or deletion of three or a multiple of three may or may not be as deleterious to the gene as multiples of one or two. In this situation, the resulting protein could include extra amino acids in the case of insertions or lose amino acids in the case of deletions. It is possible that this could hinder the protein from normally functioning.

Table 13-2
FRAMESHIFT MUTATION

FRAMESHIFT RESULTING IN ALTERED PROTEIN					
CCC Pro	GCA Ala	**TA**T Tyr	CAT His	TTT Phe	ACC Thr
		↓ Deletion of **TA**			
CCC Pro	GCA Ala	TCA **Ser**	TTT **Phe**	TAC **Tyr**	
FRAMESHIFT CAUSING PREMATURE STOP					
CCC Pro	GCA Ala	TAT Tyr	CAT His	TTT Phe	ACC Thr
		↓ Deletion of **TC**			
CCC Pro	GCA Ala	TAA **STOP**			

Many forms of fragile X syndrome are examples of **insertions in multiples of three** where the outcome is detrimental not because of altered amino acid sequence but rather a **loss of protein expression. Deletions and insertions** typically occur in highly **repetitive sequences, and this is exactly what is seen in fragile X. Fragile X syndrome is the most common form of inherited mental retardation.** It affects approximately 1 in 4000 males and 1 in 8000 females. Mutations of the X chromosome, including fragile X, may account for the higher number of male patients in mentally handicapped facilities. The **fragile X mental retardation 1 (*FMR1*) gene is highly conserved. Normal individuals have 7 to 60 (CGG) repeats** in the 5′-untranslated region, and the repeats are often broken up by AGG. **Many fragile X individuals have an increase in the number of CGG repeats, which may expand to over 230 repeats.** This is termed the *full mutation.* Repeats ranging from 60 to 230 are called a premutation. The premutation carrier has normal **fragile X mental retardation protein (FMRP)** levels, but the gene is unstable in its passage to offspring.

The fragile X phenotype is mainly caused by the **loss of FMRP resulting from CGG trinucleotide repeat expansion of the *FMR1* gene.** However, some point mutations and deletions in the *FMR1* gene show the same phenotype as the repeat expansion. The alteration arising from the trinucleotide repeat expansion within the promoter region of *FMR1* leads to enhanced local methylation and ultimately transcriptional silencing. DNA methylation causes protrusions in the DNA helix where the methylated cytosines interfere with transcription factor binding. This is a common property used for gene regulation. However, in the case of fragile X, the methylation signals to halt all expression of FMRP. Loss of FMRP gives the phenotypical characteristics commonly seen in fragile X patients.

COMPREHENSION QUESTIONS

[13.1] A 6-year-old boy visits his physician because his parents have noticed autistic behavior and speech problems. The mother's family does have a history of mental retardation. Therefore, the physician suggested a genetic screen of the fragile X mental retardation 1 (*FMR1*) gene for fragile X syndrome. Polymerase chain reaction (PCR) revealed borderline fragile X syndrome.

What situation most likely explains this result?

A. A complete loss of fragile X mental retardation protein (FMRP)
B. An *FMR1* gene CGG repeat expansion of 60 with partial DNA methylation
C. An *FMR1* gene CGG repeat expansion of 230 with minor DNA methylation
D. An *FMR1* gene CGG repeat expansion of 280 with complete DNA methylation

[13.2 to 13.3] Use the genetic code for Questions [13.2] and [13.3].

	T	C	A	G	
T	Phe	Ser	Tyr	Cys	**T**
	Phe	Ser	Tyr	Cys	**C**
	Leu	Ser	STOP	STOP	**A**
	Leu	Ser	STOP	Trp	**G**
C	Leu	Pro	His	Arg	**T**
	Leu	Pro	His	Arg	**C**
	Leu	Pro	Gln	Arg	**A**
	Leu	Pro	Gln	Arg	**G**
A	Ile	Thr	Asn	Ser	**T**
	Ile	Thr	Asn	Ser	**C**
	Ile	Thr	Lys	Arg	**A**
	Met	Thr	Lys	Arg	**G**
G	Val	Ala	Asp	Gly	**T**
	Val	Ala	Asp	Gly	**C**
	Val	Ala	Glu	Gly	**A**
	Val	Ala	Glu	Gly	**G**

Using the genetic code above, predict the type of mutation that had to occur to show these alterations in the final protein product for Questions [13.2] and [13.3]:

[13.2]

Phe	Thr	Val	Tyr	Leu	Gly	Met	→	Phe	Thr	Val	STOP
TTT	ACA	GTT	TAT	CTC	GGG	ATG					

A. Missense mutation
B. Nonsense mutation
C. Silent mutation
D. Repeat expansion

[13.3]

Phe	Thr	Val	Tyr	Leu	Gly	Met	→	Phe	Thr	Phe	Ile	STOP
TTT	ACC	GTT	TAT	CTA	GGG	ATG						

A. Deletion of G in third codon
B. Deletion of A in second codon
C. Insertion of T between second and third codons
D. Insertion of GT between AC and A of second codon

Answers

[13.1] **C.** A CGG repeat expansion of 230 base pairs is just at the limit of the number of repeats required to have the full mutation. Expansions from 7 to 60 are seen in normal patients and those exceeding 230 are positive for fragile X syndrome. Because some of the DNA was methylated, this could help to silence the expression of FMRP and cause some of the phenotype seen in the patient. Therefore, this scenario would be considered borderline fragile X.

[13.2] **B.** A nonsense mutation causes a premature stop codon on a single nucleotide substitution. The original sequence, TAT, coding for a tyrosine, must have been mutated to either TA**A** or TA**G**, leaving a stop codon in place of the tyrosine.

[13.3] **A.** Deletion of G in the third codon gives the sequence:

TTT ACC **G**TT TAT CTA GGG ATG →
TTT ACC TTT ATC TAG GGA TG

This causes a frameshift and a premature stop codon to be generated. Therefore, the final protein product would look like the one in question.

Phe	Thr	Val	Tyr	Leu	Gly	Met
TTT	ACC	GTT	TAT	CTA	GGG	ATG

→

Phe	Thr	Phe	Ile	STOP
TTT	ACC	TTT	ATC	TAG

BIOCHEMISTRY PEARLS

- Mutations are alterations in the DNA base sequence that do not get repaired.
- There are various types of mutations including point mutations, deletions, and insertions.
- Point mutations arise when one base pair is substituted by another, and they are the most common types of mutations.
- Point mutations can be further subdivided into three categories: silent, missense, and nonsense.
- Many fragile X individuals have an increase in the number of CGG repeats, which may expand to over 230 repeats.

REFERENCES

Crawford DC, Acuna JM, Sherman SL. FMR1 and the fragile X syndrome: human genome epidemiology review. Genet Med 2001 Sep–Oct;3(5):359–71.

Warren ST, Sherman SL. The fragile X syndrome. In: Scriver CR, Beaudet AL, Sly WS, et al., eds. The Metabolic and Molecular Basis of Inherited Disease, 8th ed. New York: McGraw-Hill, 2001.

❖ CASE 14

A 40-year-old female presents to the emergency department with complaints of lower back pain, fever, nausea, vomiting, malaise, chills, syncope, dizziness, and shortness of breath. Patient states that she has some burning with urination (dysuria). Her fever was as high as 39.4°C (103°F) at home earlier in the day. She has a history of non–insulin-dependent diabetes mellitus but denies any other medical problems. On exam, she is in moderate distress with a temperature of 38.9°C (102°F) degrees, pulse of 110 beats per minute, respiratory rate of 30 breaths per minute, and blood pressure of 70/30 mm Hg. Her extremities are cool to the touch with thready pulses. Her chest is clear to auscultation bilaterally, and heart is tachycardic but with regular rhythm. She has significant costovertebral tenderness on the right side. Her white blood cell (WBC) count was elevated at 20,000/mm^3. The hemoglobin and hematocrit were normal. Her glucose was moderately elevated at 200 mg/dL, and her serum bicarbonate level is low. An arterial blood gas demonstrated a pH of 7.28 and parameters consistent with a metabolic acidosis. Her urinalysis shows an abnormal number of gram-negative rods.

◆ **What is the most likely diagnosis?**

◆ **What is the biochemical mechanism of the metabolic acidosis?**

ANSWERS TO CASE 14: ANAEROBIC METABOLISM

Summary: 40-year-old female with diabetes presents with fever 38.9°C (102°F), chills, nausea, vomiting, back pain (costovertebral tenderness), chills, increased WBC count, hypotension, and metabolic acidosis.

- **Most likely diagnosis:** Septic shock and pyelonephritis.
- **Likely cause of the metabolic acidosis:** Lactic acid produced from cells undergoing anaerobic metabolism as a result of tissue hypoperfusion from shock.

CLINICAL CORRELATION

This patient developed septic shock, infection related hypotension, and low blood pressure. The decreased blood pressure then led to insufficient red blood cells and oxygen to be delivered to the various tissues in the body. Thus, the tissue had to switch from aerobic metabolism to anaerobic metabolism. Lactate accumulates, leading to acidemia. The treatment of septic shock is initially intravenous fluids, since the body is relatively volume depleted as a result of the vasodilation response to the infection. Sometimes, the blood pressure remains low despite several liters of intravenous fluids; in these cases, so-called vasoactive drugs are used such as dopamine infusion to cause vasoconstriction and therefore elevate the blood pressure. Antibiotics are also important to treat the infection. Finally, control of the source of the sepsis is critical. This may include surgery to remove an abscess or necrotic bowel, or removal of a foreign body, or simply debridement. Thus, septic shock is treated by supporting the blood pressure, antibiotics, and source control. Worsening of the acidemia and accumulation of lactate is a poor prognostic sign in septic shock.

APPROACH TO AEROBIC AND ANAEROBIC METABOLISM

Objectives

1. Be very familiar with the tricarboxylic acid (TCA) cycle.
2. Know about the differences in energy production in aerobic and anaerobic conditions.

Definitions

Acceptor control: The regulation of the rate of an enzymatic reaction by the concentration of one or more of the substrates.

Anaerobic glycolysis: The biochemical process by which glucose is converted to lactate with the production of two moles of ATP. This process is increased when the cellular demand for ATP is greater than the ability of the TCA cycle and oxidative phosphorylation to produce it because of decreased oxygen tension or heavy exercise.

Glycolysis: The biochemical process by which glucose is converted to pyruvate in the cytosol of the cell. It results in the production of 2 mol of adenosine triphosphate (ATP) and 2 mol of the reduced cofactor nicotinamide adenine dinucleotide (NADH), which transfers its reducing equivalents to the mitochondrion for the production of ATP via oxidative phosphorylation.

Krebs cycle: Citric acid cycle, TCA cycle, the mitochondrial process by which acetyl groups from acetyl-CoA are oxidized to CO_2. The reducing equivalents are captured as NADH and $FADH_2$, which feed into the electron transport system of the mitochondrion to produce ATP via oxidative phosphorylation.

DISCUSSION

Most of the energy that the body requires for maintenance, work, and growth is obtained by the **terminal oxidation of acetyl coenzyme A (acetyl-CoA) that is produced by the catabolism of carbohydrates, fatty acids, and amino acids.** The oxidation of acetyl-CoA is **achieved by mitochondrial enzymes that make up the tricarboxylic acid cycle (TCA cycle, also called the citric acid cycle or the Krebs cycle). All of the enzymes** in this metabolic pathway are **located in the matrix of the mitochondria except one, succinate dehydrogenase. Succinate dehydrogenase** is a **membrane-bound protein located on the inner mitochondrial membrane facing the matrix.** The two carbons that enter the TCA cycle as an **acetyl group** are effectively **oxidized to carbon dioxide. Oxygen** is not directly involved in this process; instead, reducing equivalents are captured by the electron carriers NAD^+ and FAD producing three (NADH + H^+) and one $FADH_2$. A high-energy phosphate bond is also produced in the form of GTP. The reduced cofactors NADH and $FADH_2$ are then reoxidized by passing their reducing equivalents to O_2 through the electron transport system (ETS) of the mitochondria with the production of water and ATP via oxidative phosphorylation.

The **TCA cycle** is shown in Figure 14-1. In the first step, **acetyl-CoA is condensed with oxaloacetate** to produce citrate with the release of free coenzyme A in a reaction catalyzed by **citrate synthase.** Citrate is then **isomerized to isocitrate by the enzyme aconitase.** The next step involves the **oxidative decarboxylation of isocitrate to produce α-ketoglutarate.** The enzyme catalyzing this reaction, **isocitrate dehydrogenase (IDH)** requires the oxidized **electron carrier NAD^+** to accept the electrons released in the oxidation and produce the first NADH + H^+. **α-Ketoglutarate is then converted to succinyl-CoA** in a second oxidative decarboxylation reaction catalyzed by **α-ketoglutarate dehydrogenase (α-KGDH).** This reaction requires the participation of two cofactors, oxidized NAD^+ and CoA, and results in the production of a second NADH + H^+. **Succinyl-CoA** is then transformed to **succinate by**

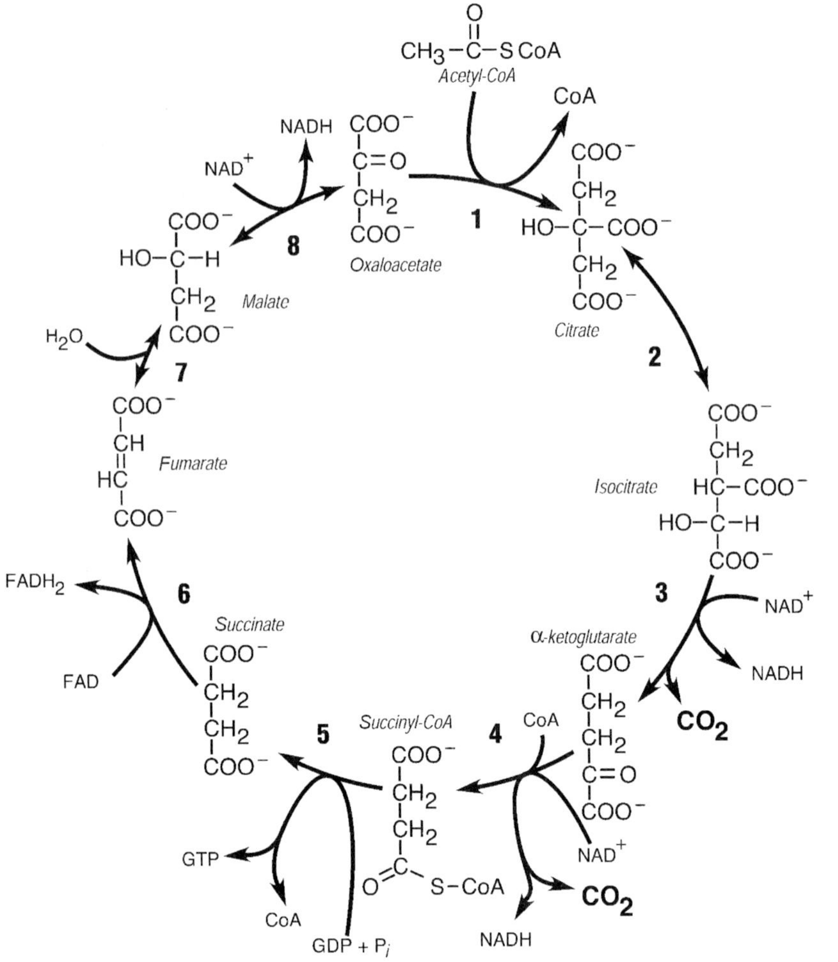

Figure 14-1. The tricarboxylic acid (TCA) cycle, also known as the citric acid cycle and the Krebs cycle. The overall reaction is

$$\text{AcCoA} + 3\text{NAD}^+ + \text{FAD} + \text{GDP} + \text{P}_\text{i} \rightarrow$$
$$2\text{CO}_2 + 3\text{NADH} + \text{FADH}_2 + \text{GTP} + \text{CoA}$$

1 Citrate synthase
2 Aconitase
3 Isocitrate
4 α-Ketoglutarate dehydrogenase
5 Succinate thiokinase
6 Succinate dehydrogenase
7 Fumarase
8 Malate dehydrogenase

succinate thiokinase with the release of free CoA. The high energy released in the hydrolysis of the thioester bond of succinyl-CoA drives the **phosphorylation of GDP to produce GTP. Succinate is then converted to oxaloacetate** in a series of reactions reminiscent of the β-oxidation of fatty acids. A double bond is introduced into succinate in an oxidation catalyzed by **succinate dehydrogenase, the only membrane-bound enzyme in the TCA cycle.** The reducing equivalents are captured as $FADH_2$. The resulting **fumarate is then hydrated to form malate by the enzyme fumarase. Oxaloacetate** is then **regenerated by the oxidation of malate by malate dehydrogenase** with the production of another NADH + H^+.

Although the TCA cycle does not directly involve the participation of molecular **oxygen** as a substrate, the reduced cofactors NADH and $FADH_2$ **must be reoxidized** by the mitochondrial ETS for continued operation of the cycle. Thus, the **rate of oxidation of acetyl-CoA by the TCA cycle** is to a large degree **regulated** by the availability of the **oxidized cofactors NAD^+ and FAD.** When tissues do not receive enough O_2 as a result of hypoperfusion, the ETS cannot regenerate these cofactors at a sufficient rate and the reactions that use them are inhibited because one of the required substrates is lacking (acceptor control). Thus, acetyl-CoA builds up in the mitochondria, depleting free CoA levels. Decreased levels of NAD^+ and CoA also inhibit the conversion of pyruvate to acetyl-CoA by pyruvate dehydrogenase.

When the ETS and oxidative phosphorylation are compromised by a **lack of O_2, cytosolic levels of ATP drop while ADP and AMP levels increase.** The flow of glucose through the glycolytic pathway is increased because of **allosteric activation of phosphofructokinase and pyruvate kinase** by the drop in ATP and rise in AMP concentrations. This results in an **increased production of NADH and pyruvate by the glycolytic pathway.** However, since electron transport is inhibited by a lack of O_2, **regeneration of NAD^+** by the **glycerol 3-phosphate or malate-aspartate shuttles,** and the **ETS is also inhibited.** To continue the production of ATP by the glycolytic pathway, **NAD^+ must be regenerated** to accept more reducing equivalents when **glyceraldehyde 3-phosphate is oxidized to 1,3-bisphosphoglycerate. Lactate dehydrogenase** is activated by increased levels of pyruvate and NADH and converts pyruvate to lactate with the regeneration of NAD^+ (Figure 14-2). **Lactate** that cannot be reused by the cells that produce it is **transported out of the cell to the bloodstream.**

Lactate has two metabolic fates, either complete combustion to CO_2 and H_2O or conversion back to glucose through gluconeogenesis. Both processes require an active ETS and oxidative phosphorylation. Reduced oxygenation of cells thus decreases the utilization of lactate and increases its production with a resulting **lactic acidosis.**

When tissues are hypoperfused, the resulting **anaerobic** conditions have energetic consequences. First, **all catabolic processes that require an active ETS and oxidative phosphorylation (e.g., β-oxidation of fatty acids and amino acid breakdown) are inhibited.** Thus, certain energy stores cannot be

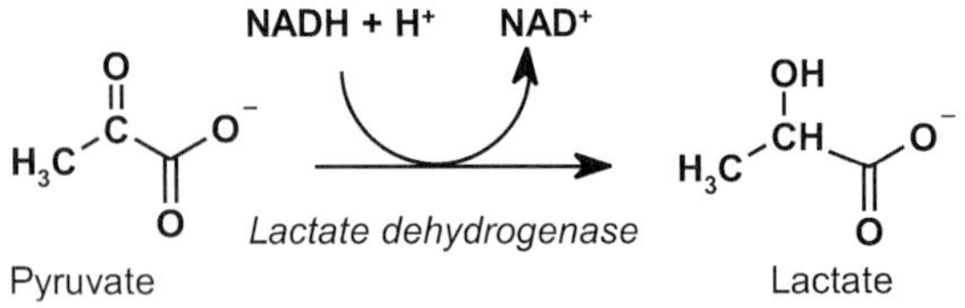

Figure 14-2. The conversion of pyruvate to lactate by the enzyme pyruvate dehydrogenase. The reaction, which is reversible, uses the reducing equivalents from NADH and regenerates NAD^+ for the continuation of the glycolytic pathway when oxygen is limiting.

used by the cell. Energy needs must be met by catabolism of carbohydrates through the glycolytic pathway. However, since under these anaerobic conditions, **pyruvate dehydrogenase, the TCA cycle and oxidative phosphorylation are compromised,** only a fraction of the chemical energy that is obtained from the oxidation of glucose under aerobic conditions is produced. Glucose that is completely oxidized to CO_2 and H_2O under aerobic conditions will produce between **36 and 38 mol of ATP per mole of glucose** depending on the shuttle used to transport cytosolic reducing equivalents to the mitochondria. **Only 2 mol of ATP are produced per mole of glucose converted to lactate through anaerobic glycolysis.** The ATP needs of the cell are met by the increased rate of the glycolytic pathway.

COMPREHENSION QUESTIONS

[14.1] Phil Hardy has decided to train for an upcoming marathon. Nearing the age of 50, Phil figures that after he trains he should be able to maintain a 9 minute-per-mile pace, which would mean that he would finish the race in approximately 4 hours. Given that he would be adequately hydrating himself at the various water stations along the way, as he is about to finish the 26 mile 385 yard course, what is the primary fuel that his leg muscles would be using?

A. Fatty acids from the blood
B. Glycerol from the blood
C. Glycogen stored in muscle
D. Glucose from the blood
E. Ketone bodies from the blood

[14.2] A postoperative patient on intravenous fluids develops lesions in the mouth (angular stomatitis). Urinalysis indicates an excretion of 15 μg riboflavin/mg creatinine, which is abnormally low. Which of the following TCA cycle enzymes is most likely to be affected?

A. Citrate synthase
B. Isocitrate dehydrogenase
C. Fumarase
D. Malate dehydrogenase
E. Succinate dehydrogenase

[14.3] After excessive drinking over an extended period of time while eating poorly, a middle-aged man is admitted to the hospital with "high output" heart failure. Which of the following enzymes is most likely inhibited?

A. Aconitase
B. Citrate synthase
C. Isocitrate dehydrogenase
D. α-Ketoglutarate dehydrogenase
E. Succinate thiokinase

Answers

[14.1] **A.** After 4 hours of heavy exercise, glycogen stored in muscle cells has been expended. Free glycerol cannot be used by the muscle cell because it does not have the enzyme (glycerol kinase) that will phosphorylate it so that it can enter the glycolytic pathway. Although glucose and ketone bodies can be taken up by the muscle cells and used for energy, fatty acid oxidation provides most of the ATP for the marathon runner at this point in the race.

[14.2] **E.** The patient has demonstrated a deficiency in riboflavin (urinary excretion of less than 30 μg/mg creatinine is considered clinically deficient). Riboflavin is a component of the cofactor FAD (flavin adenine dinucleotide), which is required for the conversion of succinate to fumarate by succinate dehydrogenase.

[14.3] **D.** This patient has exhibited symptoms of beri beri heart disease, which is a result of a nutritional deficiency in vitamin B_1 (thiamine). The active form of the vitamin, thiamine pyrophosphate, is a required cofactor for α-ketoglutarate dehydrogenase.

BIOCHEMISTRY PEARLS

- Most of the energy that the body requires for maintenance, work, and growth is obtained by the terminal oxidation of acetyl coenzyme A (acetyl-CoA), which is produced by the catabolism of carbohydrates, fatty acids, and amino acids.
- The oxidation of acetyl-CoA is achieved by mitochondrial enzymes that make up the tricarboxylic acid cycle (TCA cycle, also called the citric acid cycle or the Krebs cycle).
- When the ETS and oxidative phosphorylation are compromised by a lack of O_2, cytosolic levels of ATP drop, while ADP and AMP levels increase.
- The flow of glucose through the glycolytic pathway is increased because of allosteric activation of phosphofructokinase and pyruvate kinase by the drop in ATP and rise in AMP concentrations.
- The four carbon intermediates of the TCA cycle may be replenished or increased by metabolites of the glucogenic amino acids entering at α-ketoglutarate, succinyl-CoA, or oxaloacetate. In addition, the C_4 pool can also be increased by the carboxylation of pyruvate to oxaloacetate catalyzed by pyruvate carboxylase.

REFERENCES

Beattie DS. Bioenergetics and oxidative metabolism. In: Devlin, TM, ed. Textbook of Biochemistry with Clinical Correlations, 5th ed. New York: Wiley-Liss, 2002.

Harris RA. Carbohydrate metabolism I: major metabolic pathways and their control. In: Devlin TM, ed. Textbook of Biochemistry with Clinical Correlations, 5th ed. New York: Wiley-Liss, 2002.

CASE 15

A 59-year-old male is brought to the emergency department by the EMS after a family member found him extremely confused and disoriented, with an unsteady gait and strange irregular eye movements. The patient has been known in the past to be a heavy drinker. He has no known medical problems and denies any other drug usage. On examination, he is afebrile with a pulse of 110 beats per minute and a normal blood pressure. He is extremely disoriented and agitated. Horizontal rapid eye movement on lateral gaze is noted bilaterally. His gait is very unsteady. The remainder of his examination is normal. The urine drug screen was negative and he had a positive blood alcohol level. The emergency room physician administers thiamine.

- **What is the most likely diagnosis?**
- **What is importance of thiamine in biochemical reactions?**

ANSWERS TO CASE 15: THIAMINE DEFICIENCY

Summary: A 59-year-old male with history of heavy alcohol use presents with mental confusion, ataxia, and ophthalmoplegia.

- **Most likely diagnosis:** Wernicke-Korsakoff syndrome (thiamine deficiency) often associated with chronic alcoholics.
- **Importance of thiamine:** An important water-soluble vitamin used as a cofactor in enzymatic reactions involving the transfer of an aldehyde group. Without thiamine, individuals can develop dementia, macrocytic anemia (folate deficiency), gastritis, peptic ulcer disease, liver disease, depression, nutritional deficiencies, cardiomyopathy, and pancreatitis.

CLINICAL CORRELATION

Thiamine, also known as vitamin B_1, is fairly ubiquitous. Thiamine deficiency is uncommon except in alcoholics as a result of nutritional deficiencies and malabsorption. The classic clinical triad of dementia, ataxia (difficulty with walking), and eye findings may be seen, but more commonly, only forgetfulness is noted. Sometimes, thiamine deficiency can lead to vague symptoms such as leg numbness or tingling. Because thiamine is water soluble, it can be added to intravenous fluids and administered in that way. Other manifestations include beri beri, which is cardiac involvement leading to a high cardiac output, and vasodilation. Affected patients often feel warm and flushed, and they can have heart failure.

APPROACH TO THIAMINE PYROPHOSPHATE

Objectives

1. Know about the role of thiamine pyrophosphate (TPP) in pyruvate dehydrogenase.
2. Understand the role of TPP in α-ketoglutarate dehydrogenase and the role of TPP in transketolase in pentose phosphate pathway.
3. Be familiar with how thiamine deficiency results in decreased energy generation and how it results in decreased ribose and NADPH production.

Definitions

Carbanion: A carbon within a molecule that has a negative charge because of the removal of a proton (hydrogen ion).

Decarboxylation: The process of removing a carboxyl group (-COOH) from a molecule. Frequently this is achieved by oxidizing the compound in a process known as oxidative decarboxylation.

Dehydrogenase: An enzyme that oxidizes a molecule by the removal of a pair of electrons and one or two protons.

Nucleophilic addition: The process of forming a bond between an electron-rich group (nucleophile) and an electron-deficient atom (electrophile).

Oxidation: The removal of electrons from an atom or compound.

Thiamine: Vitamin B_1, a water-soluble vitamin containing a thiazolium ring. The active form of the vitamin is thiamine pyrophosphate, which is an important coenzyme for many biochemical reactions.

DISCUSSION

Thiamine (vitamin B_1) is an important water-soluble vitamin that, in its **active form of thiamine pyrophosphate,** is used as a **cofactor in enzymatic reactions that involve the transfer of an aldehyde group.** Thiamine can be synthesized by plants and some microorganisms, but not usually by animals. Hence, **humans must obtain thiamine from the diet,** though small amounts may be obtained from synthesis by intestinal bacteria. Because of its importance in metabolic reactions, it is present in large amounts in skeletal muscle, heart, liver, kidney, and brain. Thus, it has a widespread distribution in foods, but there can be a substantial loss of thiamine during cooking above 100°C (212°F).

Thiamine is absorbed in the intestine by both active transport mechanisms and passive diffusion. The active form of the cofactor, **thiamine pyrophosphate (thiamine diphosphate, TPP), is synthesized by an enzymatic transfer of a pyrophosphate group from ATP to thiamine** (Figure 15-1). The resulting TPP has a reactive carbon on the thiazole ring that is easily ionized to form a carbanion, which can undergo nucleophilic addition reactions.

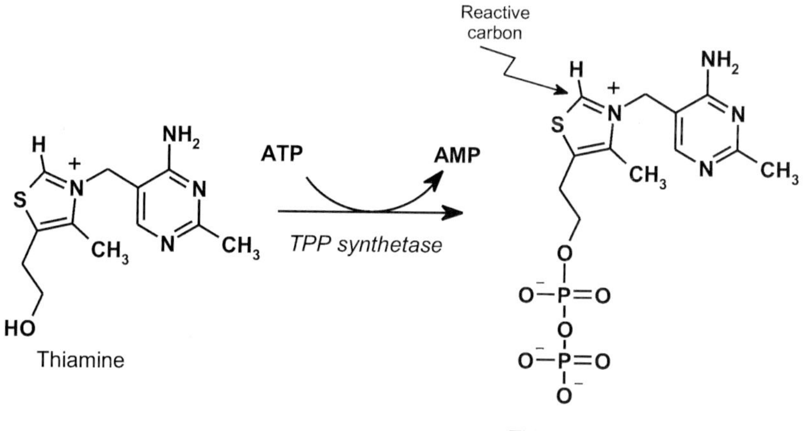

Figure 15-1. Activation of thiamine (vitamin B_1) to the active cofactor, thiamine pyrophosphate (TPP) by the enzyme TPP synthetase.

Thiamine pyrophosphate is an essential cofactor for enzymes that catalyze the **oxidative decarboxylation of α-keto acids to form an acylated coenzyme A (acyl CoA).** These include **pyruvate dehydrogenase (pyruvate → acetyl-CoA), α-ketoglutarate dehydrogenase (α-KG → succinyl-CoA) and branched-chain α-keto acid dehydrogenase.** These three enzymes operate by a similar catalytic mechanism (Figure 15-2). Each of these enzymes is a multi-subunit complex with three enzymatic activities that require the participation of several cofactors. The first activity (E_1) is the **dehydrogenase complex,** which has TPP as the enzyme-bound cofactor. The carbanion form of TPP attacks the carbonyl carbon of the α-keto acid, releasing CO_2 and leaving a hydroxyalkyl-TPP intermediate. The hydroxyalkyl group reacts with the oxidized form of lipoamide, the prosthetic group of dihydrolipoyl transacetylase (E_2), which is the second component of the complex. The resulting acyl-lipoamide reacts with coenzyme A (HSCoA) to form the product, acyl CoA, and leaving lipoamide in the fully reduced form. To regenerate the fully oxidized form of lipoamide for further rounds of the reaction, it interacts with the **third component of the complex, dihydrolipoyl dehydrogenase (E_3),** which has covalently bound **FAD.** The FAD accepts the reducing equivalents from the reduced lipoamide to form $FADH_2$ **and oxidized lipoamide.** The reducing equivalents are then transferred to **NAD^+ to form NADH,** which is **regenerated via the electron transport system with the production of ATP.**

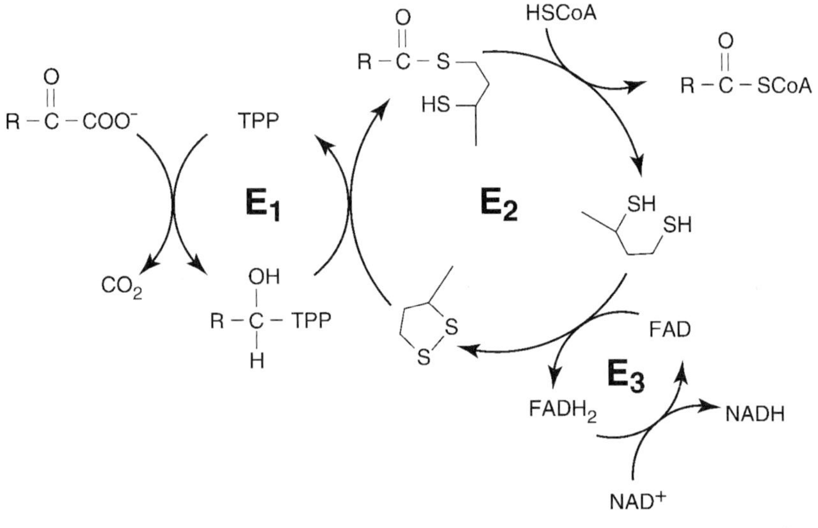

Figure 15-2. The catalytic mechanism shared by the enzymes pyruvate dehydrogenase, α-ketoglutarate dehydrogenase, and branched-chain α-ketoacid dehydrogenase. E_1 is the dehydrogenase complex; E_2 is the dihydrolipoyl transacetylase subunit, and E_3 is the dihydrolipoyl dehydrogenase component. E_1 and E_2 are specific to each enzyme, and E_3 is common to all three enzymes.

The E_1 and E_2 components are specific for each of the pyruvate dehydrogenase, α-ketoglutarate dehydrogenase, and branched-chain α-keto acid dehydrogenase complexes. However, the E_3 **component is identical** for all of the enzymes.

Thiamine pyrophosphate is also an important cofactor for the **transketolase** reactions in the **pentose phosphate pathway of carbohydrate metabolism** (Figure 15-3). These reactions are important in the **reversible transformation of pentoses into the glycolytic intermediates fructose 6-phosphate and glyceraldehyde 3-phosphate.** Again, it is the reactive carbon on the thiazole ring of TPP that reacts with a ketose phosphate (xylulose 5-phosphate) to cause the release of an aldose phosphate with two fewer carbons (glyceraldehyde 3-phosphate). The TPP-bound glycoaldehyde unit is then transferred to a different aldose phosphate (ribose 5-phosphate or erythrose 4-phosphate) to produce a ketose phosphate that has two carbons more (sedoheptulose 7-phosphate or fructose 6-phosphate).

A **deficiency in thiamine will decrease the efficiency of the enzymes for which TPP is required as a cofactor.** Thus, the rate of conversion of pyruvate to acetyl-CoA and the flow of acetyl-CoA through the tricarboxylic acid cycle will be depressed as a result of the inefficiency of the TPP-requiring enzymes pyruvate dehydrogenase and α-ketoglutarate dehydrogenase. The production of the reduced electron carrier, NADH, and the ATP produced from it via oxidative phosphorylation will be decreased as a consequence. Because **nervous tissue and heart use at high rates ATP** synthesized from the **oxidation of NADH produced from pyruvate conversion to acetyl-CoA** and from the TCA cycle, **these tissues are most affected by a deficiency in thiamine.** When deficient in thiamine, the brain can no longer efficiently metabolize pyruvate through the TCA cycle to produce ATP and thus must convert it to lactate to produce ATP.

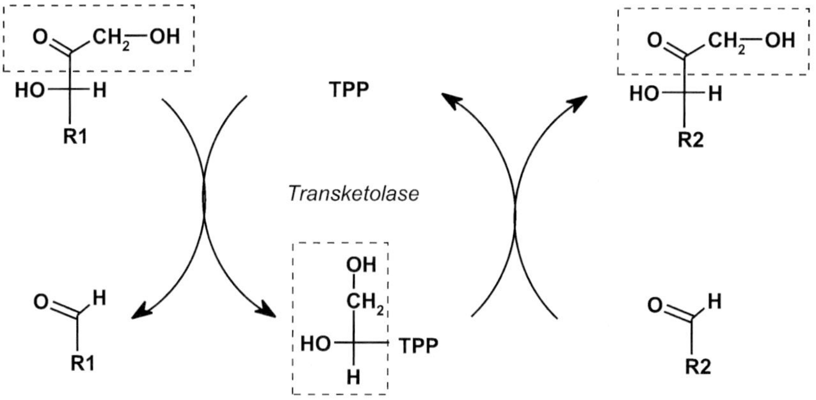

Figure 15-3. The reaction catalyzed by the enzyme transketolase, which transfers a glycoaldehyde group from a ketose to an aldose.

This increased conversion of pyruvate to lactate decreases the pH in areas of the brain that have rapid ATP turnover rates and leads to cellular destruction.

A **deficiency in thiamine also adversely affects the flux of glucose metabolized by the pentose phosphate pathway. Transketolase activity requires the cofactor TPP to transfer the glycoaldehyde unit from a ketose to an aldose** in the remodeling reactions of the pathway. When these reactions cannot proceed, precursor metabolites build up, and the flow through the pathway is decreased. This results in a **decreased production of NADPH and decreased conversion of glucose to pentose, including ribose.** This can lead to decreased regeneration of reduced glutathione and **susceptibility to oxidative stress.**

Thiamine turnover is rapid because of the ubiquitous presence of **thiaminase enzymes** that **hydrolyze thiamine into its pyrimidine and thiazole components.** Thus, **symptoms of thiamine deficiency can appear within 2 weeks of a diet depleted in thiamine.** In Western societies, severe thiamine deficiency is most frequently found in **alcoholics.** Patients who chronically misuse alcohol are prone to thiamine deficiency arising from a number of factors including poor nutrition and poor absorption and storage, as well as an increased breakdown of TPP. Alcohol is known to inhibit the active absorption of thiamine.

Thiamine deficiency is most frequently assessed by assaying erythrocyte transketolase activity in the presence and absence of added TPP. If the red blood cells have sufficient thiamine, the transketolase will be fully saturated with TPP, and no increase in activity will be observed when TPP is added to the assay system. An increase in transketolase activity indicates that the patient is thiamine deficient.

COMPREHENSION QUESTIONS

[15.1] A full-term female infant failed to gain weight and showed metabolic acidosis in the neonatal period. A physical examination at 6 months showed failure to thrive, hypotonia, small muscle mass, severe head lag, and a persistent acidosis (pH 7.0 to 7.2). Blood lactate, pyruvate, and alanine were greatly elevated. Treatment with thiamine did not alleviate the lactic acidosis.

Which of the following enzymes is most likely deficient in this patient?

A. Alanine aminotransferase
B. Phosphoenolpyruvate carboxykinase
C. Pyruvate carboxylase
D. Pyruvate dehydrogenase
E. Pyruvate kinase

[15.2] A 3-month-old male infant developed seizures and progressively worsened, showing hypotonia, psychomotor retardation, and poor head control. He had lactic acidosis and an elevated plasma pyruvate level, both more than seven times the normal amount. Pyruvate carboxylase activity was measured using extracts of fibroblasts and was found to be less than 1 percent of the normal level. Oral administration of which of the following amino acids would you recommend as the best therapy for this patient?

A. Alanine
B. Glutamine
C. Leucine
D. Lysine
E. Serine

[15.3] A deficiency in thiamine (vitamin B_1) would most likely lead to which of the following clinical manifestations?

A. A decrease in carboxylase enzyme activity
B. A decrease in serum lactate concentrations
C. A decrease in red blood cell transketolase activity
D. An increase in urinary methylmalonate
E. An increase in prothrombin time

Answers

[15.1] **D.** The increased concentrations of pyruvate, lactate, and alanine indicate that there is a block in the pathway leading from pyruvate toward the TCA cycle. A deficiency in pyruvate dehydrogenase would lead to a buildup of pyruvate. Pyruvate has three fates other than conversion to acetyl-CoA by pyruvate dehydrogenase: conversion to oxaloacetate by pyruvate carboxylase, reduction to lactate by lactate dehydrogenase, and transamination to the amino acid alanine. Thus, because pyruvate builds up, an increase in lactate and alanine would be expected if pyruvate dehydrogenase was deficient.

[15.2] **B.** A deficiency in pyruvate carboxylase results in a diminution of oxaloacetate, the C_4 acid that acts as the acceptor for an acetyl group from acetyl-CoA. In order for the TCA cycle to continue efficiently, C_4 acids must be replenished. Amino acids whose carbon skeletons feed into the TCA cycle and increase the C_4 pool will accomplish this. Glutamine, which is converted to α-ketoglutarate, will lead to an increase in all of the C_4 acids (succinate, fumarate, malate, and oxaloacetate). Alanine and serine are converted to pyruvate, which as a result of the deficiency in pyruvate carboxylase will not increase the C_4 pool. Lysine and leucine are ketogenic amino acids and thus also do not increase the C_4 pool.

[15.3] **C.** In addition to being an important cofactor for the enzymes involved in the oxidative decarboxylation of pyruvate, α-ketoglutarate, and branched-chain α-ketoacids, thiamine is also a cofactor for the enzyme transketolase, the enzyme that transfers a glycoaldehyde group from a ketose sugar to an aldose sugar in the pentose phosphate pathway. One of the diagnostic tools in determining a thiamine deficiency is determination of the activity of red blood cell transketolase in the presence and absence of added thiamine. A thiamine deficiency would be expected to increase blood lactate concentrations. A deficiency of biotin would lead to decreased carboxylase activity, whereas an increased methylmalonate concentration would be observed with a deficiency in vitamin B_{12}. A deficiency in vitamin K would lead to an increase in prothrombin time.

BIOCHEMISTRY PEARLS

- Thiamine (vitamin B_1) is an important water-soluble vitamin that, in its active form of thiamine pyrophosphate, is used as a cofactor in enzymatic reactions that involve the transfer of an aldehyde group.
- Thiamine deficiency is uncommon except in alcoholics who, as a consequence of nutritional deficiencies and malabsorption, may become deficient.
- The classic clinical triad of dementia, ataxia (difficulty with walking), and eye findings may be seen, but more commonly, only forgetfulness is noted.
- Thiamine pyrophosphate is also an important cofactor for many dehydrogenase reactions as well as the transketolase reactions in the pentose phosphate pathway of carbohydrate metabolism.

REFERENCES

Murray RK, Granner DK, Mayes PA, et al. Harper's Illustrated Biochemistry, 26th ed. New York: Lange Medical Books/McGraw-Hill, 2003.

Wilson JD. Vitamin deficiency and excess. In: Fauci AS, Braunwald E, Isselbacher KJ, et al., eds. Harrison's Principles of Internal Medicine, 14th ed. New York: McGraw-Hill, 1998.

CASE 16

A 68-year-old female in a hypertensive crisis is being treated in the intensive care unit (ICU) with intravenous nitroprusside for 48 hours. The patient's blood pressure was brought back down to normal levels; however, she was complaining of a burning sensation in her throat and mouth followed by nausea and vomiting, diaphoresis, agitation, and dyspnea. The nurse noticed a sweet almond smell in her breath. An arterial blood gas revealed a significant metabolic acidosis. A serum test suggests a metabolite of nitroprusside, thiocyanate, is at toxic levels.

- **What is the likely cause of her symptoms?**
- **What is the biochemical mechanism of this problem?**
- **What is the treatment for this condition?**

ANSWERS TO CASE 16: CYANIDE POISONING

Summary: A 69-year-old female with new onset burning sensation in mouth and throat, nausea and vomiting, agitation, and diaphoresis after a medication error was noted. Metabolic acidosis is seen on the arterial blood gas. A thiocyanate level is in the toxic range.

- **Diagnosis:** Cyanide poisoning from toxic dose of nitroprusside.
- **Biochemical mechanism:** Cyanide inhibits mitochondrial cytochrome oxidase, blocking electron transport and preventing oxygen utilization. Lactic acidosis results secondary to anaerobic metabolism.
- **Treatment:** Supportive therapy, gastrointestinal (GI) decontamination, oxygen, and antidotal therapy with amyl nitrite, sodium nitrite, and sodium thiosulfate.

CLINICAL CORRELATION

Hypertensive emergencies are defined as episodes of severely elevated blood pressure, such as systolic levels of 220 mm Hg and/or diastolic blood pressures exceeding 120 mm Hg *with* patient symptoms of end-organ dysfunction. These symptoms may include severe headache, neurological deficits, chest pain, or heart failure symptoms. Hypertensive emergencies require immediate lowering of the blood pressure to lower (but not necessarily to normal) levels. In contrast, hypertensive urgencies are circumstances of markedly elevated blood pressures in the absence of patient symptoms; lowering the blood pressure over 24 to 48 hours is reasonable in these cases.

One hazard of abruptly lowering the blood pressure is causing hypotension and subsequent ischemia to the brain or heart. In other words, the very treatment designed to prevent end-organ disease may cause the problem. To avoid precipitous hypotension, agents that induce a smooth fall in blood pressure are preferable, such as sodium nitroprusside, a titratable intravenous agent used for malignant hypertension. Its desirable properties include the ability to precisely increase or decrease the infusion to affect the blood pressure. One side effect of sodium nitroprusside is that its metabolite is thiocyanate, and with prolonged use, cyanide poisoning may result, which inhibits the electron transport chain. Thus, in clinical practice, short-term nitroprusside is used, or serum thiocyanate levels are drawn.

APPROACH TO ELECTRON TRANSPORT SYSTEM (ETS) AND CYANIDE

Objectives

1. Know about the function of the electron transport chain (ETC).
2. Understand what factors may inhibit the ETC.

3. Be familiar with the biochemical process by which the therapy for cyanide poisoning works. (Nitrates convert the hemoglobin to methemoglobin which has a higher affinity for cyanide and promotes dissociation from cytochrome oxidase. Thiosulfate reacts with cyanide which is slowly released from cyanomethemoglobin to form thiocyanate. Oxygen reverses the binding of cyanide to cytochrome oxidase.)
4. Recognize other ETC sites and agents of inhibition.

Definitions

Oxidative phosphorylation: The mitochondrial process whereby electrons from NADH or reduced flavin bound in enzymes are transferred down the electron transport chain to oxygen forming water and providing energy through the formation of an hydrogen ion gradient across the inner mitochondrial membrane. The hydrogen ion gradient is used to drive the formation of ATP from ADP and inorganic phosphate (P_i). This process is also called *coupled oxidative phosphorylation* to emphasize that ATP formation from ADP and P_i is coupled to and linked with electron transport such that inhibition of one also inhibits the other.

Hydrogen ion gradient: A situation developed across the inner mitochondrial membrane wherein the concentration of hydrogen ions outside the mitochondrion is higher than the concentration inside. Hydrogen ions are extruded from the mitochondrion by the transfer of electrons from complex I to coenzyme Q, from coenzyme Q to complex III, and from complex III to complex IV. The gradient is discharged by ATP synthase, which admits hydrogen ions into the mitochondrion thereby driving the phosphorylation of ADP by P_i.

Electron transport chain: Present in the mitochondrial membrane, this linear array of redox active electron carriers consists of NADH dehydrogenase, coenzyme Q, cytochrome c reductase, cytochrome c, and cytochrome oxidase as well as ancillary iron sulfur proteins. The electron carriers are arrayed in order of decreasing reduction potential such that the last carrier has the most positive reduction potential and transfers electrons to oxygen.

Reduction potential: The tendency of an electron carrier to give up electrons, stated in electron volts, is called reduction potential. In any reduction–oxidation reaction electrons flow from the species with the more negative reduction potential to the more positive reduction potential.

Cytochrome: A heme (protoporphyrin IX) containing electron transfer protein. Some heme moieties are covalently attached to the protein components (cytochrome c), whereas others have isoprenoid side chains (cytochromes a and a_3).

Iron sulfur proteins: These carry one electron and contain centers that chelate iron with organic and inorganic sulfur. Some centers contain a single iron atom chelated by four cysteine sulfurs; others contain two iron atoms chelated through four cysteine sulfurs and two inorganic sulfurs; yet others contain four iron atoms chelated by four cysteine sulfurs and four inorganic sulfurs.

Coenzyme Q (ubiquinone): A two electron accepting quinone that can accept and transfer one electron at a time allowing it to exist in a semi-quinone state as well as the fully oxidized quinone or fully reduced dihydroxy state. It is bound to multiple isoprenoid units (ubiquinone has ten units), allowing it to bind to the membrane.

Flavin mononucleotide (FMN): An isoalloxazine ring bound to ribosyl monophosphate in an *N*-glycosidic bond. FMN can accept two electrons or donate one at a time to another electron acceptor.

Flavin adenine dinucleotide (FAD): An isoalloxazine ring bound to ribosyl monophosphate in an *N*-glycosidic bond which is attached to adenosine monophosphate. Like FMN, FAD can accept or donate two electrons one at a time to another electron acceptor.

DISCUSSION

The **electron transport chain (ETC) or electron transport system (ETS)** shown in Figure 16-1 is located on the **inner membrane of the mitochondrion** and is responsible for the **harnessing of free energy** released as **electrons travel from more reduced (more negative reduction potential, E'_0) to more oxidized (more positive E'_0) carriers** to drive the phosphorylation of ADP to ATP. Complex I accepts a pair of electrons from NADH ($E'_0 = -0.32$ V)

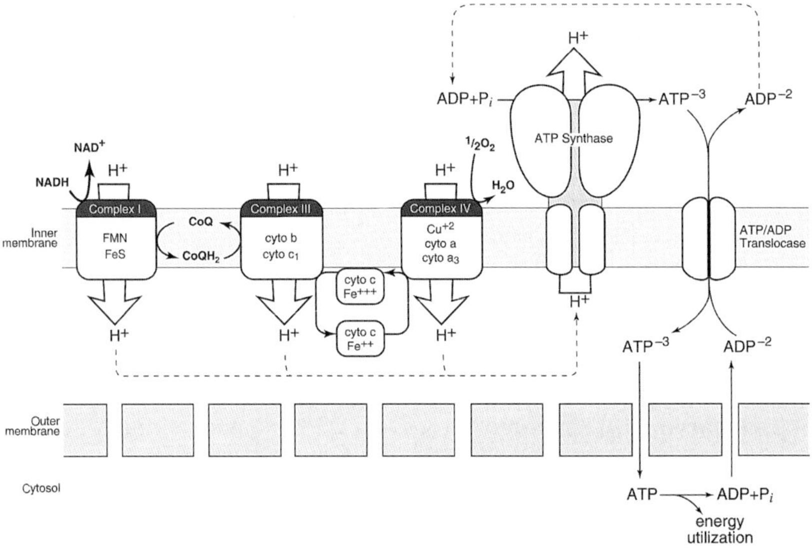

Figure 16-1. Schematic diagram of electron transport chain ATP synthase and ATP/ADP translocase.

and passes the electron pair through the intervening carriers to **complex IV**, which passes the electrons to one atom of molecular oxygen ($E_0' = +0.82$ V) to form water with hydrogen ions (H^+) from the medium.

The transport of electrons through the carriers is **highly coupled to the formation of ATP from ADP and P_i** through the formation and relaxation of the **proton gradient formed across the inner mitochondrial membrane by electron transport.** Each time **electrons are transported between complexes I and III,** between **complexes III and IV,** or **between complex IV** and **oxygen, protons are extruded from the mitochondrial matrix** across the inner membrane to the intermembrane space/cytosol. (The outer membrane poses no barrier to proton passage.) In other words, the energy gained from these electron transfers is used to pump protons from the mitochondrial matrix side to the cytosol side. Because the mitochondrial membrane is impermeable to protons, there is a gradient that develops with a higher concentration of protons outside the matrix. The protons then come through the ATP synthase complex through proton pores, and as they come back into the mitochondrial matrix, ADP is phosphorylated to ATP. Thus, because the process of electron transport is tightly coupled to ADP phosphorylation, ADP must be present for electron transport to proceed, and therefore the ADP/ATP translocase must be able to exchange a molecule of ADP in the cytosol for a molecule of ATP (newly made) in the matrix of the mitochondria. When these **various processes operate in concert the mitochondria** are said to exhibit **coupled respiration.**

The components of the electron transport chain have various **cofactors. Complex I, NADH dehydrogenase,** contains a **flavin cofactor and iron sulfur centers,** whereas **complex III, cytochrome reductase,** contains cytochromes b and c_1. **Complex IV, cytochrome oxidase,** which transfers electrons to **oxygen,** contains **copper ions** as well as **cytochromes a and a_3**. The general structure of the cytochrome cofactors is shown in Figure 16-2. Each of the cytochromes has a heme cofactor but they vary slightly. The b-type cytochromes have protoporphyrin IX, which is identical to the heme in hemoglobin. The c-type cytochromes are covalently bound to cysteine residue 10 in the protein. The a-type

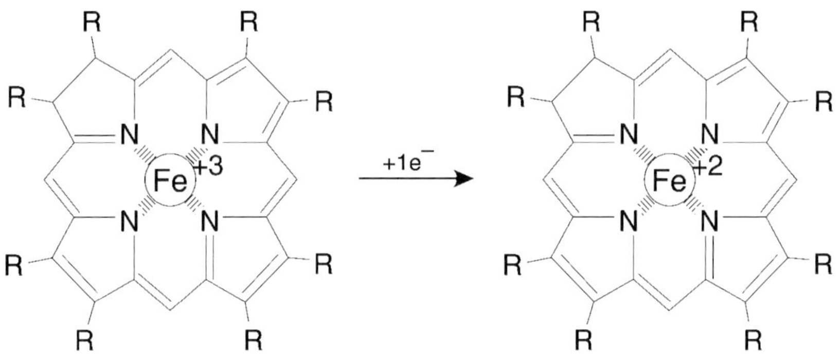

Figure 16-2. Heme active center of cytochromes a, b, and c components of electron transport chain.

cytochromes have a long isoprenoid $[(CH_2–CH=C(CH_3)–CH_2)_n]$ tail bound at one side-chain position.

Inhibition of the electron transport chain in coupled mitochondria can occur at any of the three constituent functional processes; electron transport per se, formation of ATP, or antiport translocation of ADP/ATP (Table 16-1). **The best known inhibitor of the ADP/ATP translocase is atractyloside** in the presence of which **no ADP for phosphorylation is transported across the inner membrane to the ATP synthase** and **no ATP is transported out.** In the absence of ADP phosphorylation the proton gradient is not reduced allowing other protons to be extruded into the intermembrane space because of the elevated $[H^+]$, and thus electron transfer is halted. Likewise the antibiotic oligomycin directly inhibits the ATP synthase, causing a cessation of ATP formation, buildup of protons in the intermembrane space, and a halt in electron transfer. Similarly, a blockade of complex I, III, or IV that inhibits electron flow down the chain to O_2 would also stop both ATP formation and ADP/ATP translocation across the inner mitochondrial membrane.

Cyanide ion (CN^-) is a potent inhibitor of complex IV the cytochrome c oxidase component of the electron transport system in the oxidized state of the heme (Fe^{3+}). It can be delivered to tissue electron transport systems as a dissolved gas after breathing HCN or ingested as a salt such as KCN or as a medication leading to the formation of CN^- such as nitroprusside. **Cyanide ion competes effectively with oxygen for binding to cytochrome c oxidase at the oxygen-binding site.** Cyanide binding and therefore cyanide poisoning is reversible if treated properly and early. The treatment strategy depends on dissociation of cyanide from cytochrome a/a_3 (Fe^{3+}). **Increasing the percentage of oxygen breathed will increase the competition of oxygen over cyanide for the cytochrome a/a_3 (Fe^{3+}).** Two other medications foster this competition. Nitrite ion (NO_2^-) is administered to convert some oxyhemoglobin $[HbO_2(Fe^{2+})]$ to methemoglobin

Table 16-1
RESPIRATORY CHAIN SITES AND INHIBITORS

RESPIRATORY CHAIN SITES	INHIBITOR
Complex I: NADH: CoQ reductase	Piericidin A Amytal Rotenone
Complex III: Cytochrome c reductase	Antimycin A
Complex IV: Cytochrome oxidase	Cyanide ion Carbon monoxide
ATP synthase	Oligomycin
ADP/ATP translocase	Atractyloside, Bongkrekate

$$Hb(Fe^{+2})O_2 + NO_2^- \longrightarrow MetHb(Fe^{+3})OH + NO_3^-$$

$$MetHb(Fe^{+3})OH + cyt\ a/a_3(Fe^{+3})\text{-}CN^- \longrightarrow cyt\ a/a_3(Fe^{+3}) + MetHb(Fe^{+3})\text{-}CN^-$$

$$MetHb(Fe^{+3})\text{-}CN^- + S_2O_3^{-2} \xrightarrow[\text{rhodanese}]{} SCN^- + SO_3^{-2} + MetHb(Fe^{+3})OH$$

Figure 16-3. Strategy for reversal of cyanide binding to cytochrome oxidase (cyt a/a_3 Fe^{3+}).

[met-HbOH(Fe^{3+})], another competitor for cyanide binding (Figure 16-3). The cyanide adduct of methemoglobin is formed releasing cytochrome oxidase in the Fe^{3+} form ready to bind oxygen and disinhibit the electron transport chain. To remove the cyanide adduct in a nontoxic fashion, thiosulfate ion is administered. The mitochondrial enzyme rhodanese catalyzes the conversion of cyanide and thiosulfate to thiocyanate and sulfite. Thiocyanate is incapable of inhibiting cytochrome oxidase and is excreted. The methemoglobin can be reconverted to oxyhemoglobin by NADH and methemoglobin reductase.

Other sites of the electron transport chain can be targets of inhibitors based on similarity in structure to enzyme components or to substrates of the various components. For instance, the fish poison rotenone resembles the isoalloxazine ring of the FMN cofactor of complex I, the NADH CoQ reductase. Rotenone binds the enzyme quite avidly and prevents transfer of electrons from NADH to coenzyme Q through the iron sulfur centers and thus inhibits oxidation of NADH and subsequent reduction of oxygen to water. On the other hand, carbon monoxide resembles molecular oxygen and binds with a higher affinity than oxygen to complex IV, the cytochrome oxidase component, inhibiting transfer of electrons to oxygen.

COMPREHENSION QUESTIONS

A 16-month-old girl was found to have ingested approimately 30 mL of an acetonitrile-based cosmetic nail remover when she vomited 15 minutes postingestion. The poison control center was contacted, but no treatment was recommended because it was confused with an acetone-based nail polish remover. The child was put to bed at her normal time, which was 2 hours postingestion. Respiratory distress developed sometime after the child was put to bed, and she was found dead the next morning.

[16.1] Inhibition of which of the following enzymes was the most likely cause of this child's death?

A. Cytochrome c reductase
B. Cytochrome oxidase
C. Coenzyme Q reductase
D. NADH dehydrogenase
E. Succinate dehydrogenase

[16.2] Which of the following best describes the reason for the latency of acetonitrile toxicity and why prompt treatment would have prevented this child's respiratory distress and death?

A. Acetonitrile crosses the mitochondrial membrane slowly.
B. Acetonitrile induces hemolysis by inhibiting glucose 6-phosphate dehydrogenase.
C. Acetonitrile is only poorly absorbed by the intestinal system.
D. Complex IV of the electron transport system binds acetonitrile weakly.
E. Cytochrome P450 enzymes oxidize acetonitrile and slowly release cyanide.

[16.3] Inhibition of oxidative phosphorylation by cyanide ion leads to increases in which of the following?

A. Gluconeogenesis to provide more glucose for metabolism
B. Transport of ADP into the mitochondria
C. Utilization of fatty acids substrates to augment glucose utilization
D. Utilization of ketone bodies for energy generation
E. Lactic acid in the blood causing acidosis

[16.4] Which of the following procedures best describes the emergency intervention for cyanide poisoning?

A. Decrease the partial pressure of oxygen
B. Treatment with nitrites to convert hemoglobin to methemoglobin.
C. Treatment with thiosulfate to form thiocyanate
D. Use of Mucomyst (*N*-acetylcysteine) taken orally

An unskilled worker in a water garden/plant nursery was sent to sweep up a spill of a white powder in the storage shed. Later he was found with labored breathing and convulsions. On further examination, the white powder was identified as rotenone.

[16.5] Respiratory distress is induced on rotenone exposure because it inhibits the complex that catalyzes which of the following?

A. Electron transfer from NADH to coenzyme Q
B. Oxidation of coenzyme Q
C. Reduction of cytochrome c
D. Electron transfer from cytochrome c to cytochrome a_1/a_3
E. Electron transfer from cytochrome a_1/a_3 to oxygen

[16.6] The major metabolic consequence of perturbation of the electron transfer in mitochondria is which of the following?

A. Increased production of NADPH
B. Increased oxidation of NADH
C. Increased reduction of O_2 to H_2O
D. Decreased regeneration of NAD^+
E. Decreased reduction of FAD

Answers

[16.1] **B.** The culprit here is cyanide produced from acetonitrile. Cyanide inhibits the electron transport chain of cytochrome oxidase.

[16.2] **E.** Acetonitrile itself is not the toxicant but undergoes metabolism and produces cyanide, which is the toxic agent here.

[16.3] **E.** Gluconeogenesis requires ATP, which is in short supply, turning up the catabolism of glucose to lactate in the absence of an intact electron transport chain. ADP cannot be transported into the mitochondrion because ATP, its antiporter partner, isn't made by oxidative phosphorylation as a result of cyanide inhibition of cytochrome oxidase. Metabolism of fatty acids and ketone bodies requires a functional electron transport chain for their metabolism, and these possibilities are also ruled out.

[16.4] **B.** Increased oxygen competes with cyanide bound to cytochrome oxidase displacing it. Nitrites bind to hemoglobin converting it to methemoglobin, which binds cyanide more tightly than cyanohemoglobin and pulls cyanide from cyanohemoglobin to form cyanomethemoglobin. Thiosulfate is used to displace cyanide from cyanomethemoglobin to form thiocyanate, which can be excreted, a happy ending for cyanide poisoning. *N*-acetylcysteine is used for acetaminophen toxicity and not cyanide toxicity.

[16.5] **A.** Rotenone binds avidly to the flavoprotein NADH CoQ reductase, complex I (also called NADH dehydrogenase). The central portion of the rotenone structure resembles the isoalloxazine ring of the FMN molecule, and when it binds to complex I, rotenone prevents the transfer of electrons from NADH to coenzyme Q.

[16.6] **D.** Inhibition of the electron transport chain shuts down the major pathway of regenerating NAD^+ from the NADH produced in intermediary metabolism. This forces the cytosolic conversion of pyruvate to lactate to regenerate NAD^+ so that glycolysis can continue in the absence of a functioning electron transport system.

BIOCHEMISTRY PEARLS

- The electron transport chain (ETC) or electron transport system (ETS) is located on the inner membrane of the mitochondrion and is responsible for the harnessing of free energy. The chain consists of a series of carriers arranged along the inner membrane of mitochondria that transport electrons from NADH and reduced flavin carriers to molecular oxygen.
- Energy is released as electrons travel from more reduced (more negative reduction potential, E'_0) to more oxidized (more positive E'_0) carriers to drive the phosphorylation of ADP to ATP.
- The components of the electron transport chain have various cofactors, which form vital complexes.
- The disruption of various complexes may interfere with the ETC, leading to inability to manufacture ATP.
- The energy gained from electron transfer is used to drive protons out of the inner mitochondrial matrix to the cytosol, establishing a gradient of protons. As the protons come back into the matrix via the ATP synthase complex, ADP is phosphorylated to ATP.
- The normal mitochondria produce ATP as they transport electrons to oxygen; any interference with ATP synthesis or translocation across the mitochondrial membrane will inhibit electron transfer.
- Rotenone binds avidly to the flavoprotein NADH CoQ reductase, complex I (also called NADH dehydrogenase).

REFERENCES

Davidson VL, Sittman DB. National Medical Series–Biochemistry. Philadelphia, PA: Harwol, 1994:381–3.

Devlin TM, ed. Textbook of Biochemistry with Clinical Correlations, 5th ed. New York: Wiley-Liss, 2002:563–82.

Goodman AG, Gilman LS, eds. The Pharmacological basis of Therapeutics, 10th ed. New York: McGraw-Hill, 2001:1642.

McGilvery RW. Biochemistry: A Functional Approach. Philadelphia, PA: W.B. Saunders, 1979:397–400.

CASE 17

An elderly couple is taken by ambulance to the emergency department after their son noticed that they were both acting "strangely." The couple had been in good health prior to the weekend. Their son usually visits or calls them daily, but because of a terrible blizzard was not able to make it to their house. They had been snowed in at their house until the snowplows cleared the roads. They had plenty of food and were kept warm by a furnace and blankets. When the son was able to see them for the first time in 2 days, he noticed that they both were complaining of bad headaches, confusion, fatigue, and some nausea. On arrival to the emergency department, both patients were afebrile with normal vital signs and O_2 saturation of 99 percent on 2 L of O_2 by nasal cannula. Their lips appeared to be very red. Both patients were slightly confused but otherwise oriented. The physical examinations were within normal limits. Carboxyhemoglobin levels were drawn and were elevated.

- **What is the most likely cause of these patients' symptoms?**

- **What is the biochemical rationale for 100 percent O_2 being the treatment of choice?**

ANSWERS TO CASE 17: CARBON MONOXIDE POISONING

Summary: Two elderly patients, with no medical problems present with acute mental status change, fatigue, red lips, and nausea after being snowed in their home during a blizzard with warmth provided by a furnace. The carboxyhemoglobin level is elevated.

- **Most likely cause:** Carbon monoxide poisoning (increase carboxyhemoglobin level).
- **Rationale for treatment:** Administration of 100 percent O_2 displaces CO from hemoglobin.

CLINICAL CORRELATION

Carbon monoxide, a molecule with one carbon and one O_2 atom, binds very avidly to hemoglobin. It is a colorless and odorless gas and may arise from internal combustion engines, fossil-fueled home appliances (heaters, furnaces, stoves), and incomplete combustion of almost all natural and synthetic materials. Because it does not give warning signs, CO is considered a significant hazard. The patient generally has confusion and symptoms of O_2 deprivation, but not the symptoms of dyspnea, since the hemoglobin is saturated. The lips are a distinct red color as a result of the hemoglobin being "oxygenated." However, because CO binds so avidly to the hemoglobin, no transfer of O_2 occurs in the peripheral tissue. Carbon monoxide also disrupts the O_2-dependent steps of the electron transport chain, leading to unavailability of ATP. Treatment is thus 100 percent O_2 to displace the CO from the hemoglobin, and in severe cases, hyperbaric therapy to increase the amount of O_2 available to "drive out" the CO.

APPROACH TO ELECTRON TRANSFER CHAIN (ETC) AND CARBON MONOXIDE

Objectives

1. Understand the process by which CO causes symptoms.
2. Know how CO disrupts O_2 transport and uncouples the ETC.

Definitions

Oxidase: An enzyme requiring molecular O_2 as a substrate to produce water as a reduced product. In the case of cytochrome oxidase, water is the only product of oxygen reduction. With some other oxidases (called mixed function oxidases), one atom of molecular oxygen is converted to water, whereas the other atom of oxygen may be a hydroxylated product.

Methemoglobin: Hemoglobin in which the iron atom has been oxidized to the ferric (+3) state and is therefore incapable of binding O_2.

Carboxyhemoglobin: Hemoglobin that has bound CO is called carboxyhemoglobin. Carbon monoxide binds to hemoglobin with 200 times the affinity of O_2.

DISCUSSION

Carbon monoxide poisoning has two primary effects: **interference with O_2 transport to tissues and interference with the interaction of O_2 with cytochromes,** especially the cytochromes in the **electron transport chain.**

The transport of O_2 from the lungs to peripheral tissues and central organs is mediated by hemoglobin, a four subunit (2α- and 2β-globin chains) protein. Hemoglobin has four active sites enabling it to bind four molecules of O_2 per molecule in the lungs for transport by the red blood cell to distant tissues, where O_2 is released for direct use or for binding to other proteins such as muscle myoglobin. Although the globin portion of myoglobin is similar to hemoglobin, it acts as a monomer with only one O_2-binding site. The four O_2-binding sites of hemoglobin exhibit positive cooperativity in binding O_2. The binding of the first molecule of O_2 occurs with a modest affinity but causes a conformational change that enhances hemoglobin's binding affinity for a second molecule of O_2. Likewise, the binding of the second enhances the binding affinity for the third and the third, for the fourth molecule of O_2.

Carbon monoxide competes with O_2 for the four hemoglobin-binding sites, but **CO binds 220 times more avidly to hemoglobin than O_2.** Moreover, **CO** shows the **same positive cooperativity that O_2 does.** These features combine in favor of CO binding. Atmospheric air is 21 percent O_2, whereas the usual CO concentration is 0.1 parts per million (ppm). In heavy traffic, the CO concentration may be 115 ppm. A nearly tenfold increase, up to 1000 ppm (0.1 percent) carbon monoxide, results in a 50 percent carboxyhemoglobinemia. Although the reduction in O_2-transport capacity is proportional to the proportion of carboxyhemoglobin [COHb], the amount of O_2 available to tissues for use is further reduced by the inhibitory **effect of COHb on the dissociation of oxyhemoglobin.** Although the average person may show no symptoms at 10 percent of hemoglobin in the form of COHb, any complication such as anemia that reduces O_2 transport capacity may exhibit symptoms at a lower percentage of carboxyhemoglobin. At **10 to 20 percent COHb, headache and dilation of cutaneous blood vessels** may appear, whereas **at 20 to 30 percent, headaches may become stronger. At 30 to 40 percent carboxyhemoglobin, serious headache, dizziness, disorientation, nausea, and vomiting occur.** At levels exceeding **40 percent, patients usually collapse** and have worsening of other symptoms. These symptoms reflect failure of O_2 transport as well as direct inhibition of O_2-binding cytochromes such as cytochrome oxidase or myoglobin.

In the electron transport chain (see Figure 16-1 in Case 16) **only complex IV (cytochrome oxidase) interacts directly with O_2.** As with hemoglobin, **CO has a higher affinity for cytochrome oxidase than O_2.** Thus **CO binds tightly to cytochrome oxidase and inhibits the transfer to O_2.** In tightly coupled mitochondria the binding of CO to cytochrome oxidase also results in the **inhibition of the phosphorylation of adenosine diphosphate (ADP), reducing the production of adenosine triphosphate (ATP).** This becomes more profound as additional molecules of cytochrome oxidase are bound by CO. Similarly, myoglobin also binds CO more avidly than O_2, and the transfer of O_2 to enzymes requiring O_2 is inhibited.

The **binding of CO to hemoglobin is fully dissociable, and dissociation requires ventilation.** After removal from exposure to CO, administration of O_2 reverses CO binding to hemoglobin. **Utilization of 100 percent O_2 accelerates the washout of CO.** Use of **hyperbaric chambers** with pressures up to 2 atmospheres speeds up the CO washout process even more. Addition of 5 to 7 percent CO_2 to the O_2 is sometimes used as a prompt to ventilatory exchange. One disadvantage of the addition of CO_2 is the serious acidosis that results when the respiratory acidosis produced by CO_2 inhalation is added to the metabolic acidosis produced by O_2 deprivation in the tissues because of CO poisoning.

COMPREHENSION QUESTIONS

An elderly couple living in the suburbs of Hanover, New Hampshire, was snowbound for several days in their home during a blizzard. During periods of electrical outage they relied on an unvented gas heater to warm one room in their home where they stayed throughout the blizzard. As soon as the roads cleared, their granddaughter came to check on them and found them disoriented, complaining of headache, fatigue, and nausea and breathing and walking hesitantly and with a stumbling gait.

[17.1] Laboratory data show a remarkably increased carboxyhemoglobin level in the blood. The best explanation for this finding is that CO has which of the following effects?

A. It increases the hydrogen ion concentration causing oxyhemoglobin to precipitate
B. It changes the valence state of iron in hemoglobin
C. It competitively displaces O_2 from oxyhemoglobin
D. It converts myoglobin to carboxyhemoglobin at a rapid rate
E. It prevents transfer of O_2 across the alveolar membranes

[17.2] Which of the following treatment strategies is the most effective remediation for CO poisoning?

A. Removal from CO source
B. Removal from CO source and administration of 100 percent O_2
C. Administration of 5 to 7 percent CO_2 to stimulate respiration followed by 100 percent O_2
D. Administration of 5 to 7 percent CO_2 to stimulate respiration
E. Removal from CO source and administration of 5 to 7 percent CO_2 to stimulate respiration

[17.3] In addition to forming carboxyhemoglobin, the toxic effects of CO include inhibition of which of the following enzymes involved in oxidation-reduction reactions?

A. NADPH dehydrogenase
B. Coenzyme Q reductase
C. Cytochrome c reductase
D. Succinate dehydrogenase
E. Cytochrome oxidase

Answers

[17.1] **C.** Carbon monoxide binds to hemoglobin with 220 times the affinity of O_2. Thus CO displaces O_2 from oxyhemoglobin to form carboxyhemoglobin. Carbon monoxide has no appreciable effect on pH such as CO_2 may have. It does not change the valence state of iron or interfere with O_2 transport across membranes. CO binding does not change myoglobin to hemoglobin.

[17.2] **B.** Removal from the CO is always the first step. Increasing the O_2 concentration serves to displace the CO to hemoglobin and cytochrome oxidase. Adding CO_2 may stimulate respiration rate, but it also causes an additional pH perturbation.

[17.3] **E.** Carbon monoxide binds to cytochrome oxidase but not to the other enzymes listed.

BIOCHEMISTRY PEARLS

- Carbon monoxide poisoning interferes with O_2 transport to tissues and the interaction of O_2 with cytochromes, especially the cytochromes in the electron transport chain (ETC).
- The four O_2 binding sites of hemoglobin exhibit positive cooperativity in binding O_2, leading to the sigmoid shape of hemoglobin dissociation curve.
- The binding of the first molecule of O_2 occurs with a modest affinity but causes a conformational change that enhances hemoglobin's binding affinity for a second molecule of O_2, and so on.
- Carbon monoxide binds 220 times more avidly to hemoglobin than O_2.

REFERENCES

Devlin TM, ed. Textbook of Biochemistry with Clinical Correlations, 5th ed. New York: Wiley-Liss, 2002:577–82.

Goodman AG, Gilman LS, eds. The Pharmacological Basis of Therapeutics, 10th ed. New York: McGraw-Hill, 2001.

❖ CASE 18

A 27-year-old male presented to the emergency department with the signs and symptoms of acute appendicitis. He was promptly sent to the operating room for an emergency appendectomy. The patient was prepped and draped for the surgery, and halothane was given as an inhalation anesthetic. Two minutes after the anesthetic was given, the patient was noted to have an extremely elevated temperature, muscle rigidity, and tachypnea. An arterial blood gases test revealed a metabolic acidosis, and the serum electrolytes demonstrated hyperkalemia. A nurse quickly went to talk to the family about the events, and the family mentioned that the only other person to have surgery in their family had a similar reaction and died. The physician makes a diagnosis of malignant hyperthermia (MH).

◆ **What is the biochemical basis of this disease?**

◆ **What is the best treatment for this condition?**

ANSWERS TO CASE 18: MALIGNANT HYPERTHERMIA

Summary: 27-year-old male with acute appendicitis undergoing a halothane inhaled anesthetic with acute onset of hyperthermia, tachypnea, respiratory acidosis, hyperkalemia, and family history of similar events. The tentative diagnosis is MH.

- **Biochemical mechanism:** Uncoupling of oxidative phosphorylation.
- **Treatment:** Supportive care with attempts to bring temperature down, correct blood gas and electrolyte abnormalities, and give dantrolene.

CLINICAL CORRELATION

Malignant hyperthermia is an untoward response of susceptible individuals to preoperative anesthesia utilizing halothane and succinylcholine, although other inhalation anesthetics such as enflurane and isoflurane have been recognized as milder prompts for this complication of anesthesia. The frequency of occurrence is 1 in 15,000 children and 1 in 100,000 adults. The inheritance is dominant in 50 percent of cases and recessive in 20 percent, suggesting a more complex basis for this response. The clinical findings are muscle rigidity, hyperthermia, seizure, cardiac arrhythmias, and sometimes death. Prevention is the key, in that every patient should be queried about personal or family history of complications during surgery. Once a patient is found to have MH, the clinician should alert other family members to the same possibility. The muscle relaxant, dantrolene, is the treatment of choice. Even with prompt recognition and therapy, mortality is as high as 5 percent.

APPROACH TO OXIDATIVE PHOSPHORYLATION (UNCOUPLING)

Objectives

1. Understand how malignant hyperthermia is caused by the uncoupling of oxidative phosphorylation.
2. Understand the biochemical mechanism for the heat production.
3. Be familiar with the mechanism of how dantrolene reverses the effects.

Definitions

Malignant hyperthermia: An unusual adverse response to some inhalation anesthetics such as halothane in which there is an acute dramatic elevation in body temperature as well as tachypnea, muscle rigidity, and hyperkalemia.

Ca^{2+} signaling: Many enzymes in metabolism and signaling pathways are responsive to Ca^{2+} concentrations. Calcium is sequestered within tissues (e.g., within cisternae in the sarcoplasmic reticulum of muscle) and released on signal increasing the Ca^{2+} concentration in the cytosolic compartment and eliciting Ca^{2+}-induced responses. Increased calcium concentration stimulates the activity of muscle phosphorylase kinase and activates pyruvate dehydrogenase phosphatase to increase the metabolic flow of glucose to CO_2 and H_2O.

Ca^{2+} channel: A transport system allowing passage of the charged calcium ion through a hydrophobic membrane.

DISCUSSION

Hyperthermia is a **rare complication of anesthesia** that is not completely understood. Most of the evidence points to the **ryanodine receptor (chromosome 19q13.1) as the defective gene product.** This receptor is the **Ca^{2+} release channel of muscle sarcoplasmic reticulum.** Stimulation of this channel leads to **excessive Ca^{2+} release from the cisternae of the sarcoplasmic reticulum, and that Ca^{2+} prompts muscle contraction, an increase in body temperature, tachycardia, and subsequent metabolic acidosis.**

The scheme in Figure 18-1 summarizes the effects of Ca^{2+} release on processes that bring about the complex of symptoms seen in malignant

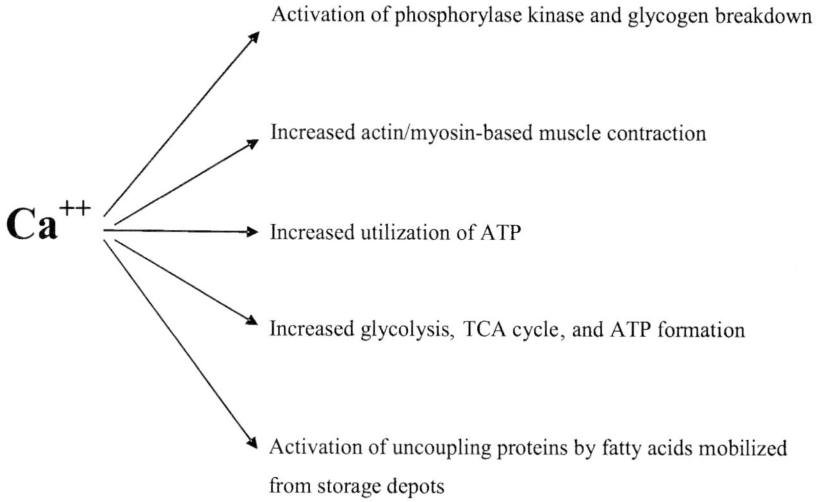

Figure 18-1. Direct and indirect effects of Ca^{2+} release in aberrant response to anesthetics halothane and succinylcholine.

hyperthermia. **Increased Ca^{2+} release triggers the binding of** adenosine triphosphate **(ATP)-charged myosin to actin to initiate muscle contraction.** This muscle contraction may be repeated to the point of **muscle rigidity with attendant muscle damage and release of myoglobin.** The sustained muscle contraction uses increased amounts of **ATP** increasing demands on **glycolysis,** tricarboxylic acid **(TCA) cycle, and oxidative phosphorylation.** Ca^{2+} release serves to activate, at least partially, **phosphorylase kinase,** which speeds the mobilization of glycogen stores for ATP generation. Sympathetic outflow triggered by these changes leads to increased activation of glycogen breakdown and lipolysis. Fatty acids serve as a signal to increase the levels of **uncoupling proteins** in various tissues. **The uncoupling proteins form channels in mitochondria allowing proton reentry into the mitochondrial matrix.** Because the phosphorylation of adenosine diphosphate (ADP) to ATP depends on the proton gradient through the ATP synthase, these channels **collapse the proton gradient allowing electron transport to occur without the phosphorylation of ADP** (Figure 18-2). Since ATP is being consumed at an elevated rate and mitochondrial production of ATP is compromised, glycolysis increases to compensate for the shortfall in ATP production. **Pyruvate and lactate, the end products of glycolysis, increase in concentration, giving rise to the metabolic acidosis** observed.

The uncoupling proteins (UCP 1 to 5) are a physiologic mechanism for maintaining body temperature through **selective uncoupling of electron transport from ATP synthesis.** The **free energy release** normally captured in the formation of the high-energy phosphate bond when phosphorylating ADP to ATP is lost as **heat** and changes body temperature. Physiologically, sympathetic outflow and catecholamine hormone release prompt new synthesis of the uncoupling proteins in response to the hypothalamus sensing lowered body temperature. In the case of malignant hyperthermia, this mechanism is triggered by anesthetic challenge to the calcium release channel and the subsequent metabolic and signaling changes assembled above unfold.

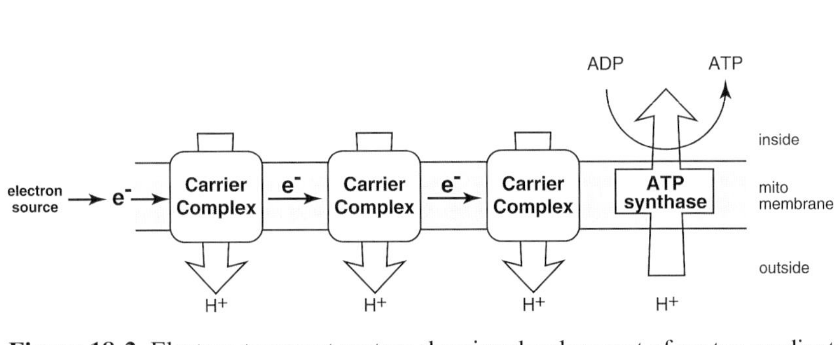

Figure 18-2. Electron transport system showing development of proton gradient outside the mitochondrial membrane by the passage of electrons down the chain.

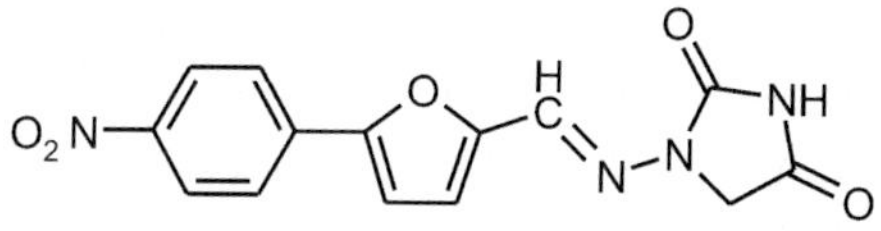

Figure 18-3. Structure of dantrolene (1-[[[5-(4-nitrophenyl)-2 furanyl]-methylene]amino]-2, 4 imidazolidinedione).

Treatment of MH depends on cessation of anesthesia, mechanical means to reduce the temperature, and the correction of blood electrolyte and gas parameters to normal. In addition, dantrolene (Figure 18-3) is infused initially intravenously and then may be given orally. **Dantrolene brings about the reduction in contraction of skeletal muscles by decreasing the amount of Ca^{2+} released from the sarcoplasmic reticulum.** A serious side effect of dantrolene use is its potential hepatotoxicity. The hepatotoxicity is fatal to 0.1 to 0.2 percent of patients treated for 60 days or longer. Thus, use of dantrolene for longer than 45 days requires monitoring of hepatic markers.

COMPREHENSION QUESTIONS

A young and otherwise healthy student undergoing preparation for a simple surgical procedure was noticed to have an elevating temperature and respiratory rate with muscle rigidity following the onset of anesthesia using halothane and succinylcholine. Laboratory findings revealed elevated levels of calcium, hydrogen ion, pyruvate, and lactate. The diagnosis was malignant hyperthermia.

[18.1] The muscle rigidity observed in this patient is most likely prompted by which of the following?

A. The patient's fear of surgery
B. Increased levels of hydrogen ion, causing the muscles to become immobile
C. Increased Ca^{2+} levels in muscle tissue, triggering muscle contraction
D. Increased pyruvate and lactate, causing precipitation of muscle protein
E. Increased body temperature

[18.2] Rapidly elevating body temperature was observed in this patient. The underlying cause for this pyretic episode is which of the following?

A. Hypothalamic upward adjustment of body temperature set point in response to cold operating room temperature
B. Muscles producing heat from exertion of contraction
C. Uncoupling proteins allowing dissipation of the mitochondrial hydrogen ion gradient releasing energy as heat
D. Metabolism of fatty acids from lipid storage depots releasing heat
E. Elevated consumption of ATP to support muscle contraction releasing heat

[18.3] Which of the following best describes the mechanism of action of dantrolene in the treatment of MH?

A. Decrease of Ca^{2+} release
B. Reduction of body temperature with hypothalamic set temperature
C. Effect on mitochondrial ATP production
D. Nuclear transcription attenuation
E. Recoupling of the sodium and ATP channels

Answers

[18.1] **C.** Release of Ca^{2+} triggers actin–myosin reaction, prompting muscle contraction. Neither fear of surgery nor hydrogen ion levels causes concerted muscle contraction. Increased pyruvate/lactate or temperature tends to counteract contraction.

[18.2] **C.** Malignant hyperthermia does not involve central brain control of temperature but is caused by metabolic alterations. Regardless of the source of energy—whether fatty acids or the utilization of ATP to support muscle contraction—the electron transport chain is involved. Uncoupling of oxidation from ADP phosphorylation is caused by uncoupling proteins that dissipate energy as heat. This causes the elevation of body temperature seen in this patient.

[18.3] **A.** Dantrolene brings about reduction in contraction of skeletal muscles by decreasing the amount of Ca^{2+} released from the sarcoplasmic reticulum. In addition, supportive care is aimed at decreasing body temperature and correcting the metabolic acidosis and electrolyte balance.

BIOCHEMISTRY PEARLS

- Malignant hyperthermia is most likely, given that anesthetic agents stimulate a calcium release channel, leading to excessive Ca^{2+} release from the cisternae of the sarcoplasmic reticulum, in turn causing muscle contraction, an increase in body temperature, tachycardia, and subsequent metabolic acidosis.
- Although the uncoupling proteins (UCP 1 to 5) are a physiologic mechanism for maintaining body temperature through selective uncoupling of electron transport from ATP synthesis, MH is a pathologic exaggeration of this process.
- The treatment of MH includes stopping the anesthesia, cooling the patient, and administration of dantrolene, which reduces muscle contractions by decreasing the amount of Ca^{2+} released from the sarcoplasmic reticulum.

REFERENCE

Goodman AG, Gilman LS, eds. The Pharmacological Basis of Therapeutics, 10th ed. New York: McGraw-Hill, 2001:295–303.

CASE 19

A 40-year-old obese female presents to the emergency center with complaints of worsening nausea, vomiting, and abdominal pain. Her pain is located in the midepigastric area and right upper quadrant. She reports a subjective fever and denies diarrhea. Her pain is presently constant and sharp in nature but previously was intermittent and cramping only after eating "greasy" foods. On examination she has a temperature of 37.8°C (100°F) with otherwise normal vital signs. She appears ill and in moderate distress. She has significant midepigastric and right upper-quadrant tenderness. Some guarding is present but no rebound. Her abdomen is otherwise soft with no distention and active bowel sounds. Laboratory values were normal except for increased liver function tests, white blood cell count, and serum amylase. Ultrasound of the gallbladder revealed numerous gallstones and a thickening of the gallbladder wall. A surgery consult was immediately sought.

- **What is the most likely diagnosis?**

- **What is the role of amylase in digestion?**

ANSWERS TO CASE 19: PANCREATITIS

Summary: 40-year-old female with history of intermittent right upper-quadrant pain worsening after "greasy" meals and now with constant midepigastric pain, nausea, and vomiting with elevated liver function tests and amylase.

- **Diagnosis:** Gallstone pancreatitis.
- **Role of amylase:** Enzyme for carbohydrate metabolism, used to digest glycogen and starch.

CLINICAL CORRELATION

Acute pancreatitis is an inflammatory process in which pancreatic enzymes are activated and cause autodigestion of the gland. In the United States, **alcohol use is the most common cause,** and episodes are often precipitated by binge drinking. The next most common cause is biliary tract disease, usually passage of a gallstone into the common bile duct. Hypertriglyceridemia is also a common cause, and that occurs when serum triglyceride levels are greater than 1000 mg/dL, as is seen in patients with familial dyslipidemias or diabetes. When patients appear to have "idiopathic" pancreatitis, that is, no gallstones are seen on ultrasound, and no other predisposing factor can be found, biliary tract disease is still the most likely cause: either biliary sludge (microlithiasis), or sphincter of Oddi dysfunction. **Abdominal pain is the cardinal symptom of pancreatitis,** and it is often severe, typically in the **upper abdomen with radiation to the back.** The pain is often relieved by sitting up and bending forward and is exacerbated by food. Patients also **commonly have nausea and vomiting,** also precipitated by oral intake. The treatment includes nothing by mouth, intravenous hydration, pain control, and monitoring for complications.

APPROACH TO AMYLASE AND CARBOHYDRATE METABOLISM

Objectives

1. Be aware of the role of amylase in carbohydrate metabolism.
2. Understand the cause for increased amylase in pancreatitis.
3. Understand why conservative treatment (intravenous [IV] fluids, NPO, pain medication, and possibly a nasogastric tube) is effective in treatment of this condition.

Definitions

α-Amylase: An endosaccharidase that catalyzes the hydrolysis of $\alpha(1{\rightarrow}4)$ glycosidic bonds present in glycogen and starch. It is present in both saliva and pancreatic digestive juice. (See Figure 25-1b in Case 25 for a diagram showing the $\alpha(1{\rightarrow}4)$ glycosidic bonds in starch.)

Endosaccharidase: An enzyme that randomly hydrolyzes glycosidic bonds within polysaccharides.

Lipase: An enzyme that hydrolyzes the ester linkage between a fatty acid and glycerol in a triglyceride.

Pancreas: A major endocrine and exocrine organ located behind the stomach. It secretes pancreatic juice into the duodenum to neutralize the effluent from the stomach and supply digestive enzymes. It also synthesizes and secretes the hormones insulin, glucagon, and somatostatin into the bloodstream from cells within the islets of Langerhans.

Zymogen: A proenzyme; an inactive precursor of an enzyme stored in secretory granules. After secretion, the zymogen is activated by cleavage of certain peptide bonds either by low pH or by other enzymes.

DISCUSSION

The **pancreas** is a **major exocrine organ** that **synthesizes and secretes digestive enzymes.** It also produces and secretes $NaHCO_3$ to neutralize the acidic effluent from the stomach. The pancreas also has an important endocrine role because it synthesizes and secretes the hormones **insulin, glucagon, and somatostatin** into the bloodstream from cells within its islets of Langerhans.

The exocrine gland is divided into small globules that are drained by an intralobular duct. The intralobular ducts feed into the interlobular duct, which is joined to the main pancreatic duct. The pancreatic duct joins with the common bile duct (usually) in the hepatopancreatic ampulla, which exits into the duodenum. The secretory unit of the pancreas consists of the acinus and the intercalated duct. The epithelial cells of the intercalated duct have high concentrations of the enzyme carbonic anhydrase, which generates HCO_3^- from CO_2 and H_2O for neutralization of stomach acid entering the duodenum. The acinus is a cluster of acinar cells that are grouped around the intercalated duct. The acinar cells are specialized epithelial cells that synthesize and secrete the 20 or so enzymes that will be used to digest the macromolecules in the lumen of the intestine. Most of the digestive enzymes, particularly those used to degrade proteins, are synthesized as zymogens or proenzymes that must be activated. These proenzymes are synthesized on ribosomes on the rough endoplasmic reticulum. They are then transported to the Golgi apparatus and are sequestered in zymogen granules until they are secreted. Storing these inactive enzymes in zymogen granules protects the acinar cell from digesting itself. Secretion of these zymogens is regulated by cholecystokinin receptors and muscarinic acetylcholine receptors. The proenzymes are activated in the intestine, usually by the action of trypsin. There are some enzymes that are synthesized and stored as the active enzymes in the zymogen granules. These include **α-amylase, carboxyl ester lipase, lipase, colipase, RNase, and DNase.**

Acute pancreatitis is a result of anatomical changes that arise from two events. The first is the autodigestion of the acinar cells by inappropriate activation of the pancreatic enzymes (especially trypsinogen) within the cell. The second is the cellular injury response that is mediated by proinflammatory cytokines. The mechanisms by which the digestive enzymes become activated within the acinar cell are unclear. However, such inappropriate activation of pancreatic enzymes leads to destruction of the acinar cell and surrounding fat deposits, and it weakens the elastic fibers of the blood vessels, resulting in leakage.

Obstruction of the main pancreatic duct as a result of a **gallstone** lodged in or near to the hepatopancreatic ampulla can result in acute pancreatitis. One theory is that obstruction increases the pressure in the main pancreatic duct. The increase in pressure causes interstitial edema, which impairs the blood flow to the acinus. The lack of blood flow leads to ischemic injury of the acinar cell, resulting in release of the digestive enzymes into the interstitial space. How this leads to premature activation of the proenzymes stored in the acinar cell is unclear.

α-Amylase and lipase are two digestive enzymes that are synthesized and stored in the acinar cell as the active enzymes. Amylase is an endosaccharidase that catalyzes the hydrolysis of the $\alpha(1 \rightarrow 4)$ glycosidic bonds that form the main polymeric backbone of the polysaccharides starch and glycogen. Present in both saliva and pancreatic juice, it is the pancreatic form of the enzyme that breaks down most of the dietary polysaccharides. α-Amylase hydrolyzes dietary starch and glycogen to glucose, maltose, maltotriose, and an oligosaccharide referred to as the α-limit dextrin.

Pancreatic lipase is the primary digestive enzyme for the breakdown of triglycerides. It acts on triglycerides to hydrolyze the fatty acyl ester bonds. Lipase is specific for the ester bonds in the 1′- and 3′-positions to produce free fatty acids and β-monoacylglycerols. Pancreatic lipase is strongly inhibited by bile acids and therefore requires the presence of colipase, a small protein that binds to the lipase and activates it.

Since both α-amylase and lipase are stored in the pancreas as the active enzymes, they are important blood markers to help diagnose acute pancreatitis. The serum level of α-amylase will increase in the first 12 hours following the onset of acute pancreatitis. During the next 48 to 72 hours, the levels will usually fall back to normal values. Serum lipase levels also rise, but they remain elevated after the α-amylase levels have returned to normal and may take 7 to 10 days to normalize.

Most cases (85 to 90 percent) of **acute pancreatitis** that are caused by **gallstones** will resolve on their own, and therefore conservative treatment modalities are appropriate. These include pain management with analgesics, administration of intravenous fluids to maintain the intravascular volume and electrolyte balance, as well as removal of oral alimentation to decrease the secretion of pancreatic juice. Nasogastric suction has also been used to decrease gastrin release from the stomach and to eliminate gastric emptying into the duodenum. However, controlled trials have not demonstrated the efficacy of nasogastric suction in the treatment of mild to moderate acute pancreatitis.

COMPREHENSION QUESTIONS

[19.1] Prior to a race, many marathon runners will try to increase their glycogen concentrations by loading up with foods with a high starch content, such as pasta. α-Amylase secreted by the pancreas will digest the starch into which of the following major products?

A. Amylose, amylopectin, and maltose
B. Glucose, galactose, and fructose
C. Glucose, sucrose, and maltotriose
D. Limit dextrins, maltose, and maltotriose
E. Limit dextrins, lactose, and sucrose

[19.2] A 3-month-old infant presents with hepatosplenomegaly and failure to thrive. A liver biopsy reveals glycogen with an abnormal, amylopectin-like structure with long outer chains. Which of the following enzymes would most likely be deficient?

A. α-Amylase
B. Branching enzyme
C. Debranching enzyme
D. Glycogen phosphorylase
E. Glycogen synthase

[19.3] A 3-year-old Caucasian female presents with chronic diarrhea and a failure to thrive. Stools were oily. History reveals that she was breast-fed and had no problems until she was weaned. Which of the enzymes would be expected to be deficient following stimulation with secretin?

A. Cholesteryl esterase
B. Gastric lipase
C. Hormone sensitive lipase
D. Lipoprotein lipase
E. Pancreatic lipase

Match the following enzymes (A–D) to the products yielded.

A. Sucrose and lactose
B. Glucose and fructose
C. Glucose and galactose
D. Glucose

[19.4] Lactase

[19.5] Sucrase

[19.6] Maltase

Answers

[19.1] **D.** α-Amylase hydrolyzes α(1→4) glycosidic bonds present in starch (amylose and amylopectin) in a random fashion leaving primarily the disaccharide maltose, the trisaccharide maltotriose, and an oligosaccharide known as the α-limit dextrin, which is composed of 6 to 8 glucose residues with one or more α(1→6) glycosidic bonds. Galactose and fructose are not present in starch.

[19.2] **B.** Amylopectin is plant starch that has some α(1→6) branch points, but not as many as normal glycogen. Glycogen, which has an amylopectin-like structure, has fewer branch points than normal glycogen and would be less soluble within the cell. A deficiency in the branching enzyme will introduce fewer α(1→6) branch points.

[19.3] **E.** Neither hormone sensitive lipase nor lipoprotein lipase is a digestive enzyme. The patient's symptoms are consistent with an inability to absorb triglycerides, which would eliminate cholesteryl esterase from consideration. Since the patient did not have any problems while being breast-fed, then the most likely enzyme to be deficient is pancreatic lipase, since gastric lipase is most active on short chain triglycerides, such as those that are found in breast milk.

[19.4] **C.** Lactase breaks down lactose into glucose and galactose.

[19.5] **B.** Sucrase breaks down sucrose into glucose and fructose.

[19.6] **D.** Maltase and isomaltase convert maltose and isomaltose into glucose.

BIOCHEMISTRY PEARLS

- ❖ The pancreas is a large exocrine organ that has a role in the digestion of food as well as an endocrine organ which secretes insulin, somatostatin, and glucagon.
- ❖ Acute pancreatitis occurs from autodigestion of the acinar cells by inappropriate activation of the pancreatic enzymes (especially trypsinogen) within the cell, leading to cellular injury mediated by proinflammatory cytokines.
- ❖ The three major breakdown products of amylase are maltose, maltotriose, and α-dextrins. Enzymes in the brush border of the intestines continue to digest the carbohydrates.

REFERENCES

Greenberger NJ, Toskes PP. Acute and chronic pancreatitis. In: Fauci AS, Braunwald E, Kasper KL, et al., eds. Harrison's Principles of Internal Medicine, 14th ed. New York: McGraw-Hill, 1998.

Hopfer U. Digestion and absorption of basic nutritional constituents. In: Devlin TM, ed. Textbook of Biochemistry with Clinical Correlations, 5th ed. New York: Wiley-Liss, 2002.

Kumar V, Cotran RS, Robbins SL. Robbins Basic Pathology, 7th ed. Philadelphia, PA: W.B. Saunders, 2003.

Marino CS, Gorelick FS. Pancreatic and salivary glands. In: Boron WF, Boulpaep EL, eds. Medical Physiology: A Cellular and Molecular Approach. Philadelphia, PA: W.B. Saunders, 2003.

CASE 20

A 21-year-old primigravid female at 35-week gestation presents to the hospital with nausea, vomiting, and malaise over the last several days. Patient has also noticed that her eyes were turning yellow in color. Her prenatal course has otherwise been unremarkable. On examination she is found to have elevated blood pressure, proteinuria, increased liver function tests, prolonged clotting studies, hyperbilirubinemia, hypofibrinogenemia, and hypoglycemia. A pelvic ultrasound identified a viable intrauterine pregnancy measuring approximately 35-week gestation. After admission, the mother underwent an emergent cesarean delivery, and she subsequently developed a worsening hypoglycemia and coagulopathy and went into hepatic coma with renal failure. After reviewing all the laboratory results and her clinical picture, the patient was diagnosed with acute fatty liver of pregnancy.

- **What is an associated biochemical disorder?**

- **What is the etiology of the hypoglycemia?**

ANSWERS TO CASE 20: ACUTE FATTY LIVER IN PREGNANCY

Summary: 21-year-old female at 35-week gestation with malaise, nausea and vomiting, jaundice, elevated blood pressures, elevated liver function tests, coagulopathy, hypoglycemia, and subsequently hepatic coma and renal failure. She has been diagnosed with acute fatty liver of pregnancy.

- **Associated biochemical defect:** Fetal deficiencies of long chain 3-hydroxyacyl-coenzyme A dehydrogenase (LCHAD).
- **Cause of hypoglycemia:** Decreased liver glycogen after liver undergoes fatty infiltration and subsequent liver failure. Histology reveals swollen hepatocytes in which the cytoplasm is filled with microvesicular fat.

CLINICAL CORRELATION

Acute fatty liver of pregnancy is a poorly understood condition affecting only pregnant women with the clinical manifestations of hypoglycemia, liver failure, metabolic acidosis, renal failure, and coagulopathy. Affected patients may become jaundiced or encephalopathic from liver failure, usually reflected by an elevated ammonia level. Profound hypoglycemia is common. The mortality rate is approximately 10 to 15 percent. Management is delivery, with supportive measures such as magnesium sulfate to prevent seizures, replacement of blood or clotting factors, and management of the blood pressure. The pathophysiology may be related to fetal deficiencies of long chain 3-hydroxyacyl-coenzyme A dehydrogenase (LCHAD).

APPROACH TO GLYCOGEN AND CARBOHYDRATE METABOLISM

Objectives

1. Know about glycogen storage, synthesis, and degradation.
2. Be familiar with the regulation of glycogen synthesis and degradation.
3. Understand the role of liver glycogen in carbohydrate metabolism.

Definitions

GLUT 2: Glucose transporter isoform 2, a facilitative glucose transporter present in the liver, the β-cells of the pancreas, and the basolateral surface of intestinal epithelial cells.

Adenylate cyclase: The enzyme that, when activated by hormones binding to receptors, catalyzes the cyclization of adenosine triphosphate (ATP) to cyclic adenosine monophosphate (cAMP) with the release of pyrophosphate.

Branching enzyme: 1,4-α-Glucan branching enzyme; an enzyme that removes an oligosaccharide of about seven glucosyl residues from the nonreducing end of a glycogen chain and transfers it to another chain, creating an $\alpha(1\rightarrow6)$ glycosidic bond.

Debranching enzyme: A bifunctional enzyme that catalyzes two reactions in the degradation of glycogen. It transfers a trisaccharide from the nonreducing end of a four glucosyl residue branch of the glycogen molecule to the nonreducing end of the same or adjacent glycogen molecule (oligo-1,4–1,4-glucantransferase activity). It also hydrolyzes the $\alpha(1\rightarrow6)$ linkage of the remaining glucosyl residue of the branch, releasing free glucose (amylo-1,6-glucosidase activity).

Glycogen: The storage form of glucose in tissues. It is a large polysaccharide composed of glucose residues in primarily $\alpha(1\rightarrow4)$ glycosidic linkages with some $\alpha(1\rightarrow6)$ branch points.

Glycogenesis: The synthesis of glycogen from glucose 1-phosphate.

Glycogenolysis: The breakdown of glycogen to glucose 1-phosphate (and some small amount of free glucose).

Glycogen phosphorylase: The enzyme that causes the release of glucose 1-phosphate from glycogen. It accomplishes this by catalyzing a phosphorolysis of glucosyl residues from glycogen; that is, it breaks the $\alpha(1\rightarrow4)$ glycosidic bonds by adding inorganic phosphate and releasing glucose 1-phosphate.

Glycogen synthase: The enzyme that causes the addition of glucosyl residues to a growing glycogen molecule using UDP-glucose and releasing inorganic pyrophosphate.

DISCUSSION

Acute fatty liver of pregnancy (AFLP) is a rare (occurring in 1 in 7000 to 16,000 deliveries) but potentially fatal disease that typically develops during the third trimester. Patients most commonly present with clinical and laboratory evidence of acute hepatic failure with decreased hepatic metabolic activity. **Hypoglycemia,** nausea and vomiting, jaundice, general malaise, elevated blood pressure, disseminated intravascular coagulation, hemorrhage, infection, and encephalopathy are the most common clinical findings. The etiology of the syndrome is not clear, although recent reports have linked some cases of AFLP with a **fetal inborn error in fatty acid metabolism.** For the majority of cases in which an inborn error in the fetus does not appear to play a role, the cause of the disease is unknown.

One of the **primary functions of the liver** is to **maintain blood glucose levels.** When blood glucose levels are high following a meal, the liver takes in glucose via the high capacity, insulin-insensitive glucose transporter GLUT 2 and converts it into glycogen for storage, metabolizes it to pyruvate by glycolysis, or uses it to produce NADPH and pentoses for biosynthetic processes via the pentose phosphate pathway. When blood glucose levels drop after fasting or vigorous exercise, the liver mobilizes its glycogen stores in response to glucagon and epinephrine and exports glucose into the blood. As glycogen levels are depleted, the liver begins to synthesize glucose via gluconeogenesis. In addition to a source of carbons to synthesize glucose, which it obtains from either lactate or the breakdown of amino acids, the liver also needs a source of energy in the form of ATP. β-Oxidation of fatty acids provides the reducing equivalents (NADH and $FADH_2$), by which ATP is synthesized through the action of the electron transport system and oxidative phosphorylation.

The **storage form of glucose is glycogen,** which is **stored in both muscle and liver.** However, the function of stored glycogen is different in these two tissues. Muscle uses glycogen as a fuel reserve to provide ATP for its own needs, whereas the liver uses stored glycogen as a reservoir for glucose to maintain blood glucose levels. When blood glucose concentrations drop, **glucagon and epinephrine** are released into the bloodstream and bind to glucagon and epinephrine receptors on hepatocytes. The binding of the hormones to the receptors activates **adenylate cyclase** producing **3′,5′-cyclic AMP (cAMP).** When cAMP binds to cAMP-dependent protein kinase (PKA), it is activated and is able to phosphorylate target proteins. This leads to **activation of glycogen phosphorylase,** the enzyme primarily responsible for mobilizing glucose from glycogen. **Phosphorylase, which is stabilized by pyridoxal phosphate (vitamin B6), catalyzes the phosphorolysis of glycogen;** it **cleaves the 1,4-glycosidic bond** of a terminal glucose residue from the **nonreducing end of the glycogen molecule** using inorganic phosphate (Figure 20-1). The products are glucose 1-phosphate and glycogen that is shorter by one glucose residue.

Mobilization of glycogen stores also requires the participation of a **debranching enzyme** because phosphorylase ceases to cleave α-1,4-glycosidic linkages four glucosyl residues from an α-1,6-branch site. The debranching enzyme has two catalytic activities: a **transferase activity and a glucosidase**

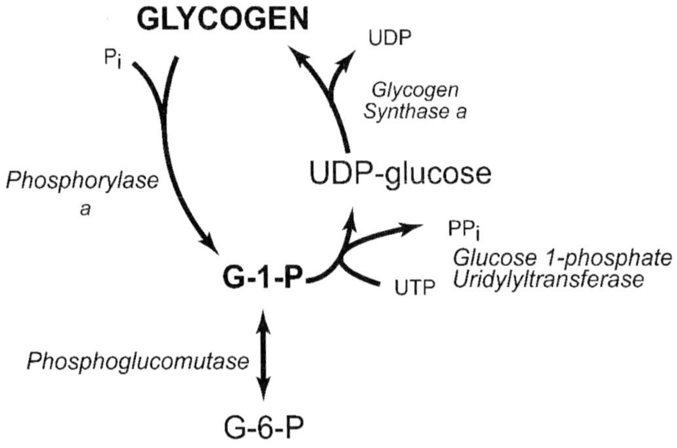

Figure 20-1. Reactions involved in the synthesis and breakdown of glycogen.

activity. The enzyme's transferase activity removes an oligosaccharide composed of the terminal three glucosyl residues from the four residue branch and transfers it to a free 4-hydroxyl group of the terminal glucosyl residue of another branch. The remaining glucosyl residue that is in an α-1,6-glucosidic linkage is then hydrolyzed by the glucosidase activity to release free glucose.

Glucose 1-phosphate released from glycogen by phosphorylase is converted to **glucose 6-phosphate by phosphoglucomutase.** Glucose-6-phosphatase, which is only present in liver and other gluconeogenic tissues, hydrolyzes the phosphate to produce free glucose. Glucose is then exported from the liver via the GLUT 2 transporter to increase the blood glucose concentration.

Following a meal, glycogen concentrations within the liver rise rapidly to high levels; this can be up to 10 percent of the wet weight of the liver. Glucose in the blood is transported into the hepatocyte by the GLUT 2 transporter and is converted to glucose 6-phosphate by glucokinase. **Phosphoglucomutase** then catalyzes the readily reversible reaction that **converts glucose 6-phosphate to glucose 1-phosphate.** The glucose 1-phosphate is further activated to **UDP-glucose** by **glucose 1-phosphate uridylyltransferase** in a reaction that consumes UTP and produces inorganic pyrophosphate. This reaction is thermodynamically favored by the hydrolysis of pyrophosphate by pyrophosphatase, which also makes the formation of UDP-glucose an irreversible reaction. Glycogen synthase catalyzes the addition of a glucosyl residue to a glycogen molecule using UDP-glucose as the substrate, forming an $\alpha(1\rightarrow4)$ glycosidic bond and releasing UDP. Since glycogen synthase cannot create an $\alpha(1\rightarrow6)$ linkage, an additional enzyme is required to form branches. When a chain of at least 11 glucosyl residues has been synthesized, 1,4-α-glucan branching enzyme removes a chain of about seven glucosyl residues and transfers it to another chain, creating an $\alpha(1\rightarrow6)$ glycosidic bond. This new branch point must be at least four glucosyl residues away from another branch point.

Since the **synthesis and mobilization of glycogen** together form a potential futile cycle, the competing processes must be **regulated to prevent waste of ATP/UTP.** This is accomplished by **hormonal as well as allosteric controls.** The enzymatic cascade, that is promulgated when glucagon or epinephrine bind to their respective receptors on the liver cells, is presented in Figure 20-2. The cAMP that is produced by activation of adenylate cyclase binds to PKA and activates it so that it can phosphorylate its target proteins. These include **phosphorylase kinase, glycogen synthase, and inhibitor 1.** Phosphorylation of glycogen synthase converts it to an inactive form, whereas phosphorylation of phosphorylase kinase and inhibitor 1 activate them. The phosphorylated inhibitor 1 is then able to bind strongly to protein phosphatase 1, but it is a poor substrate and is hydrolyzed slowly. Although the phosphorylated inhibitor 1 is bound to the phosphatase, it will inhibit it from acting on other phosphorylated proteins. Thus, while protein phosphatase 1 is inhibited, those proteins that are activated by phosphorylation remain active, and those that are inhibited by phosphorylation stay in their inactive form.

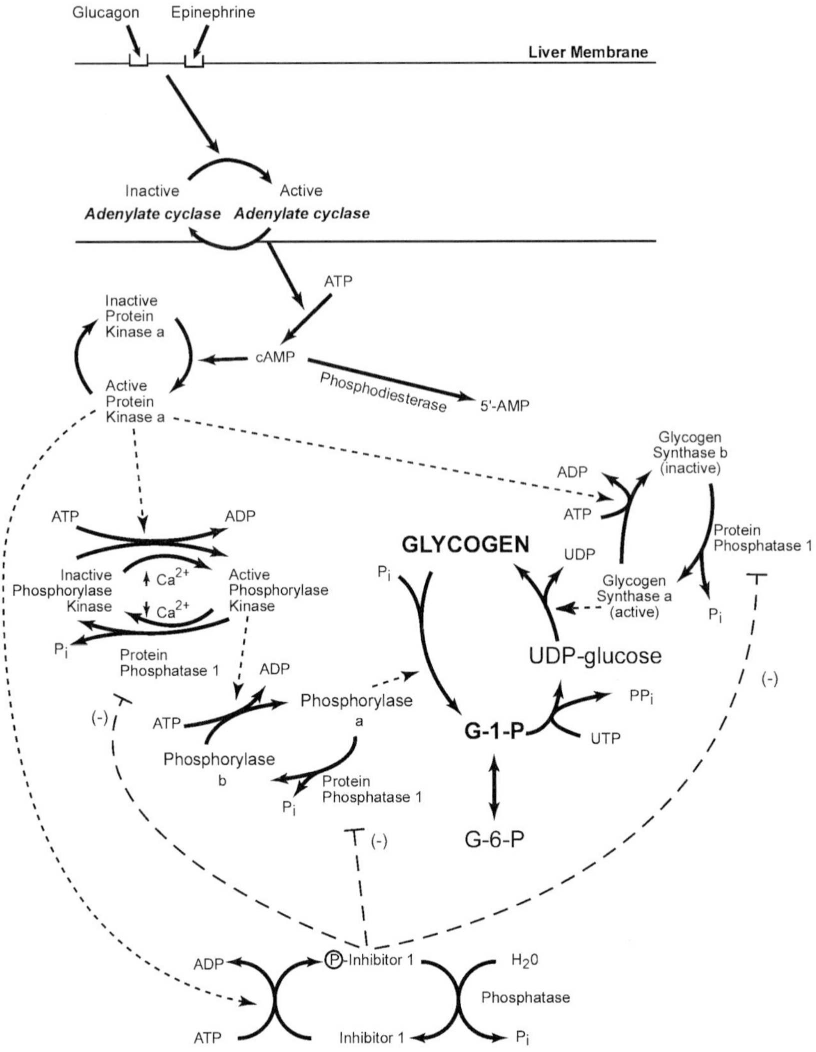

Figure 20-2. The mobilization of glycogen in the liver in response to hormonal signals. Binding of the hormones glucagon and/or epinephrine causes the activation of adenylate cyclase resulting in the production of cyclic AMP, which activates protein kinase A. By phosphorylation reactions, protein kinase A inactivates glycogen synthase, activates a cascade that results in active glycogen phosphorylase, and produces an active inhibitor of protein phosphatase 1.

Phosphorylation of phosphorylase kinase partially activates it so that it can phosphorylate **phosphorylase b to its active form.** Phosphorylase kinase is also partially activated by Ca^{2+}; full activation is obtained when it both binds Ca^{2+} and is phosphorylated. Conversion of phosphorylase b to phosphorylase a enables glucose 1-phosphate to be released from glycogen. Thus, glucagon and epinephrine start a cascade that mobilizes glucose from glycogen and at the same time inhibits the storage of glucose as glycogen.

When blood glucose levels are elevated, **insulin** is secreted from the pancreatic cells. When insulin binds to hepatic insulin receptors, it results in the activation of a **complex series of kinases** that leads to the **activation of protein phosphatase 1. Protein phosphatase 1 dephosphorylates phosphorylase kinase, phosphorylase, and inhibitor 1, thus inactivating them** and inhibiting the phosphorolysis of glycogen. It also dephosphorylates glycogen synthase, converting it to its active form and enabling the storage of glucose as glycogen. In addition, the liver form of phosphorylase a is inhibited by elevated intracellular concentrations of glucose. Thus insulin favors the storage of glycogen and inhibits its mobilization.

Although the etiology of the AFLP syndrome is unclear, it does appear to be a defect affecting mitochondrial processes. Liver biopsy usually will show mitochondrial disruption and microvesicular fat deposits, indicating decreased β-oxidation of fatty acids. The fatty acids, since they cannot be efficiently oxidized in the mitochondria, are converted to triglycerides, which build up in the hepatocyte. The fat infiltration decreases the amount of glycogen that can be stored and mobilized to maintain blood glucose levels. Gluconeogenesis is also depressed because ATP is not available from the oxidation of fatty acids. Thus, blood glucose levels decline.

As noted above, there have been reports that link some cases of AFLP with a defect in fatty acid metabolism in the fetus. These include **fetal deficiencies of long chain 3-hydroxyacyl-coenzyme A dehydrogenase (LCHAD), carnitine-palmitoyl transferase 1 (CPT 1), and medium chain acyl-coenzyme A dehydrogenase (MCAD).** The mechanism by which defective fetal fatty acid oxidation causes maternal illness is not known. However, since the fetus uses primarily glucose metabolism for its energy needs, it is likely that toxic products from the placenta, which does use fatty acid oxidation, cause the maternal liver failure.

COMPREHENSION QUESTIONS

For Questions [20.1] and [20.2] refer to the following case:

A female infant appeared normal at birth but developed signs of liver disease and muscular weakness at 3 months. She had periods of hypoglycemia, particularly on awakening. Examination revealed an enlarged liver. Laboratory analyses following fasting revealed ketoacidosis, blood pH 7.25, and elevations in both alanine transaminase (ALT) and aspartate transaminase (AST). Administration of glucagon following a carbohydrate meal elicited a normal rise in blood glucose, but glucose levels did not rise when glucagon was administered following an overnight fast. Liver biopsy revealed an increase in the glycogen content (6 percent of wet weight).

[20.1] In which of the following enzymes is a genetic deficiency most likely for this patient?

A. Branching enzyme
B. Debranching enzyme
C. Glucose-6-phosphatase
D. Glycogen synthase
E. Muscle phosphorylase

[20.2] To prevent the frequent episodes of hypoglycemia, which of the following dietary supplements would be most appropriate for this patient?

A. Casein (milk protein)
B. Fish oil
C. Fructose
D. Lactose
E. Uncooked cornstarch

[20.3] A 17-year-old male presents complaining of an inability to perform strenuous exercise without bringing on painful muscle cramps and weakness. He indicated that mild to moderate exercise resulted in no problems. When he was administered an ischemic exercise test, his serum lactate concentrations did not increase significantly. A deficiency in which of the following enzymes is most likely the cause of the patient's muscle cramps?

A. Carnitine palmitoyl transferase II
B. Glucose-6-phosphatase
C. Glycogen phosphorylase
D. Glycogen synthase
E. Very long chain acyl-CoA dehydrogenase

[20.4] A 23-year-old male has been vigorously working on the yard and begins to feel slightly light-headed from hypoglycemia. He drinks a can of soda and is aware of the competition for the glucose to be stored in his liver as glycogen versus used as energy in his muscles. What is the best explanation regarding the fate of the glucose in the soda?

A. The lower Km of hexokinase versus the Km of glucokinase will tilt the glucose toward glycolysis.
B. The bolus of glucose via the soda will lead to a higher glucose level, inducing storage of the glucose into glycogen in the liver.
C. The muscle is using high levels of glucose, leading to an increased level of glucose 6-phosphate thus inhibiting glucokinase.
D. The glucose will be equally used by muscle for metabolism and liver for glycogen storage.

Answers

[20.1] **B.** Definitive diagnosis would await analysis of the glycogen structure and enzyme activities, but the hepatomegaly, increased liver glycogen content, fasting hypoglycemia, and muscle weakness are consistent with Cori disease, glycogen storage disease type III. The increase in glycogen content results from an inability to degrade glycogen beyond the limit dextrin of phosphorylase. A deficiency in the debranching enzyme leaves glycogen with short outer branches.

[20.2] **E.** Because fasting hypoglycemia results from an inability to break down glycogen past the limit dextrin of phosphorylase, a patient with type III glycogen storage disease should be given frequent meals high in carbohydrates. Uncooked cornstarch is an effective supplement because it is slowly digested, and therefore the glucose is released slowly into the bloodstream, helping to maintain blood glucose concentrations.

[20.3] **C.** Although a deficiency in a number of enzymes can result in exercise intolerance, the lack of an increase in serum lactate following ischemic exercise points to an inability to a defect in the breakdown of glycogen in the muscle. The muscle depends on glycogenolysis for intense exercise, and fatigue rapidly ensues when glycogen is depleted. Patients with a deficiency in the muscle isoform of glycogen phosphorylase (McArdle disease) can tolerate mild to moderate exercise, but get muscle cramps with strenuous exercise as a consequence of the lack of glycogenolysis in the muscle cell.

[20.4] **A.** Hexokinase is found in most tissues, and because of the very low Km for glucose, it is designed to work maximally to provide ATP for tissue even at low levels of glucose. Hexokinase is inhibited by glucose 6-phosphate and is most active with low levels of glucose 6-phosphate. Glucokinase found in the liver has a high Km for glucose and is very active after a meal. The glucose in the soda would likely be used for ATP production.

BIOCHEMISTRY PEARLS

- The etiology of the syndrome is not clear, although recent reports have linked some cases of AFLP with a fetal inborn error in fatty acid metabolism.
- Following a meal, glycogen concentrations within the liver rise rapidly to high levels; this can be up to 10 percent of the wet weight of the liver.
- Liver insufficiency may be associated with hypoglycemia.

REFERENCES

Castro M, Fassett MJ, Telfer RB, et al. Reversible peripartum liver failure: A new perspective on the diagnosis, treatment, and cause of acute fatty liver of pregnancy, based on 28 consecutive cases. Am J Obstet Gynecol 1999;181:389–95.

Rakheja D, Bennett MJ, Rogers BB. Long-chain L-3-hydroxyacyl-coenzyme A dehydrogenase deficiency: A molecular and biochemical review. Lab Invest 2002;82:815–24.

Roe CR, Ding J. Mitochondrial fatty acid oxidation disorders. In: Scriver CR, Beaudet AL, Sly WS, et al., eds. The Metabolic and Molecular Basis of Inherited Disease, 8th ed. New York: McGraw-Hill, 2001:2297–2326.

Saudubray J-M, Charpentier C. Clinical phenotypes: diagnosis/algorithms. In: Scriver CR, Beaudet AL, Sly WS, et al., eds. The Metabolic and Molecular Basis of Inherited Disease, 8th ed. New York: McGraw-Hill, 2001:1327–1403.

Yang Z, Yamada J, Zhao Y, et al. Prospective screening for pediatric mitochondrial trifunctional protein defects in pregnancies complicated by liver disease. JAMA 2002;288:2163–66.

CASE 21

A 29-year-old male presents to the emergency department with complaints of dark-colored urine, generalized fatigue, myalgias, and weakness after completing a marathon. The patient states that this was his first marathon. He has no significant medical history and denies any medications or drug use. On examination, he appears moderately ill and is afebrile with normal vital signs. Physical exam reveals diffuse musculoskeletal tenderness. Urinalysis revealed large amounts of blood (hemoglobin and myoglobin), and serum creatine phosphokinase (CPK) was significantly elevated, as well as the potassium level on his electrolytes. The serum lactate level was markedly elevated.

- **What is the most likely diagnosis?**
- **What is the most appropriate treatment?**
- **What is the biochemical basis for the markedly elevated serum lactate level?**

ANSWERS TO CASE 21: RHABDOMYOLYSIS

Summary: 29-year-old marathon runner with acute episode of generalized myalgias, weakness, fatigue, and dark-colored urine with urine myoglobin/hemoglobin, hyperkalemia, and significantly increased CPK isoenzyme.

- **Most likely diagnosis:** Rhabdomyolysis (skeletal muscle cell lysis) after strenuous exercise.
- **Treatment:** Aggressive intravenous hydration to help clear the excess myoglobin from the serum, and correction of electrolyte abnormalities and treatment of kidney failure if present.
- **Biochemical basis for elevated lactate:** Nicotinamide adenine dinucleotide (NADH) **levels increase because of the relative lack of oxygen for muscle,** adenosine diphosphate (ADP) and adenosine monophosphate (AMP) concentrations rise in the cytoplasm, leading to an increased flux of glucose through the glycolytic pathway in the muscle, causing pyruvate levels to increase. Pyruvate is reduced by NADH to lactate in a reaction catalyzed by lactate dehydrogenase. Lactate is transported out of the muscle cell to the blood.

CLINICAL CORRELATION

Skeletal muscle has a need for oxygen and fuel (glucose and fatty acids). Short exertion allows for replenishment of these important substrates; however, long, grueling demands on muscle, such as running a marathon, can lead to relative deprivation of oxygen (because of either overexertion or dehydration and insufficient blood flow to the muscles). This lack of oxygen leads to the conversion to the glycolytic pathway versus the tricarboxylic acid (TCA) pathway for adenosine triphosphate (ATP) production. Marathon running has been shown to effect increases in the blood and urinary concentrations of a number of biochemical parameters that result from exertional muscle damage (rhabdomyolysis) and hemolysis. These include increases in serum myoglobin, CPK, as well as an increase in the anionic gap leading to a metabolic acidosis. In the case of marathon runners, this is usually caused by an increase in the serum lactate concentration. Other causes of rhabdomyolysis include cocaine intoxication, hyperthermia, convulsions, or toxins.

APPROACH TO OXIDATIVE PHOSPHORYLATION AND LACTATE

Objectives

1. Understand how hypoxemia (e.g., rhabdomyolysis) leads to reduced oxidative phosphorylation and increased lactic acid production.
2. Be familiar with the pyruvate cycle and the importance of NADH levels.
3. Know about the lactic acid pathway.

Definitions

Anion gap: A calculation of the routinely measured cations minus the routinely measured anions. Since in all fluids the sum of the positive charges (cations) must be balanced with the negative charges (anions), the anion gap is an artifact of measurement. Because the $[K^+]$ is small, it is usually omitted from the calculation. The equation most frequently used to calculate the anion gap is

$$AG = [Na^+] - ([Cl^-] + [HCO_3^-])$$

Gluconeogenesis: The series of biochemical reactions in which glucose is synthesized in the liver (and other gluconeogenic tissues) from small organic acids such as lactate, pyruvate, and oxaloacetate.

Hematin: Heme in which the coordinated iron is in the ferric (Fe^{3+}) oxidation state.

Myoglobin: A large heme-containing protein that is able to bind oxygen and release it in tissues in which the oxygen tension is low.

β-Oxidation: The series of biochemical reactions in which fatty acids are degraded to acetyl-CoA, which then enters the tricarboxylic acid cycle for the production of energy in the form of reducing equivalents and GTP. Each round of β-oxidation shortens the fatty acid by two carbons and, in addition to acetyl-CoA, produces NADH and $FADH_2$, which are fed into the electron transport system for the production of ATP.

DISCUSSION

The **exercising muscle's sources of energy are primarily glucose and fatty acids.** The muscle obtains glucose from the blood or the breakdown of stored glycogen in the muscle. Fatty acids are acquired as free fatty acids from the blood or from the breakdown of triglycerides that are stored in the muscle. For complete oxidation of these sources of energy, the **metabolic intermediate acetyl coenzyme A (acetyl-CoA)** must be **oxidized** through the **TCA cycle,** and the reducing equivalents produced (NADH and $FADH_2$) must be transferred to O_2 through the mitochondrial electron transport system. This electron transfer process produces a **proton gradient** across the inner mitochondrial matrix that drives the synthesis of ATP by ATP synthase. For this process to continue, a plentiful supply of oxygen must be supplied to the tissues.

It has been estimated that at the beginning of the marathon, a runner who is running at a reasonable pace consumes energy in a ratio of 75 percent carbohydrate to 25 percent fatty acids. This ratio will decrease as glycogen stores in the body diminish. However, as exertion continues and the muscle starts to rely more on the β-oxidation of fatty acids to provide its energy needs, its oxygen demand increases, placing a heavier demand on the heart to provide oxygenated blood. If the runner does not take steps to replace fluids lost through sweat, dehydration will occur, resulting in decreased perfusion of the muscle with adequately oxygenated blood.

If the muscle uses ATP faster than it can be produced by oxidative phosphorylation either by overexertion or because O_2 uptake is limited, **NADH levels increase in the mitochondria and in the cytoplasm.** ADP and AMP concentrations in the cytoplasm will rise as ATP is used by the muscle for contraction. This will increase the flux of glucose through the glycolytic pathway in the muscle, causing pyruvate levels to increase. To regenerate the oxidized cofactor NAD^+ that is required in the conversion of glyceraldehyde 3-phosphate to 3-phosphoglycerate, **pyruvate is reduced by NADH to lactate** in a reaction catalyzed by lactate dehydrogenase. Lactate is transported out of the muscle cell to the blood.

Lactate in the blood is taken up by the liver and used as a carbon source in the synthesis of new glucose by the gluconeogenic pathway. The liver must first reoxidize lactate to pyruvate in a reaction that is catalyzed by lactate dehydrogenase and generates NADH. Pyruvate cannot be directly converted to phosphoenolpyruvate (PEP) by pyruvate kinase because under physiologic conditions; the reaction is thermodynamically irreversible in favor of the formation of pyruvate. Instead, pyruvate must enter the mitochondria and be carboxylated by the enzyme **pyruvate carboxylase** to form **oxaloacetate** in a reaction that requires biotin as a cofactor (Figure 21-1). Oxaloacetate is then reduced to malate by malate dehydrogenase and **malate** then exits the mitochondrion. In the cytosol, malate is reoxidized back to **oxaloacetate** by cytoplasmic malate dehydrogenase. The cytoplasmic **oxaloacetate** is then converted to **PEP by PEP carboxykinase** in a reaction that requires GTP. (In a minor alternate pathway, mitochondrial oxaloacetate can be converted to PEP by the mitochondrial form of the enzyme PEP carboxykinase. PEP then exits the mitochondrion to the cytosol.) The pathway from PEP to glucose is identical to that of glycolysis except for two reactions, as shown in Figure 21-2. **Fructose 1,6-bisphosphate** is converted to fructose 6-phosphate by hydrolysis of phosphate by the **enzyme fructose-1,6-bisphosphatase.** The final reaction is hydrolysis of glucose 6-phosphate to glucose by glucose-6-phosphatase. Hepatic glucose produced via gluconeogenesis is then delivered to the blood for use by the brain and muscle. This process by which extrahepatic lactate is taken back to the liver, **converted to glucose by gluconeogenesis,** and returned to extrahepatic tissues is called the **Cori cycle.** When the rate of lactate production

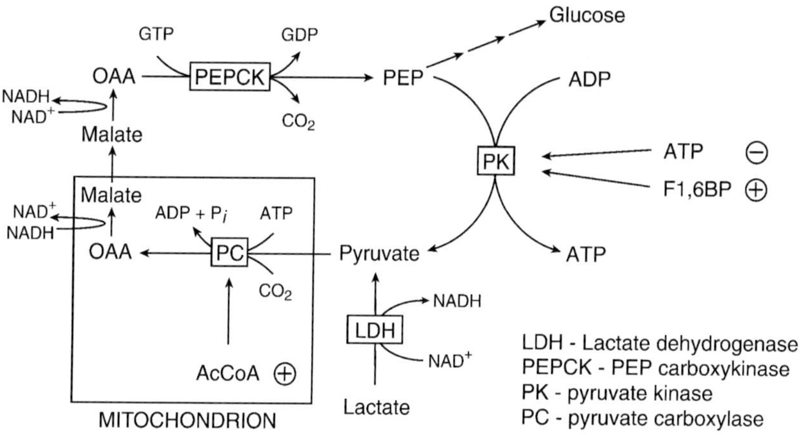

Figure 21-1. Interconversion of phosphoenolpyruvate (PEP) and pyruvate. The conversion of PEP to pyruvate is thermodynamically irreversible in the cell. To convert pyruvate back to PEP for gluconeogenesis, pyruvate must enter the mitochondrion, be carboxylated to oxaloacetate (OAA), and reduced to malate. After exiting the mitochondrion, malate is oxidized back to OAA and converted to PEP by the action of phosphoenolpyruvate carboxykinase.

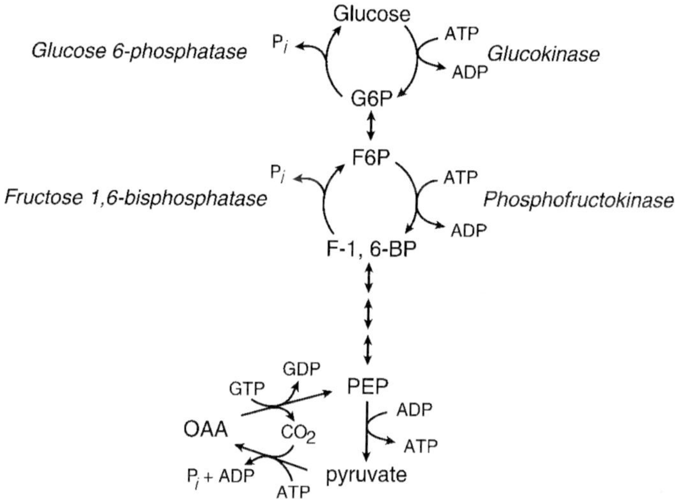

Figure 21-2. The reactions that make up the thermodynamically irreversible steps in glycolysis and gluconeogenesis. These steps make up potential "futile" cycles that must be carefully regulated.

by muscle exceeds the rate at which the Cori cycle can operate, **lactate accumulates in the blood leading to lactic acidemia.**

During a marathon run, there is constant stretching and tearing of the leg muscle fibers each time the foot hits the ground. This constant shock causes damage to the muscle cells, resulting in a release of cellular contents to the extracellular matrix and the bloodstream. The concentrations of myoglobin, which is in high concentration in slow twitch (red) muscle fibers, and K^+, which is concentrated in all cells, therefore rise in the blood. When the concentration of myoglobin increases above 0.5 to 1.5 mg/dL, it is excreted into the urine. Normally myoglobin is not toxic to the kidney; however, when the pH of the urine drops below 5.6, myoglobin undergoes oxidation to produce hematin (porphyrin bound Fe^{3+}), which is toxic to the kidney and can lead to acute renal failure. This toxic effect is exacerbated when the urine is concentrated as a result of dehydration.

COMPREHENSION QUESTIONS

[21.1] During the course of a marathon race a runner expends a large amount of energy and must use stored sources of fuel as well as oxygen. Compared to the beginning of the race (first mile), which of the following best describes the utilization of glycogen and fatty acids as fuels and amount of oxygen consumed after running for 26 miles?

	Glycogen used	Fatty acids used	Oxygen consumed
A.	↑	↑	↑
B.	↑	↓	↓
C.	↑	↓	↑
D.	↓	↑	↑
E.	↓	↓	↓

[21.2] During an extended period of exercise, the enzymes involved in the glycolytic pathway in muscle tissue are actively breaking down glucose to provide the muscle energy. The liver, to maintain blood glucose levels, is synthesizing glucose via the gluconeogenic pathway. Which of the following enzymes involved in these pathways would be most likely to exhibit Michaelis–Menten kinetics, that is, have a hyperbolic curve when plotting substrate concentration versus velocity of the reaction?

A. Fructose-1,6-bisphosphatase
B. Hexokinase
C. Lactate dehydrogenase
D. Phosphofructokinase 1
E. Pyruvate kinase

[21.3] A known alcoholic is found lying semiconscious at the bottom of a stairwell with a broken arm by his landlady, who called an ambulance to take him to the emergency room. Initial laboratory studies showed a relatively large anion gap of 34 (normal = 9 to 15). His blood alcohol was elevated at 245 mg/dL (intoxication level = 150 to 300 mg/dL), and his blood glucose was 38 mg/dL (low normal). The patient's large anion gap and hypoglycemia can best be explained by which of the following?

A. Decreased secretion of glucagon
B. Increased secretion of insulin
C. Increased urination resulting from the diuretic effect of alcohol
D. Inhibition of dehydrogenase enzymes by NADH
E. Inhibition of glycogenolysis by ethanol

Answers

[21.1] **D.** At the beginning of the race, a runner running at a reasonable pace consumes energy at a ratio of approximately 75 percent carbohydrate: 25 percent fatty acids. However, by the end of the race, glycogen stores are for the most part depleted, and the generation of ATP must come from the β-oxidation of fatty acids, which produces reducing equivalents for oxidative phosphorylation. This requires an increase in the amount of oxygen consumed.

[21.2] **C.** The activity of regulatory enzymes such as fructose-1,6-bisphosphatase, hexokinase, phosphofructokinase 1, and pyruvate kinase are frequently controlled by binding allosteric effectors. These allosteric enzymes usually exhibit sigmoidal kinetics. Lactate dehydrogenase is not controlled by allosteric effectors and therefore would be expected to exhibit Michaelis–Menten kinetics.

[21.3] **D.** Alcoholics frequently do not eat while binge drinking, so it is most likely that his liver glycogen stores became depleted and could not increase his blood glucose levels. The metabolic stress leads to the increase in secretion of epinephrine and other hormones that mobilize fatty acids from stored triglycerides in adipose cells. These fatty acids undergo β-oxidation in the liver but are converted to ketone bodies because of the inhibition of the TCA cycle by high levels of NADH produced by the oxidation of ethanol first to acetaldehyde and acetate. Key gluconeogenic dehydrogenases are also inhibited by the elevated levels of NADH, including lactate dehydrogenase, glycerol 3-phosphate dehydrogenase, and malate dehydrogenase.

BIOCHEMISTRY PEARLS

- Exercising muscle's sources of energy are primarily glucose and fatty acids.
- With insufficient oxygen to meet the muscle demands, NADH levels increase in the mitochondria and in the cytoplasm, ADP and AMP concentrations in the cytoplasm will rise, and glucose will be shunted to the glycolytic pathway in the muscle. The pyruvate is converted to lactate, which causes the metabolic acidosis.
- Muscle injury can lead to myoglobinemia and myoglobinuria (red urine) and may crystallize in the renal tubules leading to renal insufficiency.

REFERENCES

Brady HR, Brenner BM. Acute renal failure. In: Fauci AS, Braunwald E, Kasper KL, et al., eds. Harrison's Principles of Internal Medicine, 14th ed. New York: McGraw-Hill, 1998.

Kratz A, Lewandrowski KB, Siegel AJ, et al. Effect of marathon running on hematologic and biochemical laboratory parameters, including cardiac markers. Am J Clin Pathol 2002;118:856–63.

Murakami K. Rhabdomyolysis and acute renal failure. In: Glew RH, Ninomiya Y, eds. Clinical Studies in Medical Biochemistry, 2nd ed. New York: Oxford University Press, 1997.

CASE 22

A 50-year-old Hispanic female presents to your clinic with complaints of excessive thirst, fluid intake, and urination. She denies any urinary tract infection symptoms. She reports no medical problems, but has not seen a doctor in many years. On examination she is an obese female in no acute distress. Her physical exam is otherwise normal. The urinalysis revealed large glucose, and a serum random blood sugar level was 320 mg/dL.

- **What is the most likely diagnosis?**
- **What other organ systems can be involved with the disease?**
- **What is the biochemical basis of this disease?**

ANSWERS TO CASE 22: TYPE II DIABETES

Summary: 50-year-old obese Hispanic female presents with polydipsia, polyphagia, and urinary frequency and elevated random blood sugar of 320 mg/dL.

- **Diagnosis:** Type II diabetes.
- **Other organ systems involved:** Cardiovascular, eye, peripheral nerves, gastrointestinal, kidney.
- **Biochemical basis:** Insulin resistance as a result of a postinsulin receptor defect. The insulin levels are normal or increased as compared with normal individuals; however, the insulin is not "recognized," and thus the glucose levels remain elevated.

CLINICAL CORRELATION

Diabetes mellitus is characterized by elevated blood glucose levels. It is composed of two types depending on the pathogenesis. Type I diabetes is characterized by insulin deficiency and usually has its onset during childhood or teenage years. This is also called *ketosis-prone* diabetes. Type II diabetes is caused by insulin resistance and usually has elevated insulin levels, and it is diagnosed in the adult years. Type II diabetes is far more common than type I diabetes. Risk factors include obesity, family history, sedentary life style, and, in women, hyperandrogenic states or anovulation.

Diabetes mellitus is now recognized as one of the most common and significant diseases facing Americans. It is estimated that 1 of 4 children born today will become diabetic in their lifetime because of obesity and inactivity. Also, it has been noted that diabetes has a severe effect on blood vessels, particularly in the pathogenesis of atherosclerosis (blockage of arteries by lipids and plaque), which can lead to myocardial infarction or stroke. Diabetes mellitus is treated as equivalent to a prior cardiovascular event in its risk for future atherosclerotic disease. Diabetes is also associated with immunosuppression, renal insufficiency, blindness, neuropathy, and other metabolic disorders.

APPROACH TO INSULIN AND GLUCOSE

Objectives

1. Understand the role of insulin on carbohydrate metabolism.
2. Be aware of the role of glucagon on carbohydrate metabolism.
3. Know about the processes of gluconeogenesis and glycogenolysis.

Definitions

Diabetes mellitus: An endocrine disease characterized by an elevated blood glucose concentration. There are two major forms of diabetes mellitus: type I, or insulin-dependent, and type II, or non–insulin-dependent. Type I is caused by a severe lack or complete absence of insulin. Type II is caused by resistance to insulin, that is, an inability to respond to physiologic concentrations of insulin.

Fructose 2,6-bisphosphate: A metabolite of fructose 6-phosphate produced by the bifunctional enzyme 6-phosphofructokinase-2/fructose bisphosphatase-2 (PFK-2/FBPase-2). It serves as an allosteric effector that activates 6-phosphofructokinase-1 and inhibits fructose bisphosphatase-1, thus stimulating the movement of glucose through the glycolytic pathway and inhibiting gluconeogenesis.

Glucagon: A polypeptide hormone synthesized and secreted by the α-cells of the islets of Langerhans in the pancreas. Glucagon is released in response to low blood glucose levels and stimulates glycogenolysis and gluconeogenesis in the liver.

Insulin: A polypeptide hormone synthesized and secreted by the β-cells of the islets of Langerhans in the pancreas. Insulin is released in response to elevations in blood glucose and promotes the uptake of glucose into cells by increasing the number of GLUT 4 glucose transporters on cell surfaces.

Protein kinase A: An enzyme that will phosphorylate target proteins. It is activated by increased cAMP concentration in the cell that is a response to the activation of adenylate cyclase by binding of certain hormones on cell surfaces.

DISCUSSION

Every cell in the human body uses **glucose** as an energy source. Indeed, certain cells have an obligate requirement for glucose to meet their energetic demands (e.g., erythrocytes). **Neurons,** although can use alternative fuel sources under extreme conditions (e.g., **ketone bodies** during prolonged starvation), have a strong preference toward glucose utilization. Circulating levels of glucose must therefore be maintained sufficiently high to meet the energy demands of the body. Chronic elevations in blood glucose levels are also detrimental, being associated with oxidative stress and glycation of cellular proteins. It has been suggested that the latter mediate many of the complications associated with chronic hyperglycemia, such as diabetic microvascular disease and retinopathy.

Despite diurnal variations in meal times, blood glucose levels are normally maintained within a narrow range. This is made possible in large part by the counter regulatory actions of the peptide hormones **insulin and glucagon.** Insulin, secreted by the β-cells of pancreatic islets when blood glucose levels increase, promotes glucose utilization and represses endogenous glucose production (Figure 22-1a). In contrast, **glucagon,** secreted by the α-cells of pancreatic islets when blood glucose levels are low, represses glucose utilization and promotes endogenous glucose production (Figure 22-1b). A **careful balance between the actions of insulin and glucagon** therefore help maintain blood glucose levels within a normal range.

Insulin receptors are essentially expressed ubiquitously, in large part as a result of the mitogenic actions of this peptide hormone. In terms of glucose metabolism, the actions of insulin on the liver, adipose, and skeletal muscle will be the focus of this discussion, although insulin-mediated changes in satiety and blood flow undoubtedly play a role in whole body glucose homeostasis. On binding to its cell surface receptor, insulin elicits a complex cascade of cellular signaling events that have not been elucidated fully to date. This in turn increases glucose transport into the cell (skeletal muscle and adipose), promotes storage of excess carbon from glucose as glycogen (skeletal muscle and liver) and triglyceride (TAG; liver and adipose), increases glucose utilization as a fuel source (skeletal muscle, liver, and adipose), and decreases endogenous glucose production (liver) (see Figure 22-1a). These actions of insulin can be either acute (affecting activity of preexisting proteins) or chronic (altering protein levels).

Skeletal muscle and adipose express two major isoforms of **glucose transports, GLUT 1 and GLUT 4.** GLUT 1, a ubiquitously expressed glucose

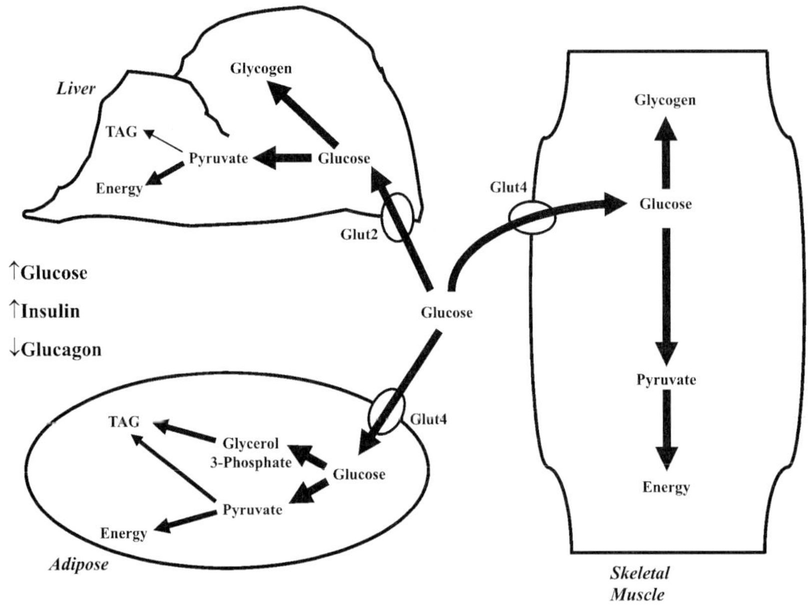

Figure 22-1a. The flow of glucose to tissues under conditions of elevated blood glucose concentration. When $[\text{glucose}]_{\text{blood}}$ is high, the insulin:glucagon ratio is high, leading to the uptake of glucose into the tissues. Abbreviation: TAG = triacylglycerol

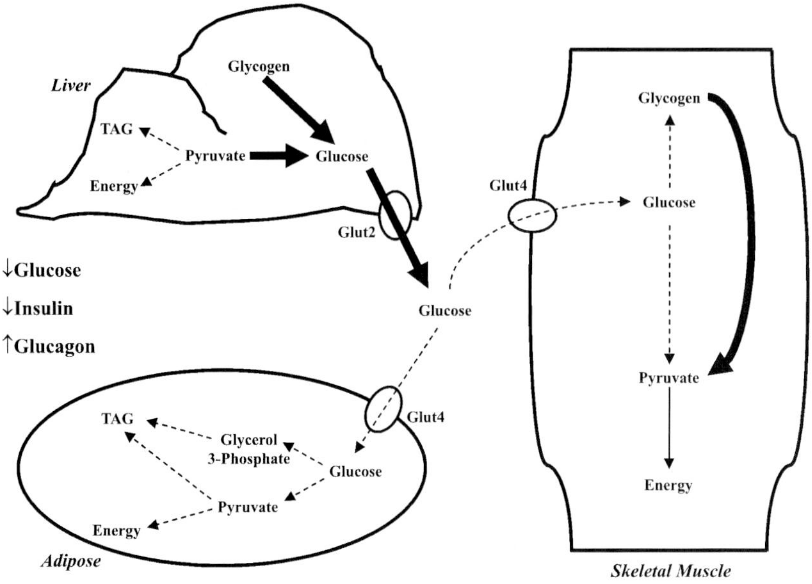

Figure 22-1b. The flow of glucose to tissues under conditions of low blood glucose concentration. When $[\text{glucose}]_{\text{blood}}$ is low, the insulin:glucagon ratio is low, leading to glycogenolysis and gluconeogenesis in the liver. Abbreviation: TAG = triacylglycerol

transporter, resides almost exclusively at the cell surface, where it facilitates a constant "basal" rate of glucose uptake into the cell. In contrast, GLUT 4, whose expression is limited to skeletal muscle, heart, and adipose, can be found both at the cell surface and within specialized intracellular vesicles. Redistribution of GLUT 4 from intracellular vesicles to the cell surface in response to insulin results in increased rates of glucose transport, thereby facilitating insulin-stimulated glucose disposal. In contrast to skeletal muscle and adipose, the liver expresses GLUT 2. This is a freely reversible glucose transporter that resides permanently at the cell surface. GLUT 2 enables glucose to pass down its concentration gradient, allowing increased hepatic glucose uptake when blood glucose levels are high and increased hepatic glucose efflux when blood glucose levels are low.

Once within the cell, glucose undergoes one of several fates. Insulin promotes incorporation of glucose moieties into **glycogen,** the storage form of glucose in mammals. This is driven in large part by insulin-mediated activation of **protein phosphatase 1** (PP1; Figure 22-2a). PP1 dephosphorylates (hydrolytic removal of regulatory phosphate groups from serine/threonine residues on target enzymes) a number of key proteins involved in glycogen metabolism. Dephosphorylation and activation of glycogen synthase (GS), with a concomitant dephosphorylation and inactivation of glycogen phosphorylase (GP),

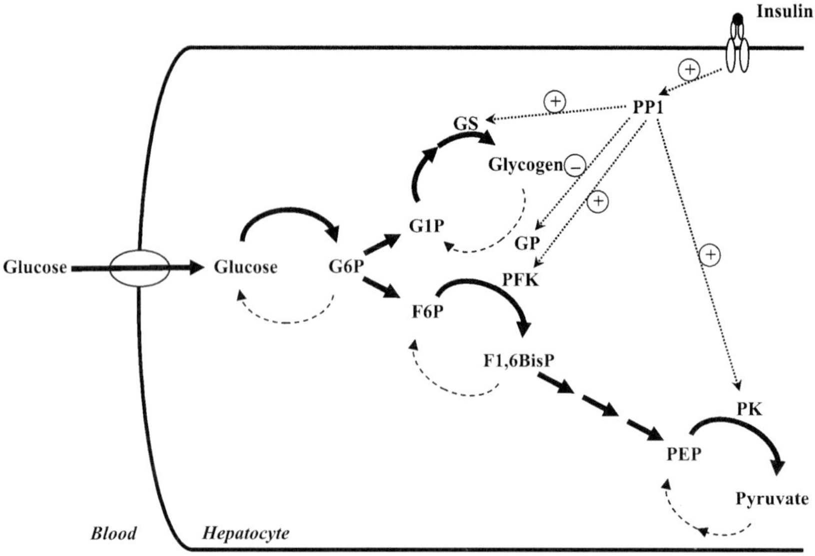

Figure 22-2a. Insulin stimulation of the flux of glucose through the glycolytic pathway. Insulin activates protein phosphatase 1, which in turn activates glycogen synthase, phosphofructokinase, and pyruvate kinase. Abbreviations: fructose 1,6-bisphosphate (F1,6BisP); fructose 6-phosphate (F6P); glucose 1-phosphate (G1P); glucose 6-phosphate (G6P); glycogen phosphorylase (GP); glycogen synthase (GS); phosphoenolpyruvate (PEP); phosphofructokinase (PFK); pyruvate kinase (PK); protein phosphatase 1 (PP1)

will stimulate net glycogen synthesis (glycogenesis) by liver and muscle, in response to insulin. One way in which insulin promotes PP1-mediated effects on glycogen metabolism is through the targeting of PP1 to the glycogen particle, a subcellular domain comprised of glycogen itself, as well as the enzymes required for glycogen synthesis and degradation. The glycogen binding subunit is a docking protein allowing PP1 association with the glycogen particle. Insulin induces tyrosine phosphorylation of this glycogen binding subunit, thereby promoting PP1 binding and therefore increased glycogenesis.

The glycogen capacity of a cell is of finite size. Once this capacity is reached, excess glucose must undergo alternative metabolic fates. Insulin promotes flux of glucose through the glycolytic pathway (glucose $\rightarrow$ 2 pyruvate), again in part through PP1 activation (see Fig. 22-2a). As glycogenolysis is the reciprocal pathway to glycogenesis, gluconeogenesis (the synthesis of glucose) is the reciprocal pathway to glycolysis. Gluconeogenesis occurs primarily in the liver and to a lesser extent in the kidney. PP1 increases glycolytic flux in the liver through activation of phosphofructokinase (PFK; indirect effect through dephosphorylation of a bifunctional enzyme, resulting in increased intracellular levels of fructose 2,6-bisphosphate, an allosteric activator of PFK) and pyruvate kinase (PK; direct effect). Glycolytically derived pyruvate could potentially undergo one of two fates in the liver, namely, full oxidation (via Krebs cycle and oxidative phosphorylation) and/or entry into the fatty acid synthesis pathway. However, humans on a Western diet, in which excess calories are more often a mixture of carbohydrate and fat, tend to use ingested carbohydrate as a fuel, while fatty acids are stored as triglyceride in adipose tissue. The latter is driven by insulin.

When blood glucose levels begin to decline (e.g., during an overnight fast), so too does insulin secretion. In contrast, circulating levels of **glucagon** increase. The latter targets primarily **hepatic glucose metabolism** in humans, increasing glucose production and decreasing glucose utilization. On binding to its cell surface receptors, glucagon increases the activity of **protein kinase A** (PKA; Figure 22-2b). In turn, PKA stimulates net glycogen breakdown (glycogenolysis) through phosphorylation of phosphorylase kinase (increases activity) and glycogen synthase (decreases activity). The former phosphorylates and activates glycogen phosphorylase. PKA further antagonizes the effects of insulin through inactivation of PP1. PKA phosphorylates the glycogen binding subunit at specific serine residues, causing the release of PP1 from the glycogen particle. Once released, PP1 binds to inhibitor 1, further inactivating PP1 activity. This PP1-inhibitor 1 association is promoted by PKA-mediated phosphorylation of inhibitor 1. Gluconeogenesis is also stimulated

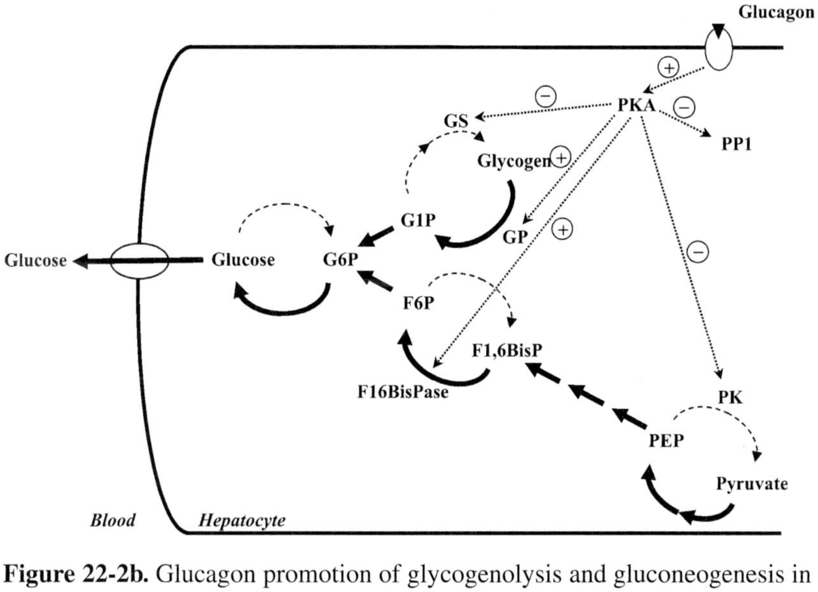

Figure 22-2b. Glucagon promotion of glycogenolysis and gluconeogenesis in the liver. Glucagon binding leads to the activation of protein kinase A, which activates glycogen phosphorylase and fructose-1,6-bisphosphatase, while inhibiting glycogen synthase, pyruvate kinase, and protein phosphatase 1. Abbreviations: protein kinase A (PKA); the rest are the same as in Figure 22-2a.

by glucagon-induced PKA activation, through activation of fructose-1,6-bisphosphatase (F1,6BisPase; indirect effect through phosphorylation of a bifunctional enzyme, resulting in decreased intracellular levels of fructose 2,6-bisphosphate, an allosteric inhibitor of F1,6BisPase) and pyruvate kinase (reverse of PP1 effects). Glycogenolysis- and gluconeogenesis-derived glucose is exported out of the liver to help maintain blood glucose levels.

Abnormalities in the above described glucose homeostatic mechanisms arise during diabetes mellitus. **Two major forms of diabetes mellitus** exist, insulin-dependent (type I) and non–insulin-dependent (type II) diabetes. **Type I diabetes** is caused by a **severe lack or complete absence of insulin.** Also known as early onset diabetes, this disease is often caused by an autoimmune destruction of pancreatic β-cells. In contrast, **type II diabetes** is caused by **insulin resistance in the face of insulin insufficiency.** Insulin resistance is defined as an inability to respond to a physiologic concentration of insulin. The pancreas initially compensates by producing more insulin. At this stage, the patient is described as glucose intolerant. As the disease progresses, the degree of insulin resistance often worsens. Type II diabetes occurs when insulin secretion is not sufficient to maintain normoglycemia. The disease is thus characterized by both hyperinsulinemia and hyperglycemia (Figure 22-3).

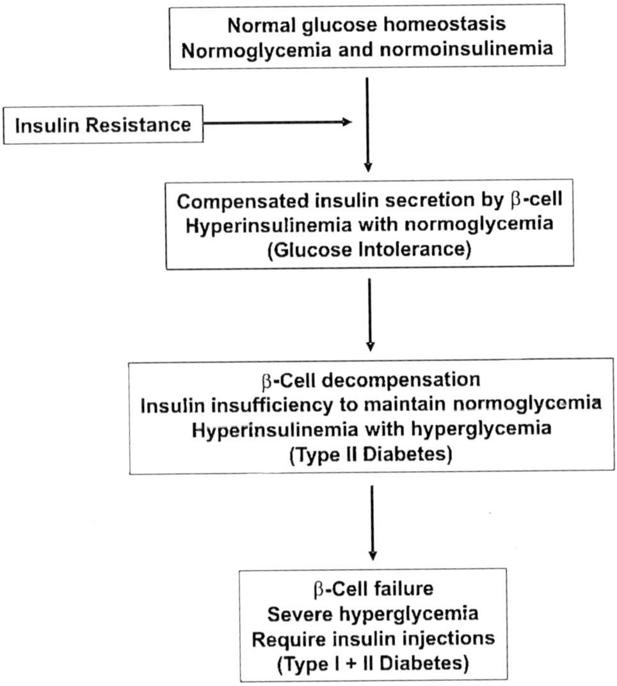

Figure 22-3. Schematic diagram showing the effects of insulin resistance leading to diabetes mellitus.

The degree at which different organs develop insulin resistance is often not uniform. Skeletal muscle insulin-mediated glucose disposal (which is normally responsible for 60 percent of whole body glucose disposal) is generally more affected, followed next by insulin suppression of hepatic glucose output. The combination of decreased peripheral glucose utilization and increased hepatic glucose production (driven by hepatic insulin resistance as well as increased circulating glucagon levels in type II diabetics) together contribute to the hyperglycemia. Insulin signaling in adipose tissue appears least affected in type II diabetics. **Hyperinsulinemia, in the face of hyperglycemia and dyslipidemia (often associated with type II diabetes), drives lipogenesis in adipose tissue** and may therefore contribute to the obesity often associated with this disease.

COMPREHENSION QUESTIONS

A 64-year-old man is presented to his family doctor with complaints of frequent episodes of dizziness and of numbness in his legs. During a routine history and physical examination, the doctor finds that the patient leads a sedentary lifestyle, is obese (body mass index of 32), and has hypertension (blood pressure of 200/120 mm Hg). The patient is asked to return to the clinic a week later in the fasting state, during which time a blood specimen is obtained, and a glucose tolerance test is performed. Humoral analysis reveals fasting hyperglycemia, hyperinsulinemia, dyslipidemia, and glucose intolerance. The diagnosis is type II diabetes mellitus.

[22.1] Alterations in substrate metabolism within which of the following organs can be a cause for the observed humoral analysis?

A. Brain
B. Kidney
C. Liver
D. Heart
E. Spleen

[22.2] A mutation, leading to decreased activity, in the gene encoding for which of these proteins is most consistent with this clinical presentation?

A. Glucagon
B. Glucose transporter isoform 1
C. Glycogen phosphorylase
D. Pyruvate carboxylase
E. Protein phosphatase 1

[22.3] Which of the following complications is less likely to occur in type II diabetics, as opposed to type I diabetics?

A. Retinopathy
B. Weight gain
C. Cardiovascular disease
D. Hypoglycemic coma
E. Neuropathy

Answers

[22.1] **C.** Of the organs listed, changes in hepatic metabolism are most likely to affect circulating glucose and lipids. This in turn influences pancreatic insulin secretion. During type II diabetes mellitus, increased hepatic glucose output contributes to the observed hyperglycemia and subsequent hyperinsulinemia, whereas complex alterations in lipid metabolism contribute to dyslipidemia. In contrast, changes in metabolic fluxes within the brain, kidney, heart, and spleen are often a consequence, rather than cause, of their environment. For example, the heart increases further reliance on fatty acids as a fuel in the diabetic milieu.

[22.2] **E.** The underlying cause of type II diabetes mellitus is insulin resistance (an inability to respond normally to physiological concentrations of insulin). Protein phosphatase 1 is an integral mediator of the metabolic effects of insulin. Failure to activate protein phosphatase 1 adequately in response to insulin would therefore attenuate the metabolic actions of this hormone. Decreased activity of glucose transporter 1, glycogen phosphorylase, and pyruvate carboxylase would influence basal glucose transport, glycogenolysis, and gluconeogenesis, respectively. Decreased glucagon levels would tend to improve effectiveness of insulin action.

[22.3] **D.** Hypoglycemia is a common complication associated with over supplementation of type I diabetics with insulin. This is less common in type II diabetics, because insulin therapy generally occurs only in the later stages of the pathogenesis of this disease. Retinopathy, cardiovascular disease, and neuropathy are common complications associated with both forms of diabetes mellitus. In contrast to type I diabetics, type II diabetics tend to be overweight. Whether weight gain is a cause or consequence of disease progression is under current debate.

BIOCHEMISTRY PEARLS

- ❖ Serum glucose levels are tightly controlled: Insulin promotes glucose utilization and represses endogenous glucose production, whereas glucagon represses glucose utilization and promotes endogenous glucose production.
- ❖ Several types of glucose transport proteins appear on specific tissues affecting the movement of glucose across cell membranes.
- ❖ Glucose is converted into glycogen, the storage form of glucose in mammals, in a process regulated by insulin and insulin-mediated activation of protein phosphatase 1 (PP1).

REFERENCES

Cohen P. Dissection of the protein phosphorylation cascades involved in insulin and growth factor action. Biochem Soc Trans 1993;21:555.

Gould GW, Holman GD. The glucose transporter family: structure, function and tissue-specific expression. Biochem J 1993;295:329.

Newsholme EA, Leech AR. Biochemistry for the Medical Sciences. New York: Wiley, 1983.

❖ CASE 23

A 2-year-old black girl is being seen by the hematologist after her pediatrician found her to be severely anemic with splenomegaly and jaundice. Her mother gives a possible history of a "blood problem" in her family but doesn't know for sure. Her hemoglobin electrophoresis was normal, and the complete blood count (CBC) revealed a normocytic anemia. The platelet and white blood cell counts are normal. On the peripheral smear, there are many bizarre erythrocytes, including spiculated cells. A diagnosis of pyruvate kinase deficiency is made.

◆ **What is the biochemical mechanism for this disorder?**

◆ **How is this disorder inherited?**

ANSWERS TO CASE 23: HEMOLYTIC ANEMIA

Summary: A 2-year-old black girl has normocytic anemia, jaundice, splenomegaly, and peripheral smear showing spiculated cells. A family history of similar symptoms is possible.

- **Biochemical mechanism:** Pyruvate kinase deficiency usually will manifest clinical symptoms on red blood cells (RBCs) with no apparent metabolic abnormalities in other cells. Insufficient adenosine triphosphate (ATP) is produced in the red cell and its membrane is affected, is rigid and removed by the spleen.

- **Inheritance:** Autosomal recessive.

CLINICAL CORRELATION

Hemolytic anemia is not a common cause of anemia, but should be considered in patients with elevated serum bilirubin or urine bilirubin levels. Lysis of the erythrocyte can occur from various mechanisms such as medications, antibodies against red blood cells, infection, coagulopathy, and mechanical processes such as abnormal heart valves, and enzyme deficiencies of the red blood cell. Patients may notice fatigue, dizziness from the anemia, and dark colored (classically "coke-colored" urine) from the bilirubinuria.

Confirmation of hemolysis can be obtained by the peripheral blood smear revealing fragmented red blood cells, or increased serum bilirubin or decreased serum haptoglobin. Immunoglobulins can cause red blood cell lysis by attacking various proteins on the surface of erythrocytes; autoimmune processes (body attacking itself), or alloimmune (immunoglobulins from outside) such as from a blood transfusion or a fetus from the mother. The Coombs tests can assess for immunoglobulin on the red blood cell or circulating in the serum. Typically, hemolysis of the erythrocyte is associated with increased levels of RBC precursors in the bone marrow and thus immature forms of the erythrocytes in the bloodstream; therefore, an increased reticulocyte concentration supports the increased destruction of red blood cells.

APPROACH TO PYRUVATE METABOLISM

Objectives

1. Understand the role of pyruvate kinase in pyruvate metabolism.
2. Be familiar with the Embden-Meyerhof pathway of RBC metabolism.
3. Know how pyruvate kinase deficiency results in anemia.

Definitions

Hemolytic anemia: A pathologic condition in which there is an abnormally lowered number of circulating RBCs caused by rupture of RBCs as a result of membrane abnormalities or deficient enzyme(s) level within the red blood cell.

Glucose 6-phosphate dehydrogenase: The enzyme that catalyzes the rate regulating step of the hexose monophosphate shunt, which produces NADPH required for inactivating oxygen radicals and thereby protects the RBC membrane from radical attack and rupture.

Pyruvate kinase: Last ATP-producing step in glycolysis and critical in the RBC for maintaining energy supply (ATP levels).

Methemoglobin reductase: Red blood cell enzyme that uses nicotinamide adenine dinucleotide (NADH) to convert the iron of oxidized hemoglobin (methemoglobin) from the ferric (Fe^{3+}) to reduced ferrous state (Fe^{2+}) hemoglobin, which alone is capable of binding O_2.

DISCUSSION

Hemolytic anemia has many causes, though this case has the marks of undersupply of RBCs caused by an enzyme deficiency within the RBC rather than a membrane abnormality or an environmental factor such as an autoantibody or mechanical trauma.

The RBC in the course of its maturation loses its mitochondria, ribosomes, and nucleus and consequently the functions associated with those organelles such as new enzyme synthesis and mitochondrial energy formation. **Thus the enzyme endowment present at maturation of the RBC cannot be replaced.** In terms of energy production in the RBC, **only the glycolytic pathway (Embden-Meyerhof pathway) is available.** This pathway together with the hexose monophosphate shunt (pentose phosphate pathway) is shown in Figure 23-1 as the only metabolic pathways that use glucose in RBCs. As a consequence of this limited metabolic capacity, glucose is the only fuel usable by the RBC to generate ATP. Glucose metabolism is required in the RBC to maintain the ionic milieu within the cell, to maintain the heme cofactor of hemoglobin in its reduced (Fe^{2+}) state, to maintain reduced sulfhydryl groups, to maintain the shape of the RBC plasma membrane and to produce 2,3-bisphosphoglycerate, a modulator of hemoglobin affinity for oxygen. To accomplish these tasks approximately 90 percent of the glucose in the RBC is metabolized directly to pyruvate/lactate while approximately 10 percent is first metabolized through the hexose monophosphate shunt before reentry into the glycolytic pathway.

Approximately 90 percent of all cases of known RBC enzyme deficiencies involve either altered protein or **decreased protein levels of pyruvate kinase** whereas **4 percent are variants of glucose 6-phosphate isomerase, which converts glucose 6-phosphate to fructose 6-phosphate.** Most of these enzyme deficiencies are inherited in an autosomal recessive pattern.

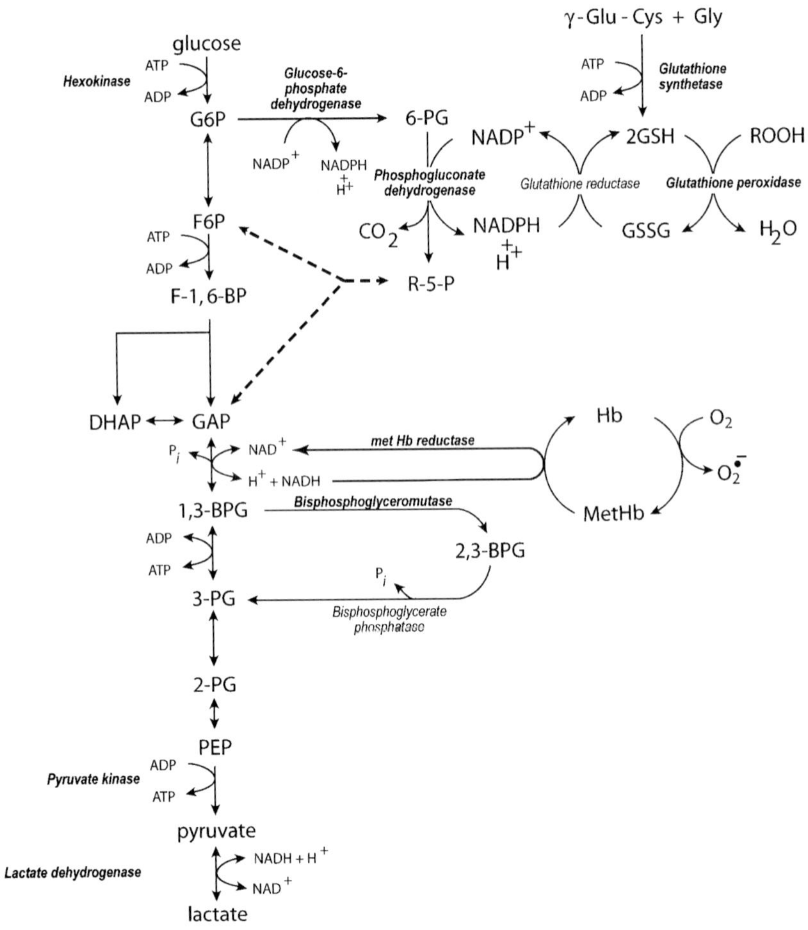

Figure 23-1. Metabolism of glucose in erythrocytes. The following abbreviations are used G6P, glucose 6-phosphate; 6-PG, 6-phosphogluconate; γ-Glu-Cys, γ-glutamyl-cysteine; Gly, glycine; GSH, reduced glutathione; GSSG, glutathione disulfide; ROOH, organic hydroperoxides; F6P, fructose 6-phosphate; R-5-P, rebulose-5-phosphate; F-1,6-BP, fructose 1, 6-bisphosphate; DHAP dehydroxyacetone phosphate; GAP, glyceraldehyde 3-phosphate; Hb, hemoglobin (Fe^{+2}); metHb, hemoglobin (Fe^{+3}); 1,3-BPG, 1,3-bisphosphoglycerate; 2,3-BPG, 2,3-bisphosphoglycerate; 3-PG, 3-phosphoglycerate; 2-PG, 2-phosphoglycerate; PEP, phosphoenolpyruvate.

As shown in Figure 23-1, 2 mol of ATP are formed per mole of glucose metabolized by the glycolytic cycle. The primary final product of glucose metabolism by glycolysis in the RBC is not pyruvate, as in other tissues most of the time, but lactate. Since RBCs have no mitochondria, NAD^+ cannot be regenerated by shuttling NADH produced in glycolysis into the mitochondrial electron transport system. Therefore, the only option to continue glycolysis is to regenerate NAD^+ from NADH by reducing pyruvate to lactate in the lactate dehydrogenase reaction. But lactate is not the sole product of RBC glycolysis. Methemoglobin reductase uses some of the NADH produced by glycolysis to reduce methemoglobin (Fe^{3+}) back to active hemoglobin (Fe^{2+}) capable of binding O_2 for transport to the tissues. Thus the final products are a mixture of lactate and pyruvate with lactate being the primary product.

In tissues other than the RBC, pyruvate has alternative metabolic fates that, depending on the tissue, include gluconeogenesis, conversion to acetyl-CoA by pyruvate dehydrogenase for further metabolism to CO_2 in the tricarboxylic acid (TCA) cycle, transamination to alanine or carboxylation to oxaloacetate by pyruvate carboxylase (Table 23-1). In the RBC, however, the restricted enzymatic endowment precludes all but the conversion to lactate. The pyruvate and lactate produced are end products of RBC glycolysis that are transported out of the RBC to the liver where they can undergo the alternative metabolic conversions described above.

How does compromise of pyruvate kinase activity lead to anemia? Pyruvate kinase lies at the end of the glycolytic pathway in RBCs followed only by lactate dehydrogenase. In any linked pathway where the product of one reaction is the substrate for the next reaction a compromise in one reaction affects the entire pathway. The RBC depends exclusively on glycolysis to produce ATP to discharge all energy-requiring tasks. Pyruvate kinase activity is critical for the pathway and therefore critical for energy production. If ATP is not produced in amounts sufficient to meet the energy demand, then those functions are compromised. Energy is required to maintain the Na^+/K^+ balance within the RBC and to maintain the flexible discoid shape of the cell. In the absence of sufficient pyruvate kinase activity and therefore ATP, the ionic balance fails, and the membrane becomes misshapen. Cells reflecting pyruvate kinase insufficiency rather than a change in membrane composition are removed from the circulation by the macrophages of the spleen. This results in an increased number of circulating reticulocytes and possibly bone marrow hyperplasia, which is a biological response to lowered RBC count as a result of hemolysis of erythrocytes.

Table 23-1

METABOLIC FATE OF PYRUVATE

$CH_3-C(=O)-COO^-$

REACTION	ENZYME	PRODUCT
Transamination	Alanine Transaminase	$CH_3-CH(NH_3^+)-COO^-$ Alanine
Carboxylation	Pyruvate Carboxylase	$^-OOC-C(=O)-CH_2-COO^-$ Oxalocetate
Decarboxylation	Pyruvate Dehydrogenase	$CH_3-C(=O)-SCoA$ Acetyl-CoA
Gluconeogenesis	Many	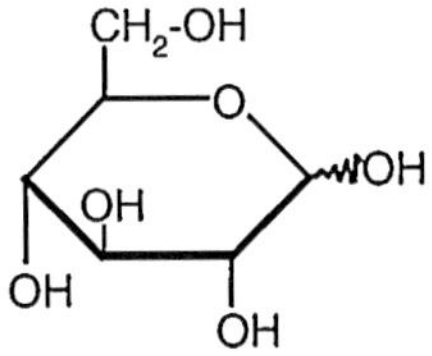 Glucose
Reduction	Lactate Dehydrogenase	$CH_3-CH(OH)-COO^-$ Lactate

COMPREHENSION QUESTIONS

A young man with normocytic anemia, jaundice, and splenomegaly was diagnosed as having RBC pyruvate kinase deficiency after a peripheral blood smear showed spiculated cells.

[23.1] Since in this patient pyruvate kinase is abnormal not only is less pyruvate made but intermediates above pyruvate in the glycolytic pathway build up slowing the pathway. Which of the following products may not be made in the appropriate amounts in the RBC because of the deficiency of pyruvate?

A. Glucose
B. Oxaloacetate
C. acetyl-CoA
D. Lactate

[23.2] In the RBCs of the patient described above, which of the following would be expected?

A. ADP to ATP ratios would be elevated above normal.
B. $NADP^+$ would increase relative to NADPH.
C. Ribulose 5-phosphate levels would decrease.
D. NADH to NAD^+ ratios would decrease.
E. Methemoglobin levels would increase.

[23.3] The glycolytic pathway is a multistep process by which glucose is broken down to a three-carbon metabolite. Some of the steps are listed below:

1. Conversion of 3-phosphoglycerate to 2-phosphoglycerate
2. Conversion of phosphoenolpyruvate to pyruvate
3. Conversion of glyceraldehyde 3-phosphate to 1,3-bisphosphoglycerate
4. Conversion of glucose to glucose 6-phosphate
5. Conversion of fructose 6-phosphate to fructose 1,6-bisphosphate

Which of the following is the correct order of these conversions?

A. $4 \rightarrow 5 \rightarrow 1 \rightarrow 2 \rightarrow 3$
B. $4 \rightarrow 3 \rightarrow 1 \rightarrow 2 \rightarrow 5$
C. $4 \rightarrow 5 \rightarrow 3 \rightarrow 1 \rightarrow 2$
D. $4 \rightarrow 1 \rightarrow 3 \rightarrow 5 \rightarrow 2$
E. $4 \rightarrow 5 \rightarrow 3 \rightarrow 2 \rightarrow 1$

Answers

[23.1] **D.** The RBC has no mitochondria so glucose cannot be made from pyruvate or acetyl-CoA or oxaloacetate. The RBC does have lactate dehydrogenase and conversion to lactate depends on pyruvate levels.

[23.2] **A.** In the RBC a deficiency of pyruvate kinase would tend to shunt glucose toward the hexose monophosphate pathway increasing ribulose 5-P levels, and the ratio of $NADP^+$ to NADPH would decrease. NADH to NAD^+ ratios would increase as a result of lower pyruvate levels making more NADH available to reduce methemoglobin and regenerate NAD^+. Because pyruvate kinase is deficient, the last ATP formation site is compromised, and so is the formation of ATP in the RBC elevating the ADP to ATP ratio.

[23.3] **C.** Glucose to glucose 6-phosphate → fructose 6-phosphate → fructose 1,6-bisphosphate → glyceraldehyde 3-phosphate → 1,3-bisphosphoglycerate → 3-phosphoglycerate → 2-phosphoglycerate → phosphoenolphosphate → pyruvate.

BIOCHEMISTRY PEARLS

- The enzyme endowment present at maturation of the RBC cannot be replaced, and only the glycolytic pathway (Embden-Meyerhof pathway) is available for energy production in the RBC.
- Glucose metabolism is required in the RBC to maintain the ionic milieu within the cell, the vast majority via conversion to lactate.
- The vast majority of RBC enzyme deficiencies involve either altered protein or decreased protein levels of pyruvate kinase.
- Insufficient pyruvate kinase activity compromises erythrocyte ATP production, leading to ionic imbalance and misshaped cell membranes. These cells are removed from the circulation by the macrophages of the spleen.
- Pyruvate kinase catalyzes one of the three irreversible steps in the glycolytic pathway, the others being the phosphorylation of glucose to glucose 6-phosphate and the phosphorylation of fructose 6-phosphate to fructose 2,6-bisphosphate.

REFERENCE

Braunwald E, Fauci AS, Kasper KL, et al., eds. Harrison's Principles of Internal Medicine, 15th ed. New York: McGraw-Hill, 2001.

CASE 24

A 3-year-old boy is brought to the emergency department after several episodes of vomiting and lethargy. His pediatrician has been concerned about his failure to thrive and possible hepatic failure along with recurrent episodes of the vomiting and lethargy. After a careful history is taken, you observe that these episodes occur after ingestion of certain types of food, especially high in fructose. His blood sugar was checked in the emergency department and was extremely low.

- **What is the most likely diagnosis?**
- **What is the biochemical basis for the clinical symptoms?**
- **What is the treatment of the disorder?**

ANSWERS TO CASE 24: FRUCTOSE INTOLERANCE

Summary: A 3-year-old boy with failure to thrive and possible hepatic failure. He presents with hypoglycemia and recurrent episodes of nausea and vomiting after ingestion of foods high in fructose.

- **Diagnosis:** Fructose intolerance.
- **Biochemical basis of disorder:** Because of a genetic disorder, the hepatic aldolase B enzyme is defective, and functions normally in glycolysis but not in fructose metabolism. Glucose production is inhibited by elevated fructose 1-phosphate. When fructose is ingested, severe hypoglycemia results.
- **Treatment:** Avoid dietary fructose.

CLINICAL CORRELATION

Individuals with a deficiency in aldolase B have the condition known as fructose intolerance. As with most enzyme deficiencies, this is an autosomal recessive disease; it does not cause difficulty as long as the patient does not consume any foods with fructose or sucrose. Frequently, children with fructose intolerance avoid candy and fruit, which should raise some eyebrows! Likewise, they usually do not have many dental caries. However, if chronically exposed to fructose-containing foods, infants and small children may have poor weight gain and abdominal cramping or vomiting.

APPROACH TO DISACCHARIDE METABOLISM

Objectives

1. Know about the metabolism of disaccharides, specifically fructose.
2. Know about the role of aldolase B in fructose metabolism.

Definitions

Disaccharide: Two sugar molecules (monosaccharides) linked together by a glycosidic bond. The major disaccharides obtained in the diet are maltose [4-(α-D-glucosido)-D-glucose], sucrose [β-D-fructofuranosyl-α-D-glucopyranoside], and lactose [4-(β-D-galactosido)-D-glucose].

Essential fructosuria: A rare, benign genetic condition in which fructose spills over to the urine because the liver, kidney, and intestine lack the enzyme fructokinase.

Fructokinase: An enzyme present in the liver, kidney, and intestine that will phosphorylate fructose to fructose 1-phosphate at the expense of an ATP.

Fructose intolerance: A genetic deficiency in the liver enzyme aldolase B. The absence of this enzyme leads to a build up of fructose 1-phosphate and depletion of liver ATP and phosphate stores.

GLUT 5: A facilitative glucose transporter isoform present in the small intestine and other tissues that will transport fructose (and glucose to a lesser extent) across the plasma membrane.

β-Glycosidase: A bifunctional, membrane-bound enzyme located on the brush-border membrane of the small intestine. This single polypeptide enzyme has two activities, lactase and glycosylceramidase, located in different domains of the protein. It will hydrolyze lactose to glucose and galactose.

SGLT1: A sodium-dependent glucose transporter located on the luminal side of the intestinal epithelial cells. It will transport glucose and galactose across the intestinal cell using a sodium ion gradient.

SGLT2: A sodium-dependent glucose transporter that has a high specificity for glucose and is specific to the kidney.

Sucrase-isomaltase complex: An enzyme complex comprised of two enzyme units. Both units have high α-1,4-glucosidase activity and will hydrolyze maltose and maltotriose to glucose. The sucrase unit will also hydrolyze sucrose to fructose and glucose, whereas the isomaltase unit will hydrolyze α-1,6 bonds found in isomaltose and the limit dextrins of starch.

DISCUSSION

The **major disaccharides** obtained in the diet are **maltose, sucrose, and lactose. Maltose** is primarily obtained from the consumption of the **plant storage polysaccharide starch.** Starch is degraded to glucose and small branched oligosaccharides called limit dextrins by exhaustive digestion by **α-amylase.** The limit dextrins are further enzymatically hydrolyzed to a branched tetrasaccharide by glucoamylase and to maltotriose and maltose and ultimately to glucose by the sucrase-isomaltase complex. Both of these enzyme complexes are located on the **brush-border membrane of the small intestine. Sucrose,** or table sugar, is hydrolyzed to **glucose and fructose by the sucrase** subunit of the sucrase-isomaltase complex. **Lactose,** or milk sugar, is enzymatically converted to **glucose and galactose by β-glycosidase,** also located on the brush-border membrane of the small intestine. This membrane-bound enzyme is a single polypeptide that has lactase and glycosylceramidase activities located in different domains of the protein.

Glucose and galactose are absorbed from the lumen of the intestine by the **sodium-dependent glucose transporter, SGLT1,** which is located on the luminal side of the intestinal epithelial cells. **Fructose** absorption is not dependent on a sodium gradient. It is transported into the intestinal cell by **facilitative diffusion** by a glucose transporter isoform, **GLUT 5.** Fructose transport is less rapid than glucose transport, and GLUT 5 does not have a high capacity. Thus, in most individuals, ingestion of fructose in amounts greater than 0.5 to 1.0 g/kg body weight can result in malabsorption. Fructose enters the bloodstream, along with glucose and galactose, via the **GLUT 2 transporter.** Fructose is taken up by the liver by the same GLUT 2 transporter, along with glucose and galactose. There is a large gradient between the extracellular and intracellular concentrations of fructose in the liver cells, indicating that the rate for fructose uptake by the hepatocyte is low.

In the liver, kidney, and intestine, **fructose** can be converted to glycolytic/gluconeogenic intermediates by the actions of three enzymes—**fructokinase, aldolase B, and triokinase** (also called triose kinase)—as shown in Figure 24-1. In these tissues, fructose is rapidly phosphorylated to fructose 1-phosphate (F1P) by fructokinase at the expense of a molecule of adenosine triphosphate (ATP). This has the effect of trapping fructose inside the cell. A deficiency in this enzyme leads to the rare but benign condition known as *essential fructosuria.* In other tissues such as muscle, adipose, and red blood cells, hexokinase can phosphorylate fructose to the glycolytic intermediate fructose 6-phosphate (F6P).

Fructose 1-phosphate is further metabolized to dihydroxyacetone phosphate (DHAP) and glyceraldehyde by the hepatic isoform of the enzyme aldolase, which catalyzes a reversible aldol condensation reaction. **Aldolase** is present in three different isoforms. Aldolase A is present in greatest concentrations in the **skeletal muscle,** whereas the B isoform predominates in the liver, kidney, and intestine. Aldolase C is the brain isoform. Aldolase B has similar activity for either fructose 1,6-bisphosphate (F16BP) or F1P; however, the A or C isoforms are only slightly active when F1P is the substrate.

Glyceraldehyde may be converted to the glycolytic intermediate, **glyceraldehyde 3-phosphate (GAP),** by the action of the enzyme triokinase. This enzyme phosphorylates glyceraldehyde at the expense of another molecule of ATP. The GAP can then enter into the glycolytic pathway and be further converted to pyruvate, or recombine with DHAP to form F16BP by the action of aldolase.

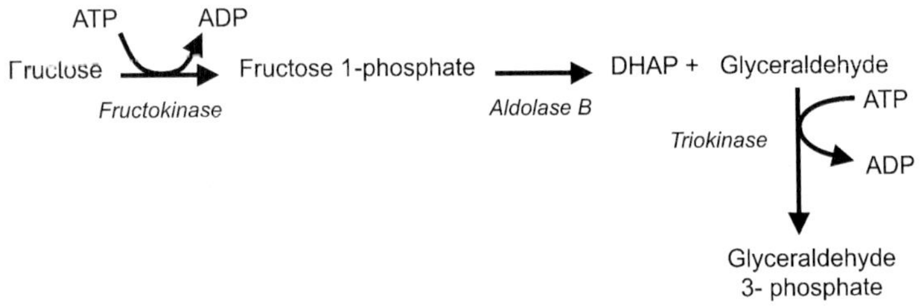

Figure 24-1. The metabolic pathway for the entrance of fructose into the glycolytic pathway. Fructokinase rapidly converts fructose to fructose 1-phosphate, which in the liver is cleaved by aldolase B to dihydroxyacetone phosphate (DHAP) and glyceraldehyde.

A **deficiency in aldolase B** leads to the condition known as **fructose intolerance.** This inherited condition is benign as long as the patient does not consume any foods with fructose or sucrose. Patients with this condition usually develop an aversion to sweets early in life and as a result frequently are without any caries. However, infants and small children if chronically exposed to fructose-containing foods usually exhibit periods of vomiting and poor feeding, as well as a failure to thrive. A defect in the aldolase B gene results in a decrease in activity that is 15 percent or less than that of normal controls. This results in a buildup of F1P levels in the hepatocyte. Because the maximal rate of fructose phosphorylation by fructokinase is so high (almost an order of magnitude greater than that of glucokinase), intracellular levels of both ATP and inorganic phosphate (P_i) are significantly decreased. The drop in ATP concentration adversely affects a number of cellular events, including detoxification of ammonia, formation of cyclic AMP (cAMP), and ribonucleic acid (RNA) and protein synthesis. The decrease in intracellular concentrations of P_i leads to a hyperuricemic condition as a result of an increase in uric acid formation. AMP deaminase is inhibited by normal cellular concentrations of P_i. When these levels drop, the inhibition is released and AMP is converted to IMP and, ultimately, uric acid (Figure 24-2).

The toxic effects of F1P can also be exhibited in patients that do not have a deficiency in aldolase B if they are parenterally fed with solutions containing fructose. Parenteral feeding with solutions containing fructose can result in blood fructose concentrations that are several times higher than can be achieved with an oral load. Since the rate of entry into the hepatocyte is dependent on the fructose gradient across the cell, intravenous loading results in increased entry into the liver and increased formation of F1P. Since the rate of formation of F1P is much faster than its further metabolism, this can lead to hyperuricemia and hyperuricosuria by the mechanisms described above.

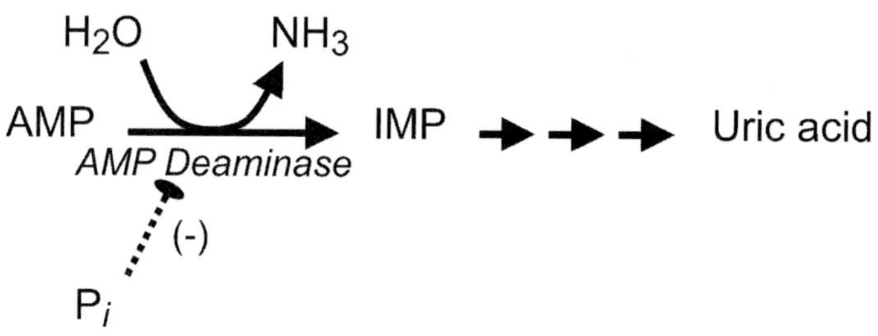

Figure 24-2. The inhibition of AMP deaminase by inorganic phosphate (P_i). A decrease in [P_i] increases the activity of AMP deaminase and leads to increased production of uric acid.

COMPREHENSION QUESTIONS

[24.1] A 22-year-old soldier collapses from dehydration during maneuvers in the desert and is sent to a military hospital. Prior to enlisting, a physician observed a high level of glucose in his urine during an examination. At first, he was not allowed to enlist because he was suspected of being a diabetic. Further tests, however, determined that his insulin level was normal. A glucose tolerance test exhibited a normal pattern. Laboratory tests following his dehydration episode repeat the previous findings, but further testing of the urine reveals that only D-glucose is elevated. Other sugars were not elevated.

This patient's elevated urinary glucose and his dehydration episode are caused by a deficiency in which of the following?

A. GLUT 2
B. GLUT 4
C. Insulin receptor
D. SGLT1
E. SGLT2

[24.2] Your patient is a 7-month-old baby girl, the second child born to unrelated parents. She did not respond well to breast-feeding and was changed entirely to a formula based on cow's milk at 4 weeks. Between 7 and 12 weeks of age, she was admitted to the hospital twice with a history of screaming after feeding, but was discharged after observation without a specific diagnosis. Elimination of cow's milk from her diet did not relieve her symptoms; her mother reported that the screaming bouts were worse after the child drank juice and that she frequently had gas and a distended abdomen. Analysis of a liver needle biopsy did not reveal any liver enzyme deficiencies. Overall, the girl is thriving (weight greater than 97th percentile) with no abnormal findings on physical examination.

If a biopsy of intestinal tissue were obtained from your patient and analyzed, which of the following would most likely be deficient or defective?

A. GLUT 2
B. GLUT 5
C. Isomaltase
D. Lactase
E. SGLT1

[24.3] A 24-year-old African-American female presents with complaints of intestinal bloating, gas, cramps, and diarrhea following a meal including dairy products. A lactose-tolerance test confirms your suspicion that she had a deficiency of lactase in her intestine. Which of the following dairy products could you recommend that would be least likely to cause her difficulties in the future?

A. Condensed milk
B. Cottage cheese
C. Ice cream
D. Skim milk
E. Yogurt

Answers

[24.1] **E.** The patient has normal levels of blood insulin and exhibits a normal glucose tolerance test. This indicates that glucose absorption from the intestine is normal as is clearance of glucose from the blood. The presence of glucose in the urine is most likely a kidney problem. Because the defect seems to involve only D-glucose and no other sugar, this points to a transporter with high specificity. The kidney has the GLUT 2, SGLT1, and SGLT2 transporters. GLUT 2 and SGLT1 are present in other tissues, and a defect in these would be expected to result in more serious sequelae. SGLT2 is a sodium-dependent glucose transporter specific to the kidney that has a high specificity for glucose. The glucose is present in the urine because of a failure to reabsorb it as a consequence of a defect in SGLT2. This leads to a loss of water also, because it is reabsorbed with glucose.

[24.2] **B.** Because the patient's liver enzymes are normal and her symptoms seem to correlate with her intake of fruit juices, most likely her problem stems from an inability to absorb fructose. Since removal of cow's milk from her diet did not eliminate the problem, a lactase deficiency can be ruled out. GLUT 5 is the primary transporter of fructose in the intestine and a deficiency in this transporter would lead to an inability to absorb fructose in the gut, making it a substrate for bacterial metabolism that produces various gases, including hydrogen, as well as organic acids.

[24.3] **E.** The microorganisms that convert milk to yogurt (*Streptococcus salivarius thermophilus* and *Lactobacillus delbrueckii bulgaricus*) metabolize most of the lactose in the milk, thus removing the source of this patient's intestinal disquietude. Yogurt is also a good source of dietary calcium.

BIOCHEMISTRY PEARLS

- ❖ The major disaccharides obtained in the diet are maltose, sucrose, and lactose.
- ❖ Sucrose, or table sugar, is hydrolyzed to glucose and fructose.
- ❖ Lactose, or milk sugar, is enzymatically converted to glucose and galactose by β-glycosidase, also located on the brush-border membrane of the small intestine.
- ❖ A deficiency in aldolase B leads to the condition known as fructose intolerance.

REFERENCES

Gitzelmann R, Steinmann B, Van den Berghe G. Disorders of fructose metabolism. In: Scriver CR, Beaudet AL, Sly WS, et al., eds. The Metabolic and Molecular Bases of Inherited Disease, 7th ed. New York: McGraw-Hill, 1995.

Semenza G, Auricchio S. Small-intestinal disaccharidases. In: Scriver CR, Beaudet AL, Sly WS, et al., eds. The Metabolic and Molecular Bases of Inherited Disease, 7th ed. New York: McGraw-Hill, 1995.

Ludueña RF. Learning Biochemistry: 100 Case-Oriented Problems. New York: Wiley-Liss, 1995.

CASE 25

A 38-year-old female presents to the clinic with complaints of alternating diarrhea and constipation. She reports some abdominal discomfort and bloating that are relieved with her bowel movement. She states that her episodes are worse in times of stress. She denies any blood in her diarrhea. She denies any weight loss or anorexia. Her physical exam is all within normal limits. She has been prescribed a cellulose-containing dietary supplement, which her doctor says will increase the bulk of her stools.

- **What is the most likely diagnosis?**

- **What is the biochemical mechanism of the dietary supplement's effect on the intestines?**

ANSWERS TO CASE 25: IRRITABLE BOWEL SYNDROME

Summary: A 38-year-old female complains of alternating constipation and diarrhea associated with times of stress and abdominal cramping and bloating relieved with bowel movements. She is prescribed a cellulose dietary supplement.

- **Diagnosis:** Irritable bowel syndrome.
- **Biochemical mechanism:** Cellulose-containing foods are not digestible but swell up by absorbing water and correlate with larger softer stools. The increase in dietary fiber also increases the intestinal transit time and decreases the intracolic pressure, thereby decreasing the symptoms of irritable bowel.

CLINICAL CORRELATION

Irritable bowel syndrome affects many individuals in Western countries, and it manifests as abdominal cramping and bloating in the absence of disease. It is thought to be caused by increased spasms of the intestines. Constipation with or without episodes of diarrhea may be seen. Weight loss, fever, vomiting, bloody stools, or anemia would be worrisome and should not be attributed to irritable bowel syndrome. Typically, affected patients are anxious and may be under stress. After ruling out other disease processes, a trial of fiber-containing foods, stress reduction, and avoidance of aggravating foods are effective therapies. Patients should be advised to avoid laxative use. Rarely antispasmodic or antiperistaltic agents can be used. Notably, increased fiber in the diet may also decrease the absorption of fats and may lower the risk of colon cancer.

APPROACH TO INDIGESTIBLE POLYSACCHARIDES

Objectives

1. Know about the indigestible polysaccharides.
2. Be aware of β-1,4 cellulose bonds.
3. Know about the major types of fiber.

Definitions

Cellulose: A polysaccharide composed of β-D-glucopyranose units joined by a β(1→4) glycosidic bond, which is not hydrolyzed by enzymes in the digestive tracts of humans.

Gums: Complex polysaccharides composed of arabinose, fucose, galactose, mannose, rhamnose, and xylose. Gums are soluble in water and, because of their mucilaginous nature, slowly digestible.

Hemicellulose: Polysaccharides with a random, amorphous structure that are components of plant cell walls. Unrelated to cellulose structurally, they are composed of a variety of monosaccharides, including some acidic sugars, with xylose being the most prevalent.

Insoluble fibers: Components of plant cell walls that are insoluble in water and not broken down by the body's digestive enzymes.

Lignins: Aromatic polymers formed by the irreversible dehydration of sugars. Because of their structure, they cannot be broken down by the digestive enzymes and make up part of the stool bulk.

Mucilaginous: Having a characteristic that is like the viscous and sticky nature of glue.

Pectins: One of the soluble fibers in the diet composed primarily of polymers of galacturonic acid with varying amounts of other hexose and pentose residues.

Soluble fibers: Mucilaginous fibers such as pectin and true plant gums that are soluble in water and digestible by the enzymes of the intestinal tract. By absorbing water and forming viscous gels, they decrease the rate of gastric emptying.

DISCUSSION

Simply stated, dietary fiber is that part of food that remains intact and not absorbed following the digestive process in humans. It consists of all of the components of the cell walls of plants that are not broken down by the body's digestive enzymes. Dietary fiber can be grouped into two main categories, those that are soluble and those that are insoluble in water. The soluble fibers include pectins, gums, some hemicelluloses, and storage polysaccharides (starch and glycogen). The insoluble fibers include cellulose, most hemicelluloses, and lignins.

Cellulose is a major structural component of plant cell walls. Cellulose is a long, linear polymer of glucose (β-D-glucopyranose) units that are joined by β(1→4) glycosidic bonds (Figure 25-1a). Cellulose molecules have an

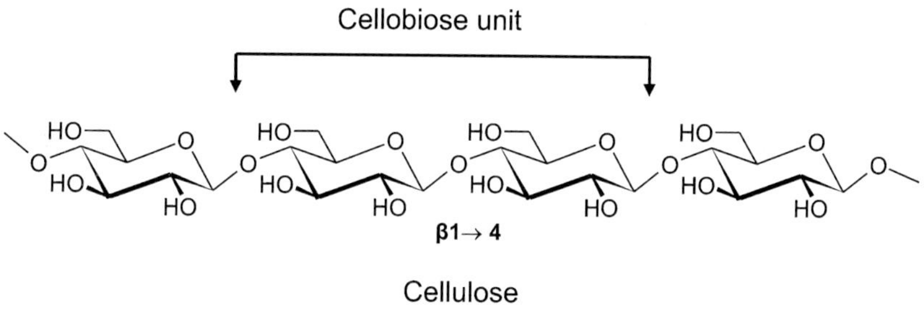

Figure 25-1a. The molecular structure of cellulose, indicating the repeating disaccharide unit, cellobiose.

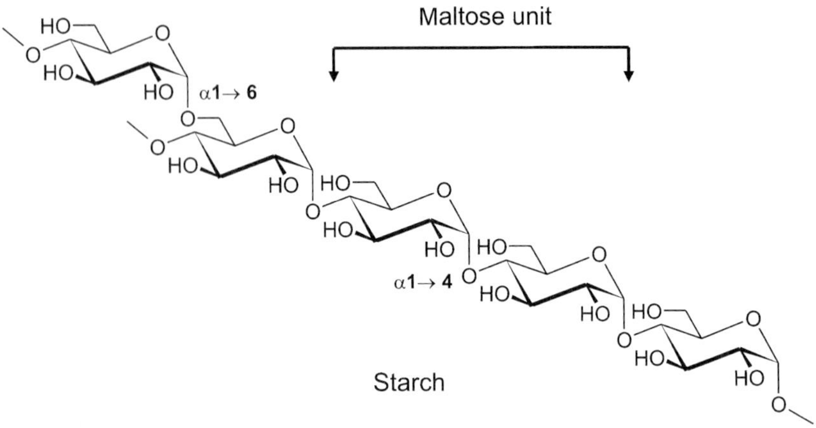

Figure 25-1b. The molecular structure of starch, indicating the repeating disaccharide unit, maltose, as well as the α-1,6-glycosidic bond present in the branch points of amylopectin.

extended, rigid structure that is stabilized by interchain hydrogen bonds. Starch, the plant storage polysaccharide, which is also a polymer of glucose, differs in its structure in that the glucose monomer units are joined by α(1→4) glycosidic bonds (Figure 25-1b). Starch is composed of two types of polymers, amylose, which has a nonbranched helical structure, and amylopectin, which is branched with α(1→6) glycosidic bonds joining the branches to the main polymer chain. Although starch is easily digested by salivary and pancreatic amylase and the disaccharidases present on the brush border of intestinal mucosal cells, cellulose cannot be hydrolyzed. The β(1→4) glycosidic bonds of the cellulose chain cannot be cleaved by the amylases present in the digestive tract.

Hemicelluloses are also polysaccharides that are structural components of plant cell walls. However, unlike what their name implies, they are unrelated to cellulose. They are polymers that are made up of a variety of sugar monomers that include glucose, galactose, mannose, arabinose, and xylose, as well as acidic forms of these monosaccharides. Xylose is the monosaccharide that is most abundant. Hemicelluloses have a random, amorphous structure that is suitable for their location in the plant cell wall matrix. Depending on their molecular structure, hemicelluloses are partially digestible.

Lignins are formed by the irreversible dehydration of sugars that result in aromatic structures. The remaining alcohol or phenol OH groups can react with each other and with aldehyde and ketone groups to form polymers. An example of a lignin molecule in the early stage of condensation is shown in Figure 25-2. These polymers cannot be broken down by the digestive enzymes and, like cellulose and the indigestible portion of hemicelluloses, form the stool bulk.

The soluble fibers such as pectin and true plant gums are mucilaginous and are digestible. Pectins are predominantly polygalacturonic acids with varying amounts of other hexose or pentose residues. True plant gums are complex polysaccharides composed of primarily arabinose, fucose, galactose, mannose, rhamnose, and xylose. The gums are soluble in water and are digestible by the enzymes in the intestinal tract. Both pectins and gums are mucilaginous; they absorb water to form viscous gels in the stomach that decrease the rate of gastric emptying.

Although cellulose and hemicellulose are insoluble, they absorb water to swell and increase the stool bulk. This results in larger, softer stools. It has been shown that diets plentiful in insoluble fiber also increase the transit time of food in the digestive tract and decrease intracolonic pressure. Lignins, in addition to increasing stool bulk, also bind organic molecules such as cholesterol and many potential carcinogens. The mucilaginous nature of the soluble fibers, pectins, and gums tends to decrease the rate at which carbohydrates are digested and absorbed, thus decreasing both the rise in blood glucose levels and the ensuing increase in insulin concentration.

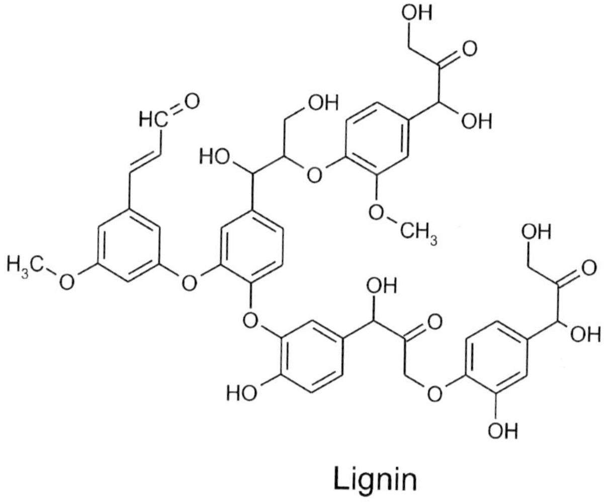

Figure 25-2. A lignin molecule in an early stage of condensation. The aromatic rings are a result of irreversible dehydration of sugar residues.

COMPREHENSION QUESTIONS

[25.1] A patient with type I diabetes mellitus has fasting and postprandial blood glucose levels that are frequently above the normal range despite good compliance with his insulin therapy. He was referred to a dietician that specialized in diabetic patients. The patient was recommended to incorporate foods high in dietary fiber. Which of the following dietary fibers would be most helpful in maintaining a normal blood glucose level?

A. Cellulose
B. Hemicellulose
C. Lignins
D. Pectins

[25.2] Some individuals complain of flatulence following a meal plentiful in beans, peas, soybeans, or other leguminous plants. All legumes contain the oligosaccharides raffinose and stachyose that contain glycosidic linkages that are poorly hydrolyzed by intestinal enzymes but are good sources of energy for intestinal bacteria that convert these sugars to H_2. Which of the following glycosidic bonds are contained in raffinose and stachyose and are not hydrolyzed by our intestinal enzymes but can be by intestinal flora?

A. Galactose (α1→6) glucose
B. Galactose (β1→4) glucose
C. Glucose (β1→2) fructose
D. Glucose (α1→4) glucose
E. Glucose (β1→4) glucose

[25.3] Cellulose is the most abundant polysaccharide and is an important structural component of cell walls. Strict vegetarians consume a large amount of cellulose, but it is not a source of energy because it is indigestible by the human intestinal tract. Cellulose is indigestible because it contains which of the following glycosidic bonds?

A. Galactose (β1→4) glucose
B. Galactose (β1→6) galactose
C. Glucose (α1→4) glucose
D. Glucose (β1→2) fructose
E. Glucose (β1→4) glucose

Answers

[25.1] **D.** Pectins and gums are soluble dietary fibers that absorb water and form mucilaginous gels. In doing so, they delay gastric emptying and decrease the rate at which monosaccharides such as glucose and fructose and disaccharides are absorbed by the intestinal tract. By decreasing the rate of sugar absorption, postprandial spikes in blood glucose concentration are avoided.

[25.2] **A.** Raffinose and stachyose are sucrose molecules that have one and two galactose residues in an $\alpha 1 \rightarrow 6$ glycosidic linkage. These bonds are not hydrolyzed by intestinal enzymes, but can be broken down by intestinal bacteria to produce CO_2 and H_2. Although they contain the glucose ($\beta 1 \rightarrow 2$) fructose link, sucrase can hydrolyze that bond. The galactose ($\beta 1 \rightarrow 4$) glucose, glucose ($\alpha 1 \rightarrow 4$) glucose, and glucose ($\beta 1 \rightarrow 4$) glucose linkages are not found in these oligosaccharides.

[25.3] **E.** Cellulose is a polymer of glucose in β1,4 glycosidic linkages. This bond is not hydrolyzed by intestinal enzymes or the flora of the human intestine. It makes up the bulk of the stool.

BIOCHEMISTRY PEARLS

- Dietary fiber comprises those components that are not digestible, which can be grouped into two main categories, those that are soluble and those that are insoluble in water.
- The soluble fibers include pectins, gums, some hemicelluloses, and storage polysaccharides (starch and glycogen). The insoluble fibers include cellulose, most hemicelluloses, and lignins.
- Cellulose is a long, linear polymer of glucose (β-D-glucopyranose) units that are joined by **β(1→4) glycosidic bonds, which cannot be broken by human enzymes.**
- Starch, the plant storage polysaccharide, which is also a polymer of glucose, differs in its structure in that the glucose monomer units are joined by α(1→4) glycosidic bonds.

REFERENCES

Chaney SG. Principles of nutrition I: macronutrients. In: Devlin TM, ed. Textbook of Biochemistry with Clinical Correlations, 5th ed. New York: Wiley-Liss, 2002:1127–8.

Mayes PA. Nutrition. In: Murray RK, Granner KK, Mayes PA, et al., eds. Harper's Illustrated Biochemistry, 26th ed. New York: Lange Medical Books/McGraw-Hill, 2003:656–7.

Pettit JL. Fiber. Clinician Reviews 2002;12(9):71–5.

CASE 26

A 56-year-old male presents to your clinic for follow-up on his diabetes. He has had diabetes since the age of 12 and has always required insulin for therapy. He reports feeling very tremulous and diaphoretic at 2 AM with the blood sugars in the range of 40 mg/dL, which is very low. He, however, notes that his morning fasting blood sugar is high without taking any carbohydrates. His physician describes the morning high sugars as a result of biochemical processes in response to the nighttime hypoglycemia.

◆ **What are the biochemical processes that govern the response to the nighttime hypoglycemia?**

ANSWER TO CASE 26: SOMOGYI EFFECT

Summary: A 56-year-old man with a long history of insulin diabetes with evening hypoglycemia and fasting morning hyperglycemia.

- **Biochemical mechanism of hypoglycemia:** The low nighttime serum blood sugar stimulates the counter-regulatory hormones to try to raise the glucose level. These include epinephrine, glucagon, cortisol, and growth hormone, which affect the glucose level and raise it by the time morning comes around.

CLINICAL CORRELATION

This individual has a classic manifestation of the Somogyi effect, which is fasting morning hyperglycemia in response to hypoglycemia in the early morning and late night hours. The danger is that if nighttime blood glucose levels are not measured, the physician may interpret the patient as having hyperglycemia and require even higher doses of insulin. This would be exactly the wrong treatment, since the hypoglycemia is leading to counter-regulatory hormone reaction, and a very low sugar level bound to the high level in the morning. The diagnosis is established by measuring a 2 AM glucose level, and when confirmed, then the bedtime NPH insulin (intermediate to long acting) needs to be decreased.

APPROACH TO GLUCOSE AND COUNTER-REGULATORY HORMONES

Objectives

1. Understand regulation of glycogen and glucose production.
2. Understand how insulin and epinephrine affect glucose levels.
3. Be familiar with the regulation of glucagons.
4. Know about diabetic ketoacidosis and biochemical mechanism.

Definitions

Epinephrine: Adrenaline; a catecholamine hormone derived from the amino acids phenylalanine or tyrosine that is synthesized and secreted by the adrenal medulla in response to stress.

GLUT 2: Glucose transporter isoform 2; a transport protein located on the plasma membrane of liver, pancreas, intestine, and kidney that will allow glucose to cross the membrane depending on the concentration gradient. GLUT 2 is the transporter that enables export of glucose from the liver.

Ketoacidosis: An elevation of the ketone body concentration that decreases the pH of the arterial blood to a pathologic condition.

Ketogenesis: The production of ketone bodies by the liver in response to increased β-oxidation with a decreased rate of the Krebs cycle as a result of shuttling C_4 acids from the mitochondrion for the synthesis of glucose via gluconeogenesis.

Ketone bodies: Acetoacetate, β-hydroxybutyrate, and acetone. Acetoacetate and β-hydroxybutyrate are formed by liver enzymes that condense molecules of acetyl-CoA, thus regenerating CoA for continual use in β-oxidation of fatty acids. Acetone is a spontaneous decomposition product of acetoacetate. Ketone bodies are exported from the liver and can be used by some extrahepatic tissues for energy generation.

Oxidative phosphorylation: The process by which adenosine triphosphate (ATP) is synthesized from a hydrogen ion gradient across the mitochondrial inner membrane. The hydrogen ion gradient is formed by the action of protein complexes in the mitochondrial membrane that sequentially transfer electrons from the reduced cofactors nicotinamide adenine dinucleotide (NADH) and $FADH_2$ to molecular oxygen. Movement of hydrogen ions back into the mitochondrion via ATP synthase drives the synthesis of ATP.

Protein phosphatase 1: An enzyme that will hydrolyze phosphate groups from target proteins such as glycogen synthase, phosphorylase, phosphorylase kinase, and the phosphorylated form of inhibitor 1. The phosphorylated inhibitor 1 is a substrate that binds well but is hydrolyzed slowly. While bound to protein phosphatase 1, the phosphorylated inhibitor 1 serves as an inhibitor of the enzyme.

DISCUSSION

The **liver is a highly specialized organ** that plays a central role in whole body **glucose metabolism.** During periods of increased glucose availability, the liver increases uptake, storage, and utilization of glucose. In contrast, when exogenous glucose availability declines (e.g., during an overnight fast), the liver increases glucose production, thereby helping to maintain blood glucose levels. The liver uses two mechanisms for **endogenous glucose production,** the mobilization of intracellular glycogen (glycogenolysis) and the **synthesis of glucose from noncarbohydrate precursors (gluconeogenesis)** (Figure 26-1). These pathways converge at **glucose 6-phosphate.** The latter is hydrolyzed to free glucose by glucose-6-phosphatase, the enzyme unique to gluconeogenic tissues, such as the liver. Once generated, glucose passes down its concentration gradient (e.g., from the cytosol of the liver cell to the blood during periods of decreased blood glucose levels), via glucose transporter isoform 2 (GLUT 2).

During periods of increased glucose availability (e.g., postprandially), glucose in excess of the energetic demands of the organism is stored as **glycogen** (the storage form of glucose in mammals). Although found within the cytosol of virtually every cell, glycogen is primarily concentrated in muscle (cardiac and skeletal) and liver. The purpose of glycogen synthesis (glycogenesis) is to

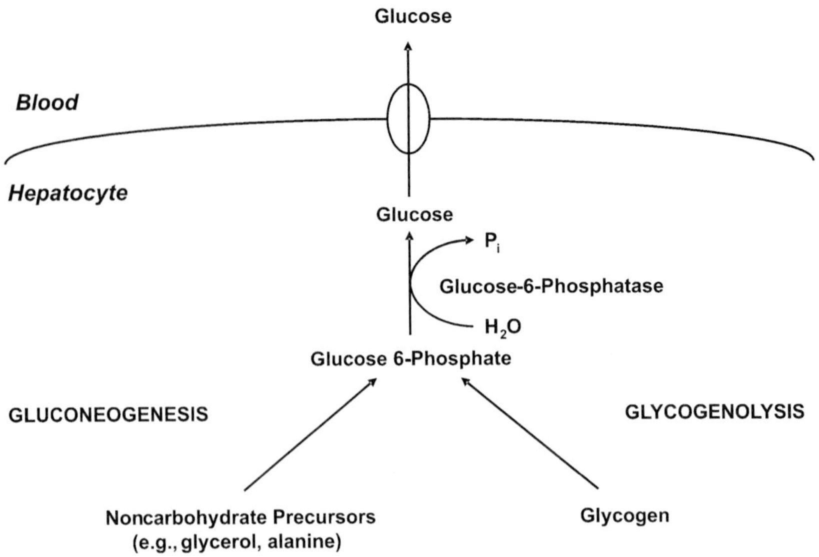

Figure 26-1. Schematic diagram showing the events that lead to the export of glucose from the liver cell during times of low blood glucose.

anticipate subsequent periods of decreased glucose availability (e.g., overnight fast). During the latter period of time, glycogen will be mobilized as a readily available source of glucose. In the case of muscle, glycogen is used selfishly, as an energy source by the myocyte only. In contrast, liver glycogen will be mobilized to help maintain blood glucose levels. The biochemical pathways of glycogenesis and glycogenolysis are illustrated in Figure 26-2.

As glycogenesis is the reciprocal pathway to glycogenolysis, gluconeogenesis is the reciprocal pathway to glycolysis. Glycolysis, the "lysis" of glucose to two pyruvate molecules, is a ubiquitous metabolic pathway, whereas gluconeogenesis occurs in only a select number of tissues, including the liver. During periods of increased glucose availability, flux through the glycolytic pathway increases, thereby utilizing this readily available fuel source. In contrast, during periods of decreased glucose availability, rates of gluconeogenesis increase, in an attempt to maintain blood glucose levels. The sources of carbon for gluconeogenesis depend on the given metabolic situation (e.g., exercise, starvation, diabetes mellitus), and include glycerol, amino acids, and lactate. Although the majority of fatty acid–derived carbon cannot be used for the net synthesis of glucose, fatty acid metabolism plays a central role in gluconeogenesis. Gluconeogenesis is an energetically demanding process, which is driven by the β-oxidation of fatty acids. When the rate of acetyl-CoA generation through fatty acid β-oxidation exceeds the rate of acetyl-CoA oxidation via the Krebs cycle and oxidative phosphorylation (ox phos), acetyl-CoA accumulates. This

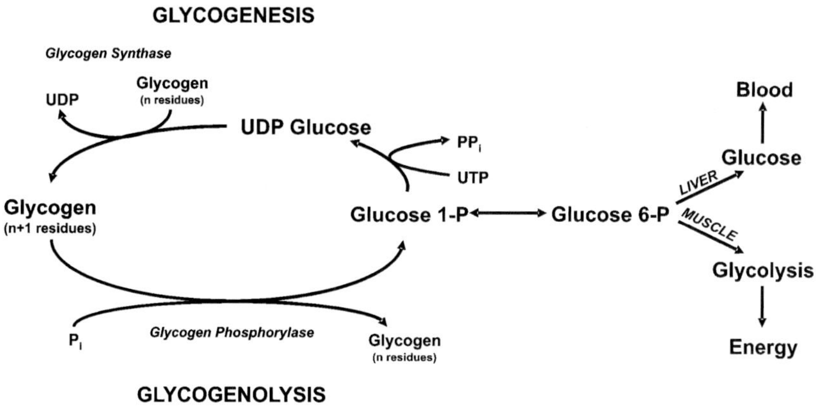

Figure 26-2. The biochemical pathways of the synthesis of glycogen (glycogenesis) and its breakdown to glucose 6-phosphate (glycogenolysis).

acetyl-CoA is shunted into the pathway of ketone body synthesis (ketogenesis), allowing continued β-oxidation of fatty acids and therefore maintenance of high rates of gluconeogenesis. The interplay between glycolysis, gluconeogenesis, β-oxidation, and ketogenesis is illustrated in Figure 26-3.

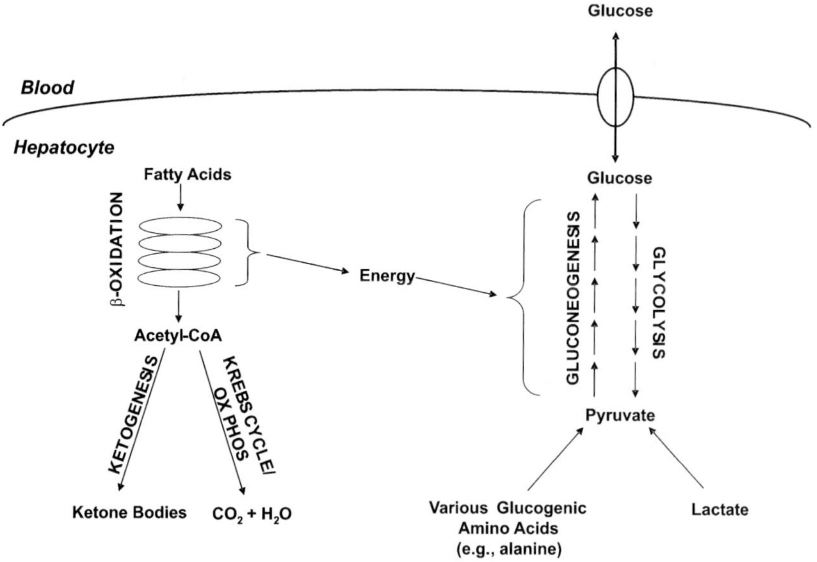

Figure 26-3. Schematic diagram of the interplay of fatty acid breakdown and ketone body formation with the synthesis (gluconeogenesis) and degradation of glucose (glycolysis). The β-oxidation of fatty acids provides the energy that drives the formation of glucose.

Flux of carbon through the pathways of hepatic glucose metabolism described above is strongly influenced by the hormones insulin, glucagon, and epinephrine. Insulin is secreted from the β-cells of pancreatic islets during periods of increased glucose availability. This peptide hormone helps to lower blood glucose back within the normal range by stimulating glycogenesis and glycolysis, and simultaneously inhibiting glycogenolysis and gluconeogenesis. The effects of insulin on hepatic glucose metabolism are mediated in large part by the enzyme protein phosphatase 1 (PP1). For a more detailed discussion of the mechanism by which PP1 affects glucose metabolism, see Case 22.

When blood glucose levels begin to decline (e.g., during an overnight fast), so too does insulin secretion. In contrast, secretion of glucagon from α-cells of pancreatic islets is stimulated. The latter targets primarily hepatic glucose metabolism, increasing glucose production (via gluconeogenesis and glycogenolysis), and decreasing glucose utilization. **Glucagon** acts in larger part by reversing the effects of insulin-mediated PP1 activation. On binding to its cell surface receptor, glucagon increases the activity of protein kinase A (PKA), which phosphorylates many of the proteins/enzymes that PP1 dephosphorylates. For a detailed discussion, of the mechanism by which PKA acts, see Case 22.

Like glucagon, **epinephrine** secretion increases during periods of decreased glucose availability. Binding of epinephrine to β-adrenergic receptors on the surface of the hepatocyte results in activation of PKA, thereby increasing further hepatic glucose production through glycogenolysis and gluconeogenesis (Figure 26-4). Epinephrine is also able to bind to a second receptor on the surface of the hepatocyte, the β-adrenergic receptor, causing elevation of intracellular Ca^{2+} levels. The latter allosterically activates a kinase called phosphorylase kinase, which in turn augments activation of glycogenolysis.

Abnormalities in the above described glucose homeostatic mechanisms arise during diabetes mellitus. Two major forms of diabetes mellitus exist, insulin-dependent (type I) and insulin-independent (type II) diabetes. Type I diabetes is caused by a severe lack or complete absence of insulin. Also known as early onset diabetes, this disease is often caused by an autoimmune destruction of pancreatic β-cells. Lack of insulin, in the face of elevated glucagon and epinephrine, leads to high rates of hepatic glucose output, being driven by β-oxidation of fatty acids. The latter results in an excessive production of ketone bodies and subsequent ketoacidosis. Treatment of type I diabetes involves regular monitoring of blood glucose levels and insulin administration as required. Insulin therapy associated with an evening meal will lower blood glucose levels. The latter triggers release of the counter-regulatory hormones glucagon and epinephrine, thereby stimulating hepatic glucose production. This inadvertently results in elevated glucose levels in the morning (the Somogyi effect). In contrast to type I diabetes, type II diabetes is caused by insulin resistance in the face of insulin insufficiency. The disease is thus characterized by both hyperglycemia and hyperinsulinemia. Ketoacidosis is a much less common complication of type II diabetes.

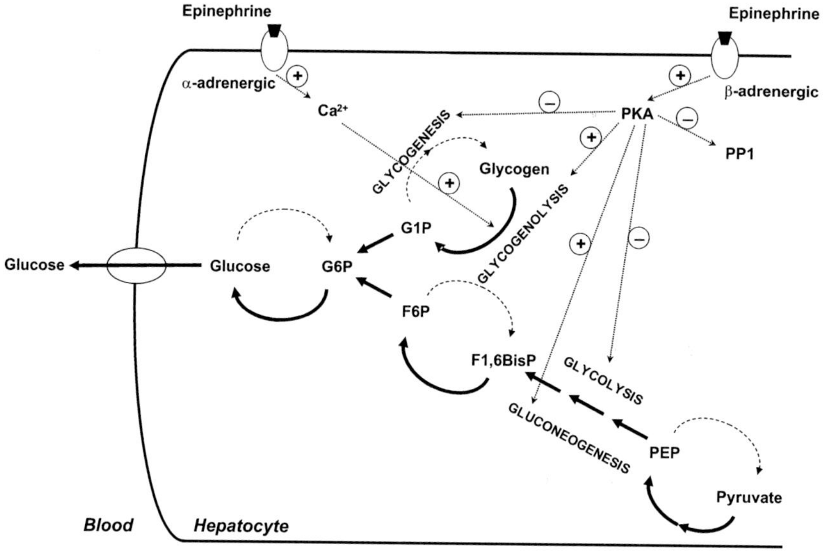

Figure 26-4. Schematic diagram showing how epinephrine leads to the breakdown of glycogen, the synthesis of glucose and export of glucose to the bloodstream. Epinephrine binds to both α- and β-adrenergic receptors on the plasma membrane of hepatocytes, leading to increased release of Ca^{2+} and activation of adenylate cyclase and protein kinase A.

COMPREHENSION QUESTIONS

A 27-year-old man has been rushed to the emergency room following his sudden collapse and entry into a state of unconsciousness. Examination of personal belongings revealed the patient is an insulin-dependent diabetic.

[26.1] A rapid decline in which of the following humoral factors likely triggered the sudden collapse of the patient?

A. Insulin
B. Glucagon
C. Fatty acids
D. Glucose
E. Triglyceride

[26.2] Which of the following is *least likely* to contribute to the hyperglycemia associated with uncontrolled type I diabetes?

A. Decreased skeletal muscle glucose uptake
B. Decreased adipose lipogenesis
C. Increased adipose lipolysis
D. Increased hepatic gluconeogenesis
E. Increased skeletal muscle glycogenolysis

[26.3] Which of the following changes in hepatic metabolism best explains the increased incidence of ketoacidosis observed in type I diabetes?

A. Increased glucose uptake
B. Increased protein synthesis
C. Increased lipoprotein synthesis
D. Increased β-oxidation
E. Increased glycogen breakdown

Answers

[26.1] **D.** Glucose is the primary source of energy for the central nervous system. A sudden decrease in circulating glucose levels will therefore impair ATP generation, in turn impeding cognitive function. If hypoglycemia persists, the patient will slip into a coma and eventually die. This unfortunately common complication in type I diabetics is a consequence of oversupplementation with insulin. In contrast, a sudden decrease in circulating insulin, glucagon, fatty acids, or triglyceride would have little immediate effect on cognitive function.

[26.2] **E.** Unlike the liver, skeletal muscle cannot export glucose into the circulation. Once glucose enters the myocyte, it is destined for use by that cell. Thus, intramyocellular glycogen is used as a fuel source by skeletal muscle and therefore cannot contribute to the hyperglycemia observed in uncontrolled type I diabetes. In contrast decreased insulin-mediated glucose utilization by skeletal muscle and adipose will contribute to hyperglycemia, as will decreased insulin-mediated suppression of hepatic glucose output. Decreased insulin-mediated suppression of lipolysis will indirectly contribute to hyperglycemia, by providing alterative, nonglucose, fuels (fatty acids and ketone bodies) for organs such as skeletal muscle and the liver.

[26.3] **D.** Decreased circulating insulin signals a need to increase hepatic glucose production. This is an energetically demanding process, driven by β-oxidation of fatty acids. However, acetyl-CoA, the major end product of β-oxidation, cannot be used for glucose production. Instead, acetyl-CoA is shunted into the pathway of ketone body synthesis (ketogenesis). In contrast, carbon from intrahepatic glycogen will contribute less to ketone body synthesis in uncontrolled type I diabetes mellitus. Net glucose uptake, protein synthesis, and lipoprotein synthesis are decreased, as opposed to increased, during uncontrolled type I diabetes.

BIOCHEMISTRY PEARLS

- ❖ The liver uses two mechanisms for endogenous glucose production, the mobilization of intracellular glycogen (glycogenolysis) and the synthesis of glucose from noncarbohydrate precursors (gluconeogenesis).
- ❖ Flux of carbon through the pathways of hepatic glucose metabolism described above is strongly influenced by the hormones insulin, glucagon, and epinephrine.
- ❖ Lack of insulin, in the face of elevated glucagon and epinephrine, leads to high rates of hepatic glucose output, being driven by β-oxidation of fatty acids. The latter results in an excessive production of ketone bodies and subsequent ketoacidosis.

REFERENCES

Cohen P. Dissection of the protein phosphorylation cascades involved in insulin and growth factor action. Biochem Soc Trans 1993;21:555.

Newsholme EA, Leech AR. Biochemistry for the Medical Sciences. New York: Wiley, 1983.

CASE 27

A 51-year-old male presents to the emergency center with chest pain. He states that he has had chest discomfort or pressure intermittently over the last year especially with increased activity. He describes the chest pain as a pressure behind his breastbone that spreads to the left side of his neck. Unlike previous episodes, he was lying down, watching television. The chest pain lasted approximately 15 minutes then subsided on its own. He also noticed that he was nauseated and sweating during the pain episode. He has no medical problems that he is aware of and has not been to a physician for several years. On examination, he is in no acute distress with normal vital signs. His lungs were clear to auscultation bilaterally, and his heart had a regular rate and rhythm with no murmurs. An electrocardiogram (ECG) revealed ST segment elevation and peaked T waves in leads II, III, and aVF. Serum troponin I and T levels are elevated.

- **What is the most likely diagnosis?**

- **What biochemical shuttle may be active to produce more adenosine triphosphate (ATP) per glucose molecule?**

ANSWERS TO CASE 27: MYOCARDIAL INFARCTION

Summary: A 51-year-old male with a history of chest pain with exertion presents with retrosternal chest pressure that radiates to the neck. He has nausea and diaphoresis while at rest. The patient has ST segment elevation and peaked T waves in the inferior ECG leads. The troponin I and T levels are elevated.

- **Likely diagnosis:** Acute myocardial infarction.
- **Biochemical shuttle:** The malate-aspartate shuttle is primarily seen in the heart, liver, and kidney. This shuttle requires cytosolic and mitochondrial forms of malate dehydrogenase and glutamate-oxaloacetate transaminase and two antiporters, the malate-α-ketoglutarate antiporter and the glutamate-aspartate antiporter, which are both localized in the mitochondrial inner membrane. In this shuttle cytosolic nicotinamide adenine dinucleotide (NADH) is oxidized to regenerate cytosolic NAD^+ by reducing oxaloacetate to malate by cytosolic malate dehydrogenase.

CLINICAL CORRELATION

The most common cause of death of Americans is coronary heart disease. The patient's symptoms in this case are very typical of myocardial infarction, that is, chest pressure or chest pain, often radiating to the neck or to the left arm. The pain is usually described as deep and "squeezing chest pain." Cardiac muscle is perfused by coronary arteries with very little redundant or shared circulation; thus, occlusion of one coronary artery usually leads to ischemia or necrosis of the corresponding cardiac muscle. Laboratory confirmation of myocardial infarction (death of cardiac muscle) includes ECG showing elevation of the ST segment and/or increase of the cardiac enzymes. When there is insufficient oxygen available for the cardiac muscle, then the glycolytic pathway must be used, which leads to a very small amount of ATP per glucose molecule. The malate-aspartate shuttle can offer two or three more times the ATP by oxidizing NADH to regenerate cytosolic NAD^+ by reducing oxaloacetate to malate by cytosolic malate dehydrogenase.

APPROACH TO GLYCOLYSIS AND THE MALATE-ASPARTATE SHUTTLE

Objectives

1. Be familiar with glycolysis.
2. Know the role of glycerol 3-phosphate and the malate-aspartate shuttle.
3. Be aware of the role of mitochondria in glycolysis.

Definitions

Myocardial infarction: An area of heart muscle is inadequately perfused as a result of cardiac vessel occlusion resulting in ischemia, cell death, and loss of cell constituents including enzymes into the circulation. Electrocardiographic changes occur as a result of damaged heart tissue.

Angina: Transient disruption of adequate blood flow to a portion of heart muscle leading to pain and a temporary shift to anaerobic glycolysis producing pyruvate and lactate that are released into the circulation.

Aerobic glycolysis: Metabolism of glucose to pyruvate. Pyruvate in the presence of sufficient oxygen can be metabolized to CO_2 via the tricarboxylic acid cycle in the mitochondrion-producing NADH and $FADH_2$, which contribute elections through the electron transfer chain to molecular oxygen producing H_2O and ATP.

Anaerobic glycolysis: Metabolism of glucose to lactate in the absence of sufficient oxygen. When oxygen is lacking, pyruvate is converted to lactate, and no further oxidative pathway is available.

Electron shuttles: Enzymatic processes whereby electrons from NADH can be transferred across the mitochondrial barrier. The **glycerol 3-phosphate shuttle** uses the reduction of dihydroxyacetone phosphate to glycerol 3-phosphate and reoxidation to transfer electrons from cytosolic NADH to coenzyme Q in the electron transport chain. The **malate-aspartate shuttle** uses malate and aspartate in a two-member transfer exchange to transfer electrons from cytosolic NADH to mitochondrial NADH (see Figures 27-2 and 27-3).

DISCUSSION

Myocardial infarction arises when perfusion of cardiac muscle is inadequate, resulting in insufficient oxygen delivery to that portion of cardiac muscle. This causes the affected muscle to rely on anaerobic metabolism for its energy supply with concomitant production of lactic acid. Even transient ischemia can lead to changes in muscle tissue but prolonged ischemia leads to breakdown of muscle cells and release of cellular proteins such as **creatine kinase, lactic acid dehydrogenase, and troponin I.** Reperfusion by thrombolytic treatment or mechanical means can restore oxygen levels and return the metabolic processes to aerobic metabolism. A secondary consequence of reperfusion is reperfusion injury in which the highly reduced state of injured cells meets increased oxygen concentration and produces reactive oxygen radicals. Most notable of these is the hydroxyl radical ($OH^{\bullet}$), which attacks tissue components such as lipids and protein sulfhydryl groups. Myocardial infarction causes changes in the pathways of energy generation triggered by oxygen insufficiency in the affected heart muscle.

The **glycolytic pathway** for all cells under conditions of adequate tissue oxygenation is shown in Figure 27-1. **Two molecules of ATP** are required to

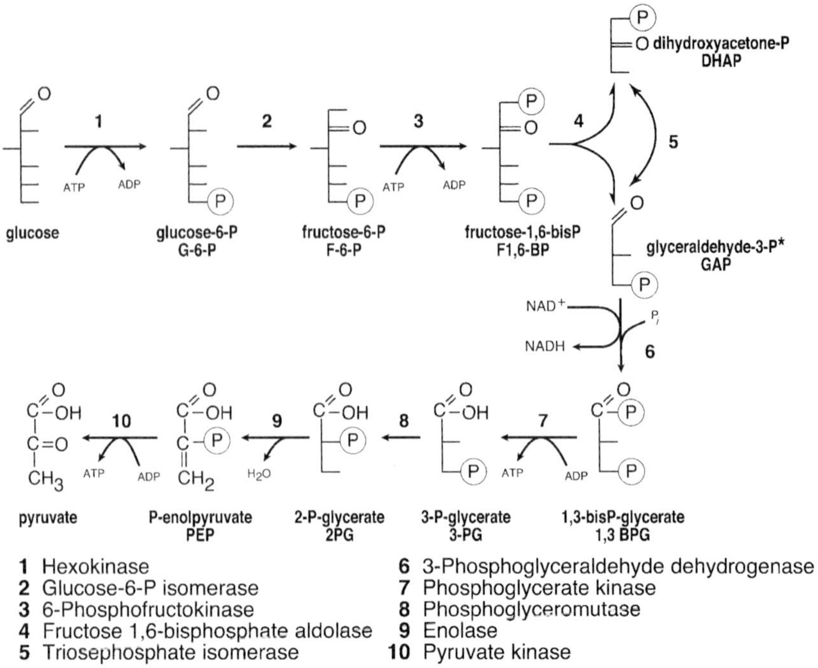

Figure 27-1. Glycolytic pathway.

*From glyceraldehyde 3-phosphate onward, two 3-carbon molecules/glucose proceed through the pathway requiring a total of $2NAD^+$ and 4ADP producing a total of 4ATP, 2NADH, and $2H_2O$.

initiate the pathway one at the **hexokinase** step to phosphorylate glucose to glucose 6-phosphate and a second to **phosphorylate fructose 6-phosphate to fructose 1,6-bisphosphate.** Cleavage of fructose 1,6-bisphosphate by aldolase yields two triose phosphates, glyceraldehyde 3-phosphate, and dihydroxyacetone phosphate, which is isomerized to glyceraldehyde 3-phosphate by triose phosphate isomerase. The oxidation of the aldehyde group to the acid level through the reduction of NAD^+ to NADH with the concomitant binding of inorganic phosphate to produce 1,3-bisphosphoglycerate is catalyzed by 3-phosphoglyceraldehyde dehydrogenase. This step requires NAD^+ regeneration to continue. Inorganic phosphate is present in tissues at such a high concentration (typically 25 mM) as not to be limiting. The first substrate level phosphorylation occurs next wherein phosphoglycerate kinase catalyzes the transfer of the acyl phosphate group of 1,3-bisphosphoglycerate to adenosine diphosphate (ADP), yielding ATP and 3-phosphoglycerate. Phosphoglucomutase shifts the phosphate of 3-phosphoglycerate to carbon position 2 producing 2-phosphoglycerate. Enolase catalyzes the removal of the elements of water between carbon positions 2 and 3 to produce phosphoenolpyruvate. Pyruvate kinase then catalyzes the second substrate level phosphorylation of ADP to produce ATP and pyruvate, the end product of aerobic glycolysis. Since fructose 1,6-bisphosphate is cleaved to two triose phosphate moieties and each triose phosphate produces two ATP molecules, the total produced by substrate level phosphorylation is four ATPs. But two ATP molecules are consumed in the activation of hexose leaving two ATPs as the net gain from glycolytic substrate level phosphorylation.

As noted above, however, NAD^+ must be regenerated from the NADH produced or the glycolytic cycle would cease. Under aerobic conditions regeneration of cytosolic NAD^+ from cytosolic NADH is accomplished by transferring electrons across the mitochondrial membrane barrier to the electron transfer chain where the electrons are transferred to oxygen. There are **two different shuttle mechanisms** whereby this transfer of electrons across the membrane to regenerate cytosolic NAD^+ can be accomplished, the glycerol 3-phosphate shuttle and the malate-aspartate shuttle.

The **glycerol 3-phosphate shuttle** (Figure 27-2) functions primarily in **skeletal muscle and brain.** The shuttle takes advantage of the fact that the enzyme glycerol-3-phosphate dehydrogenase exists in two forms, a cytosolic form that uses NAD^+ as cofactor and a mitochondrial FAD-linked form. Cytosolic glycerol-3-phosphate dehydrogenase uses electrons from cytosolic NADH to reduce the glycolytic intermediate dihydroxyacetone phosphate to glycerol 3-phosphate, thereby regenerating cytosolic NAD^+. The newly formed

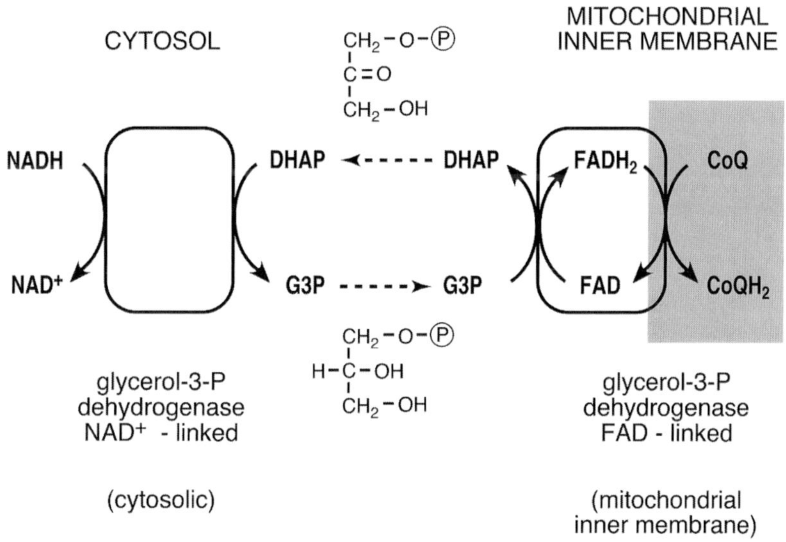

Figure 27-2. Glycerol 3-phosphate shuttle.

glycerol 3-phosphate is released from the cytosolic form of the enzyme and crosses to and is bound to the mitochondrial FAD-linked glycerol-3-phosphate dehydrogenase, which is bound to the cytosolic side of the mitochondrial inner membrane. There the mitochondrial glycerol-3-phosphate dehydrogenase reoxidizes glycerol 3-phosphate to dihydroxyacetone phosphate (preserving mass balance) reducing its FAD cofactor to $FADH_2$. Electrons are then passed to coenzyme Q of the electron transport chain and on to oxygen generating two additional ATP molecules per electron pair and therefore per glycerol 3-phosphate.

The **malate-aspartate shuttle,** however, functions primarily in the **heart, liver, and kidney** (Figure 27-3). This shuttle requires cytosolic and mitochondrial forms of malate dehydrogenase and glutamate-oxaloacetate transaminase and two antiporters, the malate-α-ketoglutarate antiporter and the glutamate-aspartate antiporter, which are both localized in the mitochondrial inner membrane. In this shuttle cytosolic NADH is oxidized to regenerate cytosolic NAD^+ by reducing oxaloacetate to malate by cytosolic malate dehydrogenase.

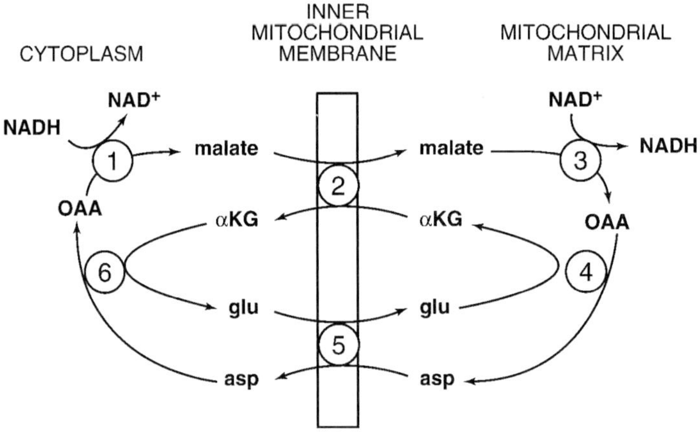

1. malate dehydrogenase (cytosolic)
2. malate - KG antiporter (mitochondrial inner membrane)
3. malate dehydrogenase (mitochondrial matrix)
4. glutamate - OAA transaminase (mitochondrial matrix)
5. glutamate - aspartate antiporter (mitochondrial inner membrane)
6. glutamate - OAA transaminase (cytosolic)

Figure 27-3. Malate-aspartate shuttle.

Malate is transported into the mitochondrial matrix while α-ketoglutarate is transported out by the malate-α-ketoglutarate antiporter, a seeming mass unbalance. Next malate is oxidized back to oxaloacetate producing NADH from NAD^+ in the mitochondrial matrix by mitochondrial malate dehydrogenase. Oxaloacetate cannot be transported per se across the mitochondrial membrane. It is, instead transaminated to aspartate from the NH_3 donor glutamate by mitochondrial glutamate-oxaloacetate transaminase. Aspartate is transported out of the matrix whereas glutamate is transported in by the glutamate-aspartate antiporter in the mitochondrial membrane, obviating the apparent mass unbalance noted above. The last step of the shuttle is catalyzed by cytosolic glutamate-oxaloacetate transaminase regenerating cytosolic oxaloacetate from aspartate and cytosolic glutamate from α-ketoglutarate both of which were earlier transported in opposing directions by the malate-α-ketoglutarate antiporter. The net effect of this shuttle is to transport electrons from cytosolic NADH to mitochondrial NAD^+. Therefore, those electrons can be presented by the newly formed NADH to electron transport system complex I thereby producing three ATPs by oxidative phosphorylation. Note that depending on which shuttle is used (i.e., which tissue is catalyzing glycolysis) either two or three extra ATPs are produced by oxidative phosphorylation per triose phosphate going through the latter steps of glycolysis.

In nonaerobic glycolysis, as in the case when a tissue is subjected to an ischemic episode (i.e., myocardial infarction), neither the extra ATP produced by the shuttle nor the ATPs produced by normal passage of electrons through the electron transport chain are produced because of oxygen insufficiency. Therefore glycolysis must increase in rate to meet the energy demand. In damaged tissue this increased rate is compromised. Moreover the shuttle mechanisms to regenerate NAD^+ from NADH formed by glycolysis are unavailable, as shown in Figure 27-4. Glycolysis under ischemic conditions satisfies the requirement for NAD^+ by reducing pyruvate, the glycolytic end product under normative conditions, to lactate with the reducing equivalents of NADH.

The new end product lactate accumulates in muscle cells under ischemic conditions and damages cell walls with its low pH causing rupture and loss of cell contents such as myoglobin and troponin I. These compounds as well as other end products combine to cause increased cell rupture and pain.

Reopening vasculature by reperfusion as rapidly as possible is a first step in treatment. Thrombolysis within an hour after infarction gives best results. Supportive measures, regulation of heart rate and pressure are required following the infarction to allow recovery from the ischemic episode and repair of tissue damage. Nutritional monitoring is required both for tissue repair and prevention of recurrence.

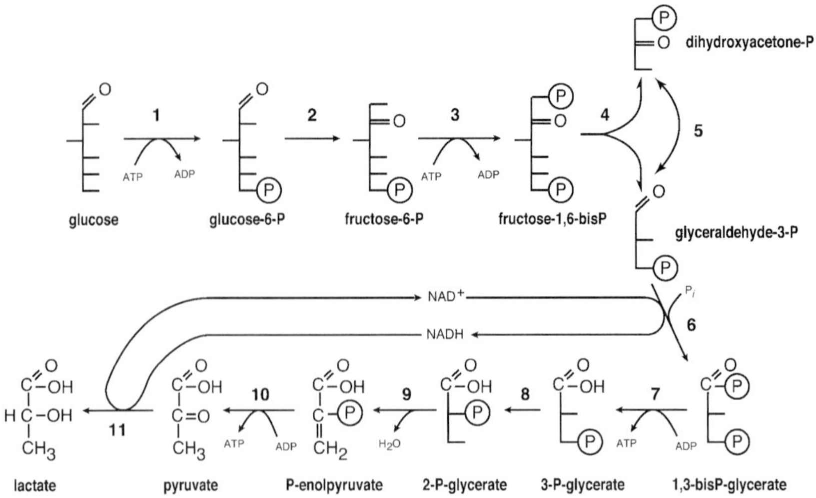

1 Hexokinase
2 Glucose-6-P isomerase
3 6-Phosphofructokinase
4 Fructose 1,6-bisphosphate aldolase
5 Triosephosphate isomerase
6 3-Phosphoglyceraldehyde dehydrogenase
7 Phosphoglycerate kinase
8 Phosphoglyceromutase
9 Enolase
10 Pyruvate kinase
11 Lactate dehydrogenase

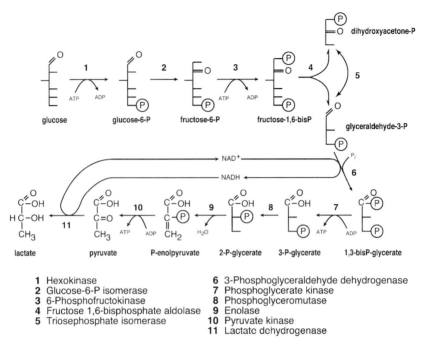

Figure 27-4. Glycolytic pathway under conditions of insufficient oxygen.

COMPREHENSION QUESTIONS

A middle-aged man, after an episode of retrosternal chest pressure with radiation to the neck and associated nausea and diaphoresis while at rest, was given a diagnosis of unstable angina and possibly myocardial infarction. Subsequent laboratory findings of increased serum levels of troponin I and cardiac enzymes were consistent with the diagnosis.

[27.1] If the patient described above did indeed have angina, which of the following change(s) in metabolism in the affected area would occur?

A. Increased oxidative phosphorylation
B. Increased rate of fatty acid oxidation
C. Increased conversion of pyruvate to acetyl-CoA
D. Increased formation of lactate
E. Increased use of ketone bodies

[27.2] In the case described above an ischemic episode very likely occurred. What changes in glucose metabolism would be observed?

A. The overall rate of glucose utilization would decrease.
B. Pyruvate kinase is allosterically inhibited.
C. The rate of ATP production in the cytosol increases.
D. NADH is reoxidized to NAD^+ via the glycerol 3-phosphate shuttle.

[27.3] If the patient described does indeed have angina, his is a focal lesion wherein only one portion of heart muscle is affected while other portions still receive adequate oxygen enabling the malate-aspartate shuttle to function. Which of the following is involved in this shuttle to transfer electrons across the mitochondrial barrier?

A. Malate-aspartate antiporter
B. Malate-α-ketoglutarate antiporter
C. Glutamate-α-ketoglutarate antiporter
D. Aspartate-α-ketoglutarate symporter
E. Malate-glutamate symporter

Answers

[27.1] **D.** In this case the problem is insufficient oxygen reaches an area of cardiac muscle. In the absence of sufficient oxygen, the use of the electron transfer chain to generate ATP from ADP is severely compromised, and therefore all products feeding into that chain accumulate reducing the formation of tricarboxylic acid products. Fatty acid oxidation, formation of acetyl-CoA from pyruvate oxidation of ketone bodies, and oxidative phosphorylation are all decreased. In glycolysis under reduced oxygen conditions, one pyruvate is converted to lactate to reoxidize NADH to NAD^+ to meet the requirement for glycolysis to continue. Therefore, there is increased formation of lactate.

[27.2] **C.** The rates of both glucose utilization and ATP generation by glycolysis increase to compensate for the absence of sufficient ATP production from oxidative phosphorylation as a result of oxygen deprivation. Under these circumstances, ATP is used as rapidly as it is made so it is not present in sufficient concentration to inhibit pyruvate kinase. Furthermore, fructose 1,6-bisphosphate tends to stimulate pyruvate kinase activity.

[27.3] **B.** The malate-α-ketoglutarate antiporter carries malate from the cytosol, where it is oxidized to oxaloacetate by mitochondrial malate dehydrogenase forming NADH. In order for the malate-α-ketoglutarate antiporter to function a molecule of α-ketoglutarate must be transported from the mitochondrial matrix to the cytosol. The only other transporter involved in this shuttle mechanism is the aspartate-glutamate antiporter. None of the other suggested transporters is required in this shuttle mechanism.

BIOCHEMISTRY PEARLS

- ❖ Coronary heart disease is the most common cause of mortality in Americans.
- ❖ Myocardial infarction is suspected with the typical chest pain, ECG findings of ST segment elevation, and increased cardiac enzymes.
- ❖ When oxygen is unavailable, the anaerobic glycolytic pathway is used.
- ❖ Two shuttles allow for more ATP to be generated under aerobic conditions: the glycerol 3-phosphate shuttle, which functions primarily in skeletal muscle and brain, and the malate-aspartate shuttle, primarily in the heart, liver, and kidney.

REFERENCES

Braunwald E, Fauci AS, Kasper KL, et al., eds. Harrison's Principles of Internal Medicine, 15th ed. New York: McGraw-Hill, 2001.

Devlin TM, ed. Textbook of Biochemistry with Clinical Correlations, 5th ed. New York: Wiley-Liss, 2002:598–613.

CASE 28

A Jewish couple of Eastern European descent presents to the clinic for prenatal counseling after their only child died early in childhood. The family could not remember the name of the disorder but said it was common in their ancestry. Their first child was normal at birth, a slightly larger than normal head circumference, an abnormal "eye finding," and a severe progressive neurologic disease with decreased motor skills and eventually death. The autopsy is consistent with Tay-Sachs disease.

- **What type of inheritance is this disorder?**

- **What is the biochemical cause of the disorder?**

ANSWERS TO CASE 28: TAY-SACHS DISEASE

Summary: A Jewish couple of Eastern European descent presents for prenatal counseling after having a child die with a progressive neurologic disorder with loss of motor skills and abnormally large head and an abnormal eye examination. An autopsy shows Tay-Sachs disease.

- **Inheritance:** Autosomal recessive; 1:30 carrier rate in Ashkenazi Jews.
- **Molecular basis of disorder:** Lysosomal storage disorder with deficiency of hexosaminidase A enzyme resulting in G_{M2} gangliosides accumulating throughout the body.

CLINICAL CORRELATION

Tay-Sachs disease is a fatal genetic disorder where harmful amounts of lipids called ganglioside G_{M2} accumulate in the nerve cells and brains of those affected. Infants with this disorder appear normal for the first several months of life, and then as the lipids distend the nerve cells and brain cells, progressive deterioration occurs; the child becomes blind, deaf, and eventually unable to swallow. Tay-Sachs disease occurs mainly in Jewish children of Eastern European descent, and death from bronchopneumonia usually occurs by age 3 to 4 years. A reddish spot on the retina also develops, and symptoms first appear around 6 months of age. It is a lysosomal storage disorder with insufficient activity of the enzyme hexosaminidase A, which catalyzes the biodegradation of the gangliosides. The diagnosis is made by the clinical suspicion and serum hexosaminidase level. Currently there is no treatment available for this disease.

APPROACH TO LYSOSOMAL STORAGE

Objectives

1. Be familiar with the lysosomal storage disorders.
2. Know about the synthesis and degradation of sphingolipids.

Definitions

Ceramide: A component of all sphingolipids that is composed of a long-chain fatty acyl group in an amide linkage to sphingosine.

Ganglioside: Complex carbohydrate-rich glycosphingolipids containing three or more monosaccharides esterified to ceramide with at least one of the sugars being *N*-acetylneuraminic acid (sialic acid). Gangliosides are membrane components found on nerve endings.

Gangliosidoses: The build-up of gangliosides in the lysosome as a result of the deficiency of one or more enzymes involved in the degradation of gangliosides.

I-cell disease: A disease in which there is a buildup of a number of different biomolecules that are normally degraded in the lysosome as a result of a deficiency in an enzyme that modifies lysosomal enzymes so that they are targeted to the lysosome.

Lysosomal storage disorders: Genetic diseases that are a result of a deficiency of one or more enzymes present in the lysosome. This leads to a buildup of the biomolecules that are normally degraded in the lysosome.

Sphingolipid: Any of a number of lipid molecules that contain sphingosine as part of its molecular structure. These include ceramide, sphingomyelin, cerebrosides, sulfatides, and gangliosides.

Tay-Sachs disease: A genetic disease that is a result of a deficiency in hexosaminidase A (β-*N*-acetylhexosaminidase), an enzyme that is involved in the degradation of gangliosides in the lysosome. The disease is prevalent in Jewish children of Eastern European descent and leads to a buildup of ganglioside G_{M2} in nerve cells of the brain and neuronal dysfunction.

DISCUSSION

Sphingolipids are a class of **lipids** found in **biological membranes** in which the **backbone of the lipid molecule is sphingosine, an 18-carbon amino alcohol.** This is in contrast to the glycerol backbone found in lipids such as phospholipids. The precursor molecules for the synthesis of sphingosine are palmitoyl-CoA and serine, which condense to form 3-dehydrosphinganine (3-ketodihydrosphingosine; Figure 28-1). This is followed by a reduction reaction using NADPH to form sphinganine (dihydrosphingosine). Ceramide is formed in a two-step reaction by the formation of an amide bond between sphinganine and a long-chain fatty acid (usually behenic acid, a saturated C-22 fatty acid). This is followed by an oxidation reaction using FAD that introduces a *trans*-double bond into the sphinganine backbone.

Ceramide is a building block for all other sphingolipids. Figure 28-2 shows the synthesis of some of these sphingolipids from ceramide. The reaction of phosphatidyl choline and ceramide (via esterification) yields sphingomyelin, a phosphate-containing subclass of sphingolipids found in the nervous tissue of higher animals. The addition of one or more sugar residues to ceramide yields the glycosphingolipids. The glycosphingolipid cerebroside contains one residue of glucose or galactose.

Gangliosides are complex carbohydrate-rich sphingolipids containing three or more sugar residues esterified to ceramide with one of the sugars being sialic acid (*N*-acetylneuraminic acid, or NANA). Gangliosides are synthesized by the stepwise addition of nucleoside-activated sugar residues to ceramide. Ganglioside nomenclature is a bit unusual. The ganglioside is

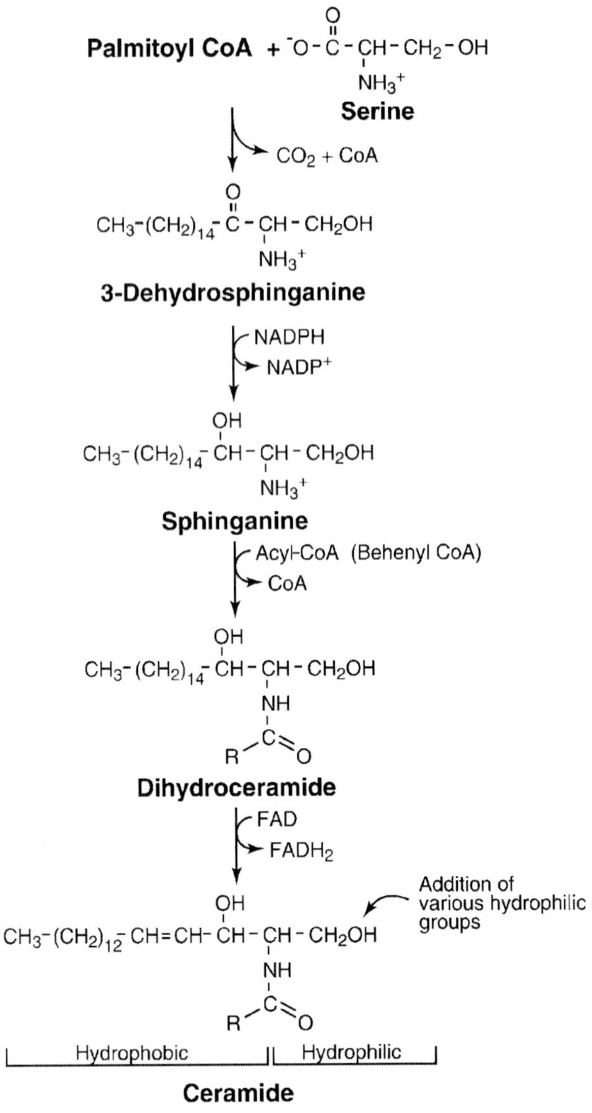

Figure 28-1. Biosynthesis of ceramide.

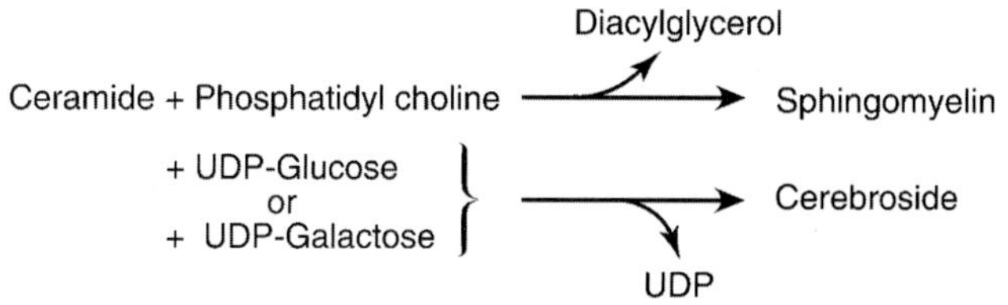

Figure 28-2. The formation of sphingomyelin and cerebrosides from ceramide.

represented by the letter G followed by a subscript letter and a number. The subscript letter indicates the number of sialic acid residues present in the molecule (M = 1, D = 2, T = 3), and the number is equal to 5 minus the number of neutral sugars. Over 60 gangliosides have been characterized, and the structure of one of these, ganglioside G_{M1}, is schematically represented in Figure 28-3. Glycosphingolipids function in cell-cell recognition and tissue immunity. Gangliosides are found at nerve endings and may be important in nerve impulse transmission.

Sphingolipids are constantly being turned over in the lysosomes of cells by specific hydrolytic enzymes that remove the sugars in a stepwise fashion. Defects in these enzymes can occur and result in disease states resulting from the accumulation of undegraded sphingolipids. Lysosomal storage diseases are a group of approximately 40 different diseases that occur in approximately 1 in 5000 live births. Many of these diseases are characterized by a deficiency in a specific lysosomal enzyme, but some diseases can be caused by the inability of enzymes to be translocated to the lysosome (I-cell disease), defective transport of small molecules out of the lysosome (cystinosis), or a deficiency in sphingolipid activator proteins, which are small molecular weight proteins that participate in the degradation of sphingolipids. The net result in all of

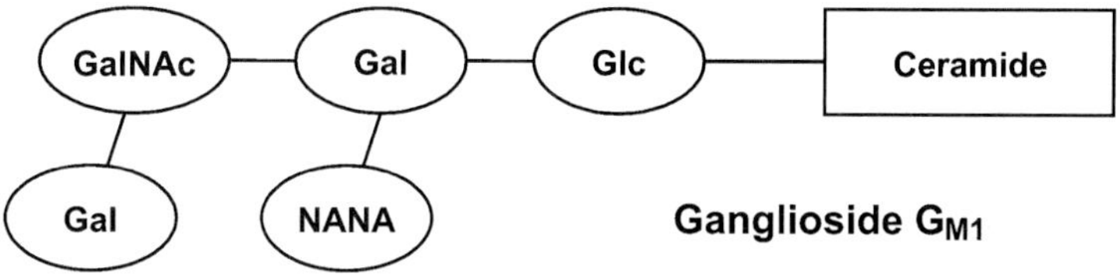

Figure 28-3. The schematic structure of ganglioside G_{M1}. The nomenclature is as follows: G stands for ganglioside; the subscript letter indicates the number of sialic acid residues; the numeral is equal to 5 − (number of neutral sugars). Abbreviations: GalNAc = *N*-acetylgalactosamine, Gal = galactose; Glc = glucose; NANA = *N*-acetylneuraminic acid (sialic acid).

these diseases is a lack of the ability of the cell to breakdown sphingolipids in the lysosome. The accumulation of these macromolecules or partial derivatives of these molecules in the lysosome results in the pathologic conditions associated with the lysosomal storage diseases known as sphingolipidoses.

Proteins destined for the lysosome must contain certain carbohydrate signals. In I-cell disease the enzyme that catalyzes the addition of a mannose-6-phosphate moiety to the protein is defective. Without the addition of this mannose residue the proteins cannot be transported to the lysosome.

Tay-Sachs disease results from a deficiency in the **enzyme hexosaminidase A** (β-*N*-acetylhexosaminidase). The human gene for this enzyme is found at **chromosome position 15q23-q24.** The deficiency in hexosaminidase A leads to the accumulation of ganglioside G_{M2} in the nerve cells of the brain. Hexosaminidase A removes a terminal *N*-acetylgalactosamine (GalNAc) residue from ganglioside G_{M2} to form ganglioside G_{M3} (Figure 28-4). The inability of patients with Tay-Sachs disease to remove these sugar residues results in the accumulation of gangliosides in the lysosome. This results in swelling of the neurons containing these lipid-filled lysosomes and the disruption of neuronal function.

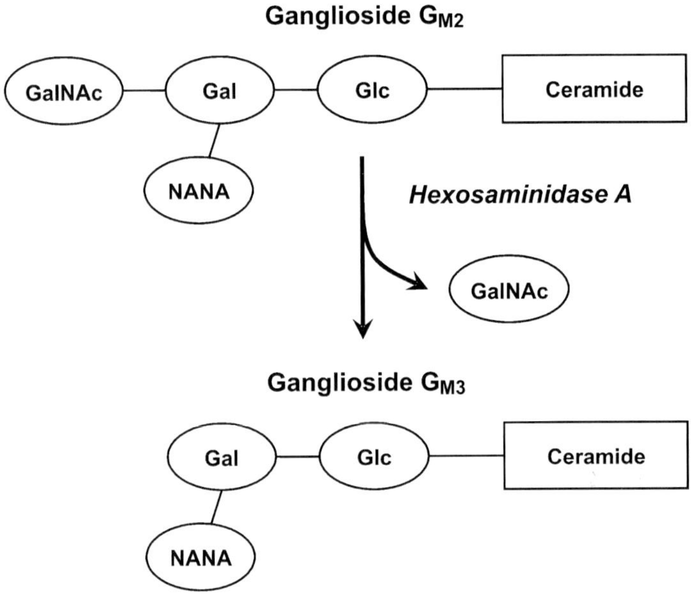

Figure 28-4. The reaction catalyzed by hexosaminidase A, the enzyme deficient in patients with Tay-Sachs disease.

Table 28-1
LYSOSOMAL STORAGE DISORDERS

DISEASE	DEFICIENT ENZYME	CONSEQUENCE
Tay-Sachs disease	β-*N*-acetylhexosaminidase	Increased ganglioside G_{M2}
Gaucher disease	β-Glucosidase	Increased glucocerebrosides
Fabry disease	α-Galactosidase A	Accumulation of glycosphingolipids with α-Galactosyl moieties
Farber disease	Ceramidase	Increased ceramide
Niemann-Pick disease	Sphingomyelinase	Accumulation of sphingomyelin
I-cell disease	*N*-Acetylglucosaminyl-1-phosphotransferase	Incorrect packaging of lysosomal enzymes impairing lysosome function

Other lysosomal storage disorders include G_{M1} gangliosidoses, G_{M2} gangliosidoses, Gaucher disease, Niemann-Pick disease, Fabry disease, fucosidosis, Schindler disease, metachromatic leukodystrophy, Krabbe disease, multiple sulfatase deficiency, Farber disease, and Wolman disease. Table 28-1 illustrates the enzyme deficiencies found in some of these disorders.

COMPREHENSION QUESTIONS

[28.1] A couple is seen in your office for genetic counseling regarding Tay-Sachs disease. They are very knowledgeable and request more information about the specific enzyme that is defective in this disease. You explain that Tay-Sachs results from the lack of an enzyme activity necessary for which of the following?

A. Removal of *N*-acetylgalactosamine from ganglioside G_{M2}
B. Addition of *N*-acetylgalactosamine to ganglioside G_{M2}
C. Removal of the disaccharide galactose-*N*-acetylgalactosamine from ganglioside G_{M2}
D. Addition of the disaccharide galactose-*N*-acetylgalactosamine to ganglioside G_{M2}
E. Removal of a galactose residue from ganglioside G_{M2}

[28.2] Tay-Sachs disease involves the metabolism of gangliosides. Gangliosides are composed of a ceramide backbone with at least which one of the following?

A. Phosphorylated sugar residue
B. Glucose residue
C. Galactose residue
D. Sialic acid residue
E. Fructose residue

[28.3] The genetic disease which results from a mutation in the gene coding for the enzyme hexosaminidase (β-*N*-acetylhexosaminidase) is called what?

A. Huntington disease
B. Lesch-Nyhan syndrome
C. Tay-Sachs disease
D. Amyotrophic lateral sclerosis
E. Neurofibromatosis

Answers

[28.1] **A.** Tay-Sachs disease is the result of the lack of the enzyme β-*N*-acetylhexosaminidase. This enzyme hydrolyzes a terminal *N*-acetylgalactosamine from the ganglioside G_{M2}. This ganglioside is found in high concentrations in the nervous system and is normally degraded in the lysosome by the sequential removal of terminal sugars. The lack of β-*N*-acetylhexosaminidase results in the accumulation of the partially degraded ganglioside in the lysosome leading to significant swelling of the lysosome. The abnormally high level of lipid in the lysosome of the neuron affects its function resulting in the disease.

[28.2] **D.** Gangliosides are carbohydrate-rich lipids in which an oligosaccharide chain is attached to ceramide. The oligosaccharide chain must contain at least one acidic sugar such as *N*-acetylneuraminate or *N*-glycosylneuraminate. These sugars are commonly referred to as sialic acid residues. Gangliosides are synthesized by the stepwise addition of sugar residues to ceramide.

[28.3] **C.** Tay-Sachs disease is the result of the lack of the enzyme β-*N*-acetylhexosaminidase. Affected infants show weakness and retarded motor skills before 1 year of age. Other abnormalities follow, and death usually occurs before age 3.

BIOCHEMISTRY PEARLS

- Sphingolipids are a class of lipids found in biological membranes in which the backbone of the lipid molecule is sphingosine, an 18-carbon amino alcohol.
- Sphingolipids are constantly being turned over in the lysosomes of cells by specific hydrolytic enzymes that remove the sugars in a stepwise fashion. Defects in these enzymes can result in accumulation of undegraded sphingolipids.
- This disease results from a deficiency in the enzyme hexosaminidase A (β-*N*-acetylhexosaminidase). The deficiency in hexosaminidase A leads to the accumulation of ganglioside G_{M2} in the nerve cells of the brain.
- Other lysosomal storage disorders include G_{M1} gangliosidoses, G_{M2} gangliosidoses, Gaucher disease, Niemann-Pick disease, Fabry disease, fucosidosis, Schindler disease, metachromatic leukodystrophy, Krabbe disease, multiple sulfatase deficiency, Farber disease, and Wolman disease.

REFERENCES

Berg JM, Tymoczko JL, Stryer L. Biochemistry, 5th ed. New York: Freeman, 2002:721–2.

Devlin TM, ed. Textbook of Biochemistry with Clinical Correlations, 5th ed. New York: Wiley-Liss, 2002:762–5.

CASE 29

A 6-year-old boy is brought to his pediatrician with profound mental retardation. Upon questioning the mother, several members of both the mother's and father's family have had mental retardation. She also has noticed that he appears very active and restless frequently getting into trouble in school. On examination, the boy appears restless and his speech is significantly delayed and incomprehensible. The boy has coarse facial features but otherwise has a normal exam. Laboratory tests reveal an elevated level of heparan sulfate.

- **What is the most likely diagnosis?**
- **What is the inheritance pattern of this disorder?**
- **What are some other causes of mucopolysaccharidoses?**

ANSWERS TO CASE 29: SANFILIPPO SYNDROME

Summary: A 6-year-old boy with severe mental retardation, family history of mental retardation, language delays, and behavioral problems.

- **Diagnosis:** Sanfilippo syndrome
- **Inheritance pattern:** Autosomal recessive
- **Other causes:** Hunter syndrome, Hurler syndrome, or Morquio syndrome

CLINICAL CORRELATION

Excessive accumulation of proteins, nucleic acids, carbohydrates, and lipids can result from deficiency of one or more lysosomal hydrolases. Lysosomal storage diseases are classified by the stored material. Accumulation of glycosaminoglycans results in mucopolysaccharidoses. Common causes of this disorder include: Hunter syndrome, Hurler syndrome, and Sanfilippo syndrome. Sanfilippo syndrome is inherited in an autosomal recessive pattern and clinically evident by profound mental retardation, lack of normal developmental milestones, and significant language delay. Sanfilippo syndrome results in an excess of heparan sulfate and can be caused by a variety of enzyme deficiencies.

APPROACH TO LYSOSOMAL DEGRADATION OF GLYCOSAMINOGLYCANS

Objectives

1. Understand the structural roles of glycosaminoglycans and proteoglycans.
2. Describe how glycosaminoglycans and proteoglycans are synthesized.
3. Describe the biochemical pathways needed for glycosaminoglycan catabolism.
4. Explain why lysosomal enzyme deficiencies and glycosaminoglycan accumulation result in clinical signs/symptoms.

Definitions

Endoglycosidase: An enzyme that hydrolyzes an interior glycosidic bond between two sugars in a polysaccharide or oligosaccharide to produce two smaller oligosaccharides.

Exoglycosidase: An enzyme that hydrolyzes the glycosidic bond between the terminal two sugars of an oligo- or polysaccharide releasing the terminal sugar from the nonreducing end of the polymer, leaving it one sugar shorter.

Glycosaminoglycan (GAG): Formerly known as mucopolysaccharide; a heteropolysaccharide composed of a repeating disaccharide of an *N*-acetylated hexosamine (glucosamine or galactosamine) and an acidic hexose (glucuronic acid or iduronic acid); this repeating disaccharide unit is frequently sulfated in one or more positions; most glycosaminoglycans are covalently linked to a core protein in structures known as proteoglycans.

Hyaluronan: Formerly known as hyaluronic acid; a glycosaminoglycan composed of alternating residues of glucuronic acid and *N*-acetylglucosamine. Hyaluronan is not sulfated nor is it covalently linked to a protein.

Mucopolysaccharidosis: A genetic disorder involving a lack of a lysosomal enzyme required for the degradation of glycosaminoglycans leading to buildup of glycosaminoglycans in the lysosome and increased excretion of glycosaminoglycan fragments in urine.

Sulfatase: An enzyme that catalyzes the hydrolysis of sulfate ester bond releasing free inorganic sulfate from the substrate.

DISCUSSION

The extracellular matrix that surrounds and binds certain types of cells is composed of a number of components including fibrous structural proteins such as various collagens, adhesive proteins like laminin and fibronectin, and proteoglycans that form the gel into which the fibrous structural proteins are embedded. Proteoglycans are very large macromolecules consisting of a core protein to which many long polysaccharide chains called glycosaminoglycans are covalently bound. Due to the high negative charge of the glycosaminoglycans, the proteoglycans are very highly hydrated, a property that allows the proteoglycans to form a gel-like matrix that can expand and contract. The proteoglycans are also effective lubricants.

The glycosaminoglycans (GAGs, formerly called mucopolysaccharides) are long, linear polymers of repeating disaccharide units containing an acidic sugar (glucuronic acid or iduronic acid) and a hexosamine (glucosamine or galactosamine, both usually *N*-acetylated). The exception to this general structure is keratan sulfate, which has galactose in place of the acidic hexose. Table 29-1 lists the various types of GAGs and the structures of their repeating disaccharide units. The GAGs are often highly sulfated, which increases their negative charge and their ability to bind water molecules. All of the GAGs except hyaluronan are covalently linked to one of approximately 30 different core proteins to form proteoglycans. The core protein is synthesized on the rough endoplasmic reticulum and transferred to the Golgi where nucleoside diphosphate–activated acidic and amino sugars are alternately added to the nonreducing end of the growing polysaccharide by glycosyltransferases, resulting in the characteristic repeating disaccharide structure common to the GAGs. Hyaluronan, which is not sulfated nor covalently linked to a core protein,

Table 29-1

GLYCOSAMINOGLYCAN COMPOSITION

Glycosaminoglycan	Repeating Disaccharide Unit *
Hyaluronan	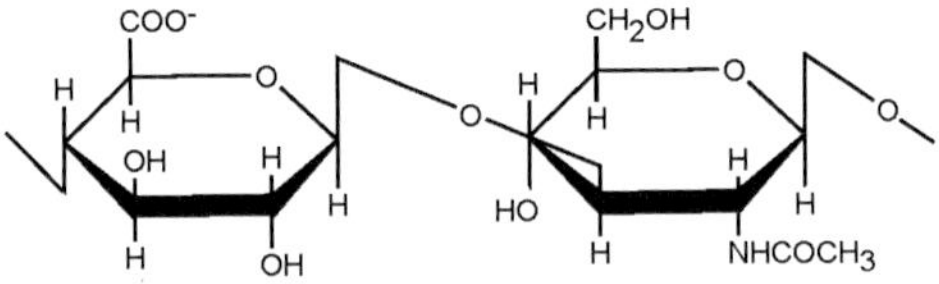 GlcA β(1→3) GlcNAc β(1→4)
Chondroitin sulfate	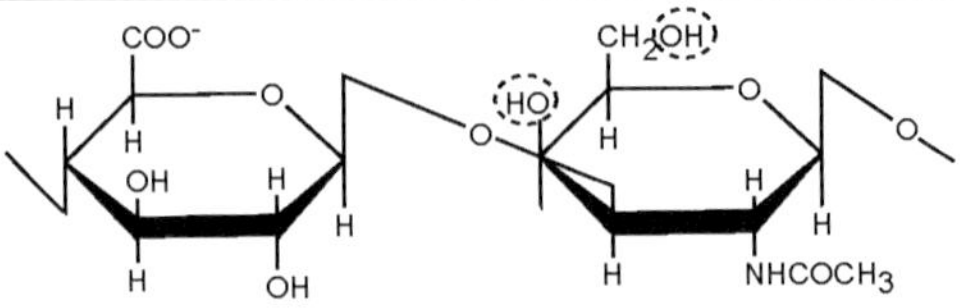 GlcA β(1→3) GalNAc β(1→4)
Dermatan sulfate	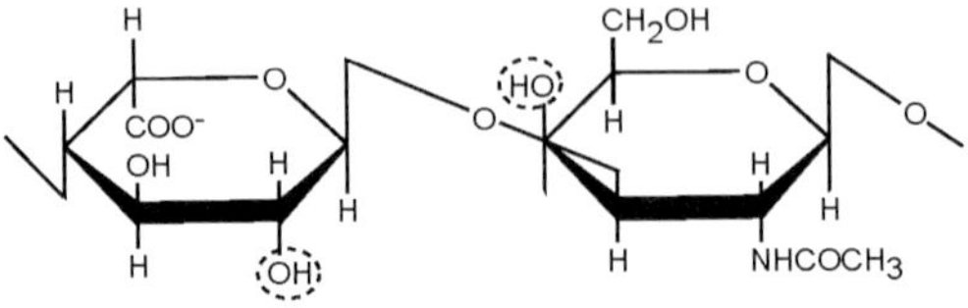 IdoA β(1→3) GalNAc β(1→4)
Heparan sulfate / Heparin **	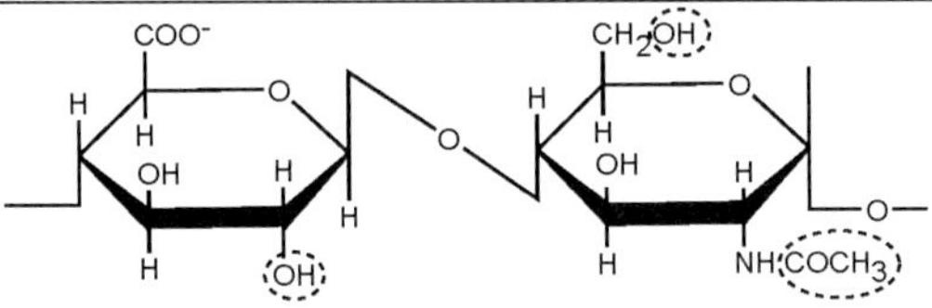 GlcA β(1→4) GlcNAc α(1→4) 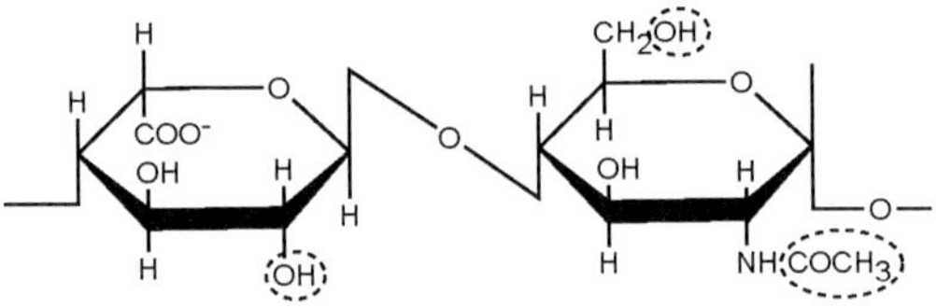 IdoA α(1→4) GlcNAc α(1→4)
Keratan sulfate I & II	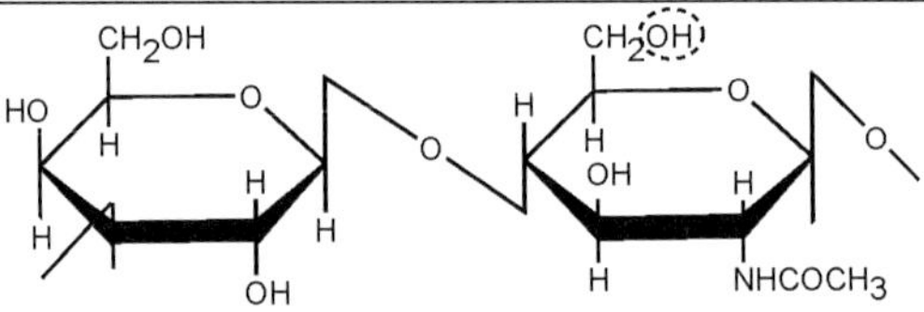 Gal β(1→4) GlcNAc β(1→3)

* Dashed ellipses indicate possible sites of sulfation

** Heparan sulfate contains primarily GlcA; heparin contains primarily IdoA

Abbreviations used: Gal = galactose; GalNAc = *N*-acetylgalactosamine; GlcA = glucuronic acid; GlcNAc = *N*-acetylglucosamine; IdoA = iduronic acid is synthesized at the plasma membrane by hyaluronan synthases. Hyaluronan synthases are integral membrane proteins that catalyze the alternate addition of UDP-glucuronate and UDP-*N*- acetylglucosamine to the reducing end of the growing hyaluronan polymer at the inner surface of the plasma membrane as it extrudes the nonreducing end of the GAG into the extracellular space. Hyaluronan can then assemble into large macromolecular complexes with other proteoglycans, which are noncovalently attached to hyaluronan by link proteins.

Degradation of proteoglycans during normal turnover of the extracellular matrix begins with proteolytic cleavage of the core protein by proteases in the extracellular matrix, which then enters the cell via endocytosis. The endosomes deliver their content to the lysosomes, where the proteolytic enzymes complete the degradation of the core proteins and an array of glycosidases and sulfatases hydrolyze the GAGs to monosaccharides. The lysosomes contain both endoglycosidases, which hydrolyze the long polymers into shorter oligosaccharides, and exoglycosidases that cleave individual acidic- or aminosugars from the GAG fragments.

Lysosomal catabolism of GAGs proceeds in a stepwise manner from the non-reducing end, as is shown with the degradation of heparan sulfate in Figure 29-1. If the terminal sugar is sulfated, the sulfate bond must be hydrolyzed by a specific sulfatase before the sugar can be removed. When the sulfate has been removed, a specific exoglycosidase then hydrolyzes the terminal sugar from the non-reducing end of the oligosaccharide leaving it one sugar shorter. Degradation continues in this stepwise fashion, alternating between removal of sulfates by sulfatases and cleavage of the terminal sugars by exoglycosidases. If removal of a sulfate leaves a terminal glucosamine residue, it must first be acetylated to *N*-acetylglucosamine because the lysosome lacks the enzyme required to remove glucosamine. This is accomplished by an acetyltransferase that uses acetyl-CoA as the acetyl group donor. When the glucosamine residue has been *N*-acetylated it can be hydrolyzed by α-*N*-acetylglucosaminidase, allowing the continuation of the stepwise degradation of the GAG.

Disease states known as mucopolysaccharidoses occur when there is a genetic deficiency of the enzymes involved in the lysosomal breakdown of the GAGs. A deficiency of any of these enzymes can lead to the accumulation of partially degraded GAGs in lysosomes and increased urinary excretion of GAG fragments. Histologic examination of affected cells shows large vacuoles, which are lysosomes engorged with partially degraded GAGs. Because GAGs are present throughout the body, deficiencies of enzymes that degrade them affect bone, connective tissues, and other organs.

The mucopolysaccharidoses are classified into seven clinical types and all are transmitted by autosomal recessive inheritance except for Hunter syndrome (MPS II, iduronate sulfatase deficiency), which is an X-linked disorder. Diagnosis of the specific disorder is made by measuring the specific enzyme activities in leukocytes or cultured skin fibroblasts. Because it takes some time

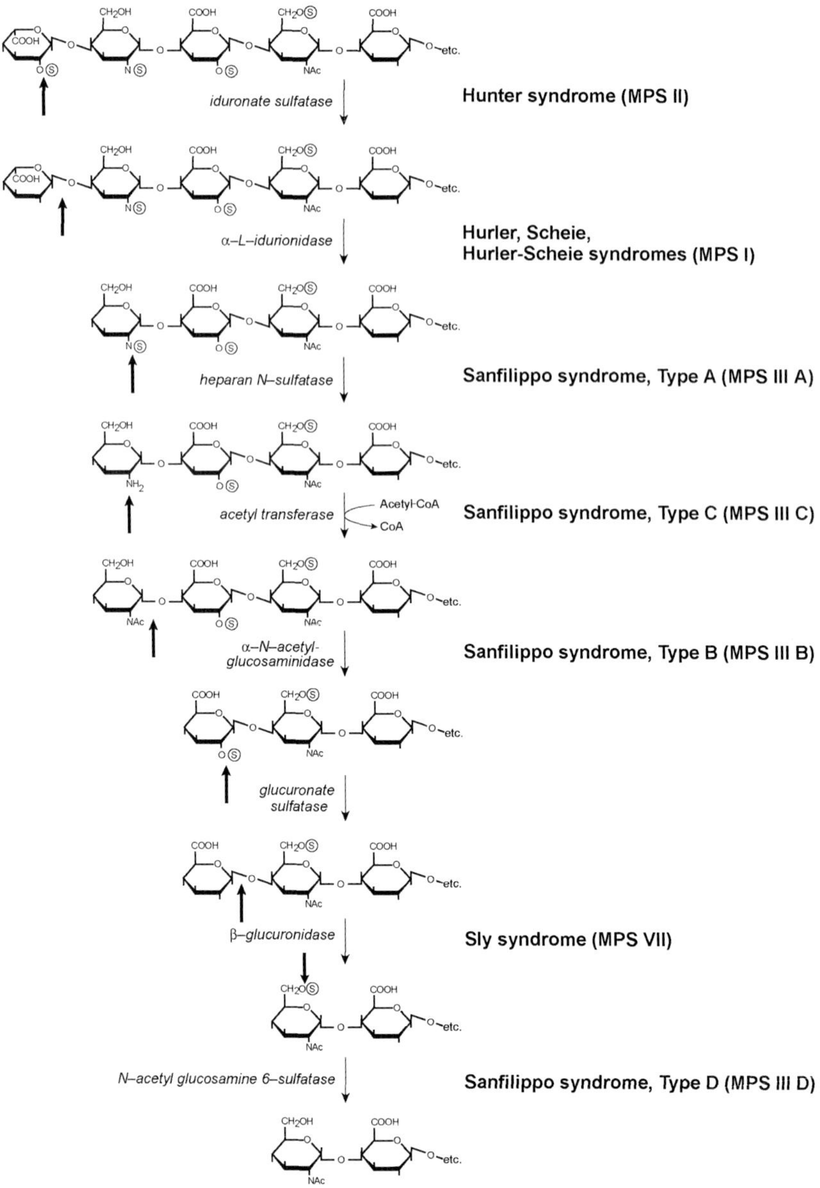

Figure 29-1. Lysosomal degradation of heparan sulfate. Dark arrows indicate the site of action of the enzyme, which is listed in italics. Mucopolysaccharidoses (MPS) due to a deficiency of the indicated enzymes are listed in bold to the right of the enzyme.

for the GAGs to accumulate, individuals with a mucopolysaccharidosis usually have a period of normal development before symptoms appear. Depending on the type and severity of the disorder, physical symptoms can include coarse facial features, dwarfism and deformities of the skeleton, cardiovascular impairments, hepatosplenomegaly, neurologic deficits, and mental retardation. As shown in Figure 29-1, Sanfilippo syndrome can be caused by a deficiency of one of four different enzymes that degrade heparan sulfate. Although the facial coarsening is mild in individuals with Sanfilippo syndrome, the accumulation of heparan sulfate in lysosomes leads to severe neurologic and mental impairment that result in death usually by the end of the second decade of life.

COMPREHENSION QUESTIONS

[29.1] A 5-year-old boy is seen by a pediatrician because his parents are concerned about his aggressive behavior, hyperactivity, and a loss of language skills. He also has recently become increasingly unsteady on his feet and has experienced a recent seizure. Slight facial feature coarsening is noted. In which of the following processes is this child most likely to have a disorder?

A. Mobilization of glycogen
B. Gluconeogenesis
C. Salvage of purine bases
D. Degradation of glycosaminoglycans
E. Cholesterol metabolism

[29.2] A 15-month-old white female is brought to the pediatrician because of recurrent upper-respiratory tract infection. During physical exam the girl is noted to have a short stature, some clouding of the corneas, coarse facial features, and an enlarged tongue. She also appears to have some hearing loss and other developmental delays. The pediatrician suspects the child has a mucopolysaccharidosis. Which of the following is she least likely to have?

A. Hurler syndrome (MPS I)
B. Hunter syndrome (MPS II)
C. Morquio syndrome (MPS IV)
D. Sly syndrome (MPS VII)
E. Sanfilippo syndrome (MPS III)

[29.3] A 3-year-old male with coarse facial features, progressive loss of motor skills, hepatosplenomegaly and chronic diarrhea is suspected of having Hunter syndrome (MPS II). Which of the following monosaccharide residues would be expected to be found at the nonreducing end of glycosaminoglycans in this patient's urine?

A. *N*-Acetylglucosamine
B. *N*-Acetylgalactosamine
C. Glucuronate
D. Iduronate
E. Iduronate 2-sulfate

Answers

[29.1] **D.** The patient is exhibiting the classic symptoms of Sanfilippo syndrome, which is a deficiency in one of four different lysosomal enzymes that breakdown glycosaminoglycans leading to the buildup of heparan sulfate and dermatan sulfate in lysosomes.

[29.2] **B.** All of the mucopolysaccharidoses are transmitted by autosomal recessive inheritance except Hunter syndrome (MPS II), a deficiency in iduronate sulfatase that is X-linked recessive. Since Hunter syndrome is X-linked, it is almost exclusively seen in males. Since our patient is female, she would not be expected to have an X-linked disorder.

[29.3] **E.** Since this patient is suspected of having Hunter syndrome, a deficiency in iduronate sulfatase, iduronate 2-sulfate would be expected to be present at the nonreducing end of glycosaminoglycans found in this patient's urine. A deficiency of iduronate sulfatase would prevent the sulfate ester bond of iduronate 2-sulfate residues from being hydrolyzed and further degradation of the glycosaminoglycan would be halted.

BIOCHEMISTRY PEARLS

❖ Often enzyme deficiencies are inherited as autosomal recessive disorders, so that both chromosomes are defective for the individual to be affected.

❖ Mucopolysaccharidoses occur when there is a genetic deficiency of the enzymes involved in the lysosomal breakdown of the glycosaminoglycans.

❖ In Sanfilippo syndrome, the accumulation of heparan sulfate in lysosomes leads to severe neurologic and mental impairment that result in death usually by the end of the second decade of life.

REFERENCES

Neufeld EF, Muenzer J. The Mucopolysaccharidoses. In: Scriver CR, Beaudet AL, Sly WS, et al., eds. The Metabolic and Molecular Bases of Inherited Disease, 8th ed. New York: McGraw-Hill, 2001.

Bittar T, Washington ER III. Mucopolysaccharidosis. http://www.emedicine.com/orthoped/topic203.htm.

Nash D, Varma S. Mucopolysaccharidosis type III. http://www.emedicine.com/ped/topic2040.htm

Fenton CL, Rogers W. Mucopolysaccharidosis type II. http://www.emedicine.com/ped/topic1029.htm.

CASE 30

A 48-year-old male presents to the clinic because of concerns about heart disease. He reports that his father died from a heart attack at age 46, and his older brother has also had a heart attack at age 46 but survived and is on medications for elevated cholesterol. The patient reports chest pain occasionally with ambulation around his house and is not able to climb stairs without significant chest pain and shortness of breath. The physical exam is normal, and the physician orders an electrocardiogram (ECG), exercise stress test, and blood work. The patient's cholesterol result comes back as 350 mg/dL (normal 200). The physician prescribes medication, which he states is directed at the rate-limiting step of cholesterol biosynthesis.

- **What is the rate-limiting step of cholesterol metabolism?**
- **What is the class of medication prescribed?**

ANSWERS TO CASE 30: HYPERCHOLESTEROLINEMIA

Summary: A 48-year-old male with strong family history of heart disease and now angina and exertional dyspnea presents with a significantly elevated cholesterol level. A medication is prescribed that is directed at the rate-limiting step of cholesterol biosynthesis.

- **Rate-limiting step:** The enzyme hydroxymethylglutaryl-CoA reductase (HMG-CoA reductase) catalyzes an early rate-limiting step in cholesterol biosynthesis.
- **Likely medication:** HMG-CoA reductase inhibitor, otherwise known as "statin" medications.

CLINICAL CORRELATION

Hyperlipidemia is one of the most treatable risk factors of atherosclerotic vascular disease. In particular, the level of the low-density lipoprotein (LDL) correlates with the pathogenesis of atherosclerosis. Exercise, dietary adjustments, and weight loss are the initial therapy of hyperlipidemia. If these are not sufficient, then pharmacologic therapy is required. The exact LDL targets depend on the patient's risk of cardiovascular disease. For example, if an individual has had a cardiovascular event previously (heart attack or stroke), the LDL target is 100 mg/dL; 1 to 2 risk factors without prior events = 130 mg/dL; and no risk factors = 160 mg/dL.

APPROACH TO LIPID METABOLISM

Objectives

1. Know about cholesterol metabolism.
2. Understand the role of serum lipoproteins.
3. Be aware of the types of hereditary hyperlipidemias.
4. Know why the LDL level is increased with familial hypercholesterolemia.

Definitions

Apolipoprotein: The protein component of a lipoprotein; in addition to being a structural component of a lipoprotein, apolipoproteins also serve as activators of enzymes and ligands for receptors.

Chylomicron: A lipoprotein synthesized by the intestine to transport dietary lipids to peripheral tissues and the liver.

HDL: High-density lipoprotein; synthesized in the liver, the HDL serves as a source of apolipoproteins for other lipoproteins, as the site of action for the conversion of cholesterol to cholesterol ester in the plasma by the enzyme lecithin-cholesterol acyltransferase (LCAT), and delivers cholesterol esters derived from peripheral membranes to the liver. It is commonly called the "good cholesterol."

LDL: Low-density lipoprotein; a product of the degradation of very-low-density lipoproteins (VLDLs) by the action of lipoprotein lipase. LDLs are taken up by a receptor-mediated endocytosis by both peripheral tissues and the liver. It is commonly called the "bad cholesterol."

Lipoprotein: A macromolecular particle composed of varying quantities of protein, triacylglycerol, phospholipids, cholesterol, and cholesterol esters. The lipoprotein structure has a phospholipid and free cholesterol skin surrounding a core composed of triacylglycerol and cholesterol esters, with the proteins imbedded on the surface. They serve as carriers of lipid in the circulation.

VLDL: Very-low-density lipoprotein; synthesized by the liver to transport triacylglycerol from the liver to peripheral tissues.

DISCUSSION

Although cholesterol can be synthesized in almost all cells, the **liver, intestine, and the steroidogenic tissues such as the adrenal glands and reproductive tissues** are the **primary sites.** In the **liver,** cholesterol is synthesized in the **endoplasmic reticulum (ER)** along with phospholipids, triacylglycerides, and apoproteins. Prelipoprotein particles are assembled in the ER and then transferred to the Golgi. Further processing and addition of **cholesterol esters occur in the Golgi,** culminating in the formation of secretory granules containing **lipoprotein particles.** These vesicles then fuse with the plasma membrane and export cholesterol from the cell via exocytosis in the form of VLDLs (see below), which then enter the circulation. As the lipoprotein complexes are transported through the bloodstream they are converted from VLDL to IDL to LDL by the removal of triacylglycerides by lipoprotein lipase, which is located on the surface of capillary epithelial cells.

Cholesterol and triacylglycerides (TAGs) are transported as complexes in the form of lipoprotein particles. These **lipoprotein particles** contain a core of TAGs and cholesteryl esters surrounded by a monolayer of phospholipids, cholesterol, and specific proteins called apoproteins. The **apoproteins,** specific to each type of lipoprotein, enable the hydrophobic lipids to be transported in the aqueous environment of the bloodstream. They also contain signals that target the lipoprotein particles to the cells or activate enzymes. The lipoprotein particles vary in density depending on the lipid/protein ratio and are named based on these densities. The higher the lipid/protein ratio, the lower the density of the particle.

Table 30-1
MAJOR APOPROTEINS INVOLVED IN LIPID TRANSPORT BY LIPOPROTEINS

APOPROTEIN	LOCATION	FUNCTION
A	HDL	Activates LCAT (lecithin-cholesterol acyltransferase)
B-48	Chylomicrons, chylomicron remnants	Structural, binds to B-48:E receptor in liver
B-100	VLDL, IDL, LDL	Structural, binds to LDL receptor in liver and extrahepatic tissues
C-II	Chylomicrons, VLDL, (IDL), HDL	Binds to and activates lipoprotein lipase
E	Chylomicrons, chylomicron remnants, VLDL, (IDL), HDL	Structural, binds to B-48:E receptor in liver

The major classes of lipoproteins (Table 30-1) and some of their properties are

Chylomicrons
- Lowest density
- Transport dietary lipids from the intestine to target tissues

Very-low-density lipoproteins (VLDLs)
- Produced by liver cells to transport endogenously synthesized lipids to target cells
- Production controlled by lipid availability
- Contain five different apoproteins

Intermediate density lipoproteins (IDLs)

Low-density lipoproteins (LDLs)
- Primarily cholesteryl esters
- Apoprotein B-100

High-density lipoproteins (HDLs)
- Secreted by liver and intestine
- Contains apoprotein A-1
- "Good cholesterol," high levels associated with low incidence of atherosclerosis

The LDLs (containing cholesteryl esters) are taken up by cells by a process known as **receptor-mediated endocytosis.** The LDL receptor mediates this endocytosis and is important to cholesterol metabolism. LDLs bind to these receptors, which recognize apoprotein B-100. After LDL binding to the LDL receptor, the ligand-receptor complexes cluster on the plasma membrane in coated pits, which then invaginate forming coated vesicles. These coated vesicles are internalized and clathrin, the protein composing the lattice in membrane coated pits, is removed. These vesicles are now called endosomes and these endosomes fuse with the lysosome. The LDL receptor–containing membrane buds off and is recycled to the plasma membrane. Fusion of the lysosome and endosome releases lysosomal proteases that degrade the apoproteins into amino acids. Lysosomal enzymes also hydrolyze the cholesteryl esters to free cholesterol and fatty acids. The free cholesterol is released into the cell's cytoplasm, and this free cholesterol is then available to be used by the cell. Excess cholesterol is reesterified by acyl-CoA:cholesterol acyltransferase (ACAT), which uses fatty acyl-CoA as the source of activated fatty acid. Free cholesterol affects cholesterol metabolism by inhibiting cholesterol biosynthesis. Cholesterol inhibits the enzyme β-hydroxy-β-methylglutaryl-CoA reductase (HMG-CoA reductase), which catalyzes an early rate-limiting step in cholesterol biosynthesis. HMG-CoA reductase is the target of the statin drugs in wide use for treating patients with elevated cholesterol levels. In addition, free cholesterol inhibits the synthesis of the LDL receptor, thus limiting the amount of LDLs that are taken up by the cell.

Hyperlipidemia is defined as elevated lipoprotein levels in the plasma, which may be primary or secondary. Several different types of hereditary hyperlipidemias have been defined.

- Type I—A relatively rare inherited deficiency of either lipoprotein lipase activity or the lipoprotein lipase-activating protein apo C-II. This results in the inability to effectively remove chylomicrons and VLDL triglycerides from the blood.
- Type II—Includes familial hypercholesterolemia described in detail below.
- Type III—Associated with abnormalities of apolipoprotein E (apo E) and defective conversion and removal of VLDL from the plasma.
- Type IV—A common disorder characterized by variable elevations of plasma triglycerides contained predominantly in VLDL. This leads to a possible predisposition to atherosclerosis and often has a familial distribution.
- Type V—An uncommon disorder, sometimes familial, associated with defective clearance of exogenous and endogenous triglycerides and the risk of life-threatening pancreatitis.

The disease ***familial hypercholesterolemia*** results from a mutation in the LDL-receptor gene found on **chromosome 19.** The overall phenotype of the inability to internalize the LDL receptor can be caused by three different types of defects. In the first type, the LDL receptor is not produced. The second type is a result of a mutation in the terminal region of the receptor that results in an LDL receptor unable to bind LDL. The third type is caused by a mutation in the C-terminal region that prevents the LDL-receptor complex from undergoing endocytosis. In the absence of a functioning LDL receptor, LDL cholesterol levels are greatly elevated in individuals with this disease. This elevation results in premature atherosclerosis of the coronary arteries. Genetic defects in apoprotein B-100, the protein in the LDL recognized by the LDL receptor, also exist and lead to elevated LDL because the LDL complex is not recognized by the LDL receptor. A diet low in fat and cholesterol, an exercise regimen, and anticholesterol medications are used in the treatment of this disease.

COMPREHENSION QUESTIONS

[30.1] A patient presents in your office with very high levels of serum cholesterol. After a series of tests, you conclude that the patient has high circulating levels of LDL cholesterol, but has normal levels of the liver LDL receptor. One possible explanation for this observation is which of the following?

A. The patient has a mutated form of apoprotein B-100.
B. The inability to selectively remove cholesterol from the LDL complex.
C. The absence of the enzyme lipoprotein lipase.
D. Decreased levels of acyl-CoA: cholesterol acyltransferase.
E. Altered phosphorylation of the LDL receptor.

[30.2] A patient with hereditary type I hyperlipidemia presents with elevated levels of chylomicrons and VLDL triglycerides in the blood. The main function of the chylomicrons in circulation is to do which of the following?

A. Transport lipids from the liver
B. Transport dietary lipids from the intestine to target tissues
C. Transport cholesterol from IDL to LDL
D. Act as a receptor for triacylglycerols in the liver
E. Bind cholesterol esters exclusively

[30.3] Free cholesterol can affect cholesterol metabolism in the body by inhibiting cholesterol biosynthesis. The step at which free cholesterol inhibits its biosynthesis is by inhibiting which of the following processes?

A. Cyclizing of squalene to form lanosterol
B. Reduction of 7-dehydrocholesterol to form cholesterol
C. Formation of mevalonate from hydroxymethylglutaryl-CoA
D. Kinase that phosphorylates hydroxymethylglutaryl-CoA reductase
E. Condensation of acetyl-CoA and acetoacetyl-CoA to form hydroxymethylglutaryl-CoA

[30.4] A patient presents in your office with very high levels of serum cholesterol. He states that he has tried to follow the diet and exercise regimen you gave him last year. You decide that this patient would benefit from a drug such as Lipitor (atorvastatin). This class of drugs is effective in treating hypercholesterolemia because it has what effect?

A. Stimulates phosphorylation of the β-hydroxy-β-methylglutaryl-CoA reductase enzyme
B. Decreases the stability of the β-hydroxy-β-methylglutaryl-CoA reductase protein
C. Binds cholesterol preventing it from being absorbed by the intestine
D. Directly prevents the deposition of cholesterol on artery walls
E. Inhibits the enzyme β-hydroxy-β-methylglutaryl-CoA reductase

Answers

[30.1] **A.** A common genetic mutation leading to high circulating LDL cholesterol levels is caused by mutations in the LDL receptor. The lack of a functional receptor prevents the removal of LDL cholesterol from circulation. Apoprotein B-100 is found in the LDL-cholesterol complex and is the protein recognized by the LDL receptor. In this patient, the LDL receptor is normal so it is reasonable to conclude that the reason that the LDL cholesterol remains in circulation is because it is not recognized by the normal LDL receptor. A mutation in the apoprotein B-100 such that it is not recognized by the receptor would lead to elevated LDL-cholesterol levels.

[30.2] **B.** The liver and intestine are the main sources of circulating lipids. Chylomicrons carry triacylglycerides and cholesterol esters from the intestine to other target tissues. VLDLs carry lipids from the liver into circulation. Lipoproteins are a mix of lipids and specific proteins and these complexes are classified based on their lipid/protein ratio. Lipoprotein lipases degrade the triacylglycerides in the chylomicrons and VLDLs with a concurrent release of apoproteins. This is a gradual process which converts the VLDLs into IDLs and then LDLs.

[30.3] **C.** The major regulatory enzyme of cholesterol metabolism, β-hydroxy-β-methylglutaryl-CoA reductase, is regulated by three distinct mechanisms. The first is phosphorylation by a cAMP dependent protein kinase. Phosphorylation of β-hydroxy-β-methylglutaryl-CoA reductase inactivates the enzyme. The other two mechanisms involve the levels of cholesterol. The degradation of the enzyme is controlled by cholesterol levels. The half-life of β-hydroxy-β-methylglutaryl-CoA reductase is regulated by cholesterol levels with high concentrations of cholesterol leading to a shorter half-life. The final regulatory mechanism involves control of the expression of the β-hydroxy-β-methylglutaryl-CoA reductase gene. High levels of cholesterol lead to a decrease in the mRNA levels coding for β-hydroxy-β-methylglutaryl-CoA reductase.

[30.4] **E.** The statin class of drugs—Lipitor (atorvastatin), Mevacor (lovastatin), and Zocor (simvastatin)—is used to treat hypercholesterolemia. This class of drugs lowers cholesterol levels by inhibiting the biosynthesis of cholesterol. Specifically, these drugs inhibit the enzyme β-hydroxy-β-methylglutaryl-CoA (HMG-CoA) reductase, which catalyzes the reaction that converts HMG-CoA to mevalonate. This is the rate-limiting step of cholesterol biosynthesis.

In addition to the statin drugs which inhibit HMG-CoA reductase a number of other drugs are used to lower cholesterol levels. The first are resins which are also referred to as bile acid sequestrants such as cholestyramine. The resins work by binding to the bile acids followed by excretion of the resin-bile complex. To make up for the loss of the bile acids the body converts cholesterol into bile acids thus reducing the cholesterol levels.

Another type of drug used is the fibrates such as gemfibrozil. These compounds work by lowering the levels of triglycerides and increasing the levels of the "good" high-density lipoproteins (HDLs). Niacin is also effective in lowering cholesterol levels when used in large doses (more than that required for niacin as a vitamin). Niacin acts to lower levels of triglycerides and low-density lipoproteins (LDLs) and increasing the levels of the "good" high-density lipoproteins (HDLs).

Drugs such as ezetimibe which inhibit the absorption of cholesterol in the intestine are effective in lowering cholesterol levels. This drug is often given in combination with a statin and this combination therapy is very effective in lowering cholesterol levels.

BIOCHEMISTRY PEARLS

- Cholesterol can be synthesized in almost all cells, but the liver, intestine, and the steroidogenic tissues such as the adrenal glands and reproductive tissues are the primary sites.
- The major classes of lipoproteins are chylomicrons (lowest density), very-low-density lipoproteins (VLDLs), intermediate-density lipoproteins (IDLs), low-density lipoproteins (LDLs), and high-density lipoproteins (HDLs) (considered the "good cholesterol").
- Familial hyperlipidemia is subdivided into five types, with type V being most common, characterized by variable elevations of plasma triglycerides contained predominantly in VLDL.
- The rate-limiting step in cholesterol biosynthesis is the enzyme hydroxymethylglutaryl-CoA reductase (HMG-CoA reductase), which is the target of the statin drugs.

REFERENCES

Berg JM, Tymoczko JL, Stryer L. Biochemistry, 5th ed. New York: Freeman, 2002:722–31.

Devlin TM, ed. Textbook of Biochemistry with Clinical Correlations, 5th ed. New York: Wiley-Liss, 2002:742–52.

CASE 31

A 45-year-old female presents to the clinic with concerns over occasional midepigastric discomfort and nausea/vomiting after eating "greasy meals." The symptoms gradually disappear, and she has no further discomfort. She denies any hematemesis, and pain is worse after eating. She further mentions that she had elevated cholesterol levels in the past and was on an exercise program, but not anymore. On exam, she is afebrile with normal vital signs. Her physical exam is completely normal with no evidence of abdominal pain. An abdominal ultrasound is performed and revealed a few gallstones in the gallbladder but no thickening of the gallbladder wall.

- **What factors would you need to consider to assess the need for cholecystectomy?**
- **What are gallstones made of?**
- **Can gallstones be seen on abdominal x-ray?**

ANSWERS TO CASE 31: GALLSTONES

Summary: A 45-year-old female presents with hypercholesterolinemia, ultrasound evidence of gallstones, and recurrent symptoms of gallbladder disease.

- **Surgical candidates:** Frequent and severe attacks, previous complications from gallstones, presence of underlying condition predisposing the patient to increased risk of gallbladder disease.
- **Components of gallstones:** Cholesterol, calcium bilirubinate, and bile salts.
- **Abdominal x-rays and diagnosis:** Mixed stones much easier to see on plain film secondary to calcifications, comprising approximately 10 percent of gallstones.

CLINICAL CORRELATION

This individual fits the "classic" patient with gallbladder disease: female, middle-aged, overweight. The gallbladder acts to store bile salts produced by the liver. The gallbladder is stimulated to contract when food enters the small intestine; the bile salts then travel through the bile duct to the ampulla of Vater into the duodenum. The bile salts act to emulsify fats, helping with the digestion of fat. Gallstones form when the solutes in the gallbladder precipitate. The two main types of stones are cholesterol stones and pigmented stones. Cholesterol stones are usually yellow-green in appearance and account for approximately 80 percent of gallstones. Pigmented stones are usually made of bilirubin and appear dark in color. Patients may have pain from the gallstones, usually after a fatty meal. The pain is typically epigastric or right upper quadrant and perhaps radiating to the right shoulder. If the gallbladder becomes inflamed or infected, cholecystitis can result. The stones can also travel through the bile duct and obstruct biliary flow leading to jaundice (yellow color or the skin), or irritate the pancreas and cause pancreatitis.

APPROACH TO BILE SALT METABOLISM

Objectives

1. Know about bile salt metabolism.
2. Be able to identify where bile salts are synthesized.
3. Know where bile salts emulsify dietary fats.

Definitions

Bile salts: Cholesterol derivatives with detergent-like properties used to solubilize cholesterol, assist in intestinal absorption of fat-soluble vitamins, and emulsify dietary lipids passing through the intestine to enable fat digestion and absorption by exposing fats to pancreatic lipases.

Bile acids: Neutral, protonated form of bile salts.

Primary bile acids: Synthesized from cholesterol as cholic acid and chenodeoxycholic acid. They are secreted as taurine and glycine conjugates.

Secondary bile acids: Products of deconjugated and reduced primary acids. Bacteria in the intestine remove the 7α-hydroxyl group, leaving the secondary bile acids.

Cholesterol 7α-hydroxylase (*CYP7A1*): The mixed-function oxidase cytochrome P450 enzyme catalyzing the initial, rate-limiting step for conversion of cholesterol to bile acids.

β-hydroxy-β-methylglutaryl-coenzyme A (HMG-CoA) reductase: The rate-limiting enzyme in cholesterol biosynthesis.

DISCUSSION

Bile salts are derivatives of cholesterol that make up the major component of bile. They are very efficient detergents when conjugated to amino acids because of the presence of both polar and nonpolar regions. This property helps to emulsify dietary lipids in the intestine, which aids in fat digestion and absorption by making fat vulnerable to pancreatic lipases. Other functions of **bile salt** (ionized, deprotonated form) and **bile acid** (neutral, protonated form) are to solubilize cholesterol thus preventing precipitation of cholesterol crystals and facilitating cholesterol excretion. The only significant removal of excess cholesterol from the body is achieved through the excretion of bile salts. Finally, they assist in intestinal absorption of fat-soluble vitamins.

The conversion of cholesterol to bile acids is a multienzyme process. **The initial and rate-limiting step of bile acid synthesis is oxidation of cholesterol to 7α-hydroxycholesterol by a mixed function oxidase from the cytochrome P450 superfamily, cholesterol 7α-hydroxylase (*CYP7A1;*** Figure 31-1). The remaining steps include reduction of the Δ^5-double bond, side chain shortening, and oxidation. The products of this pathway, cholic acid and chenodeoxycholic acid, are termed the primary bile acids because they have been synthesized de novo from cholesterol. To increase the pH range over which the bile salts remain ionized and serve as good detergents, they may be conjugated via amide bonds with either of the amino acids glycine or taurine (Figure 31-2).

Bile salts/acids are synthesized in the liver, stored and concentrated in the gallbladder, and **secreted into the intestine** where they may undergo **deconjugation** and reduction by intestinal bacteria to produce secondary bile acids (see Figure 31-1). Once formed in the liver, the bile salts/acids are

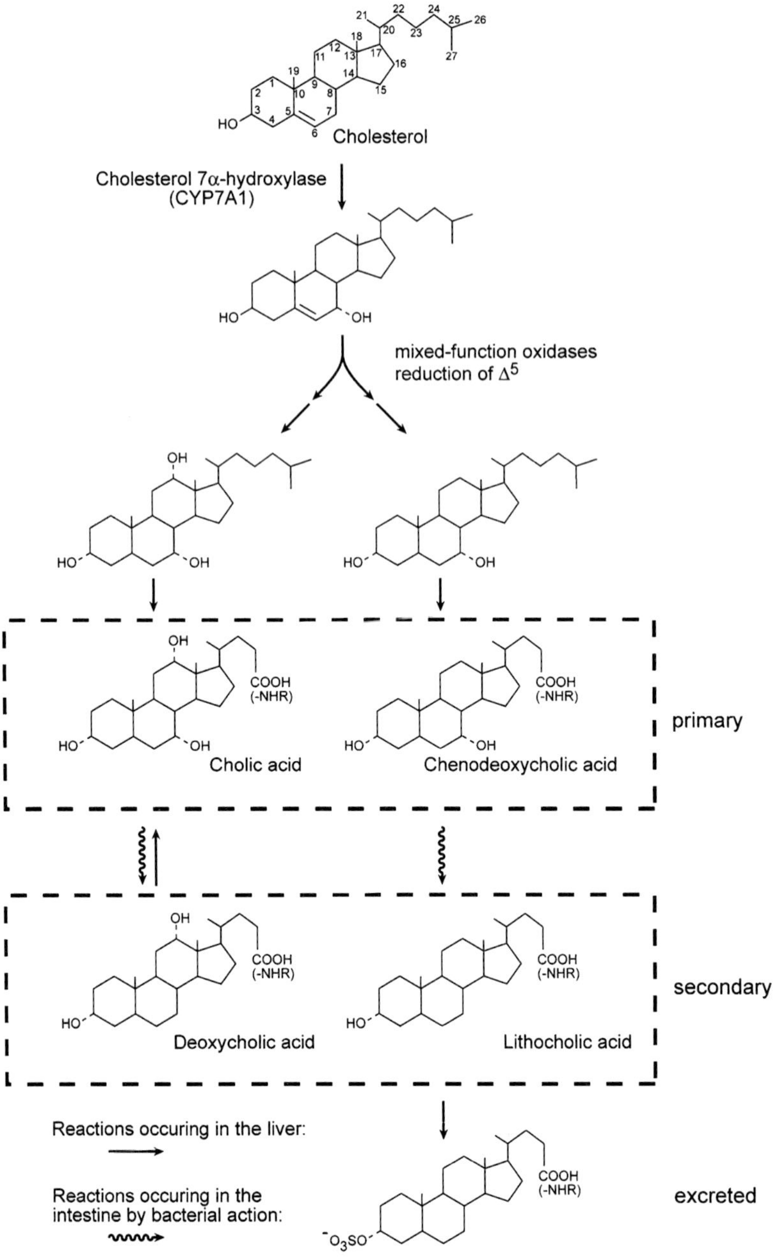

Figure 31-1. Simplified biosynthetic pathway of bile salts.

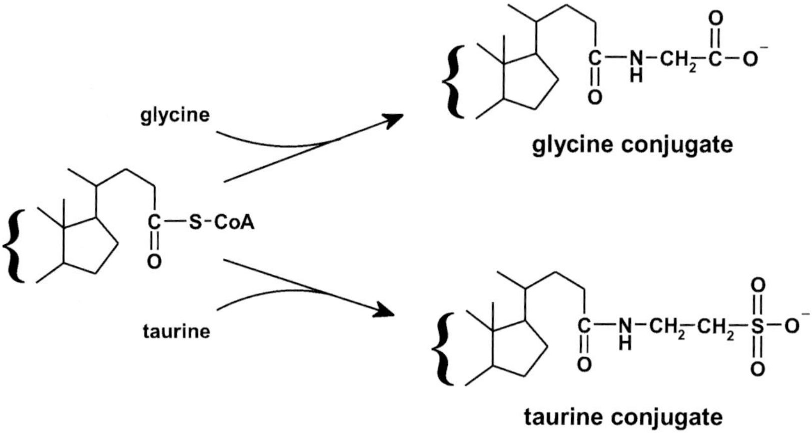

Figure 31-2. Conjugation of bile salts.

secreted through the bile ducts, which are made up of the bile canaliculi and bile ductules, and passed on to the gallbladder for storage as bile. Ultimately, they are passed on to the intestine. Most of the bile acids/salts are deconjugated in the intestinal ileum and reduced to secondary bile acids by bacteria, which remove the 7α-hydroxyl group by a dehydroxylation reaction. Although some are lost by excretion, approximately 90 percent of the bile acids and salts are reabsorbed in the terminal ileum and returned to the liver. The liver cannot provide sufficient amounts of newly synthesized bile acids for the body's daily needs; therefore, the body relies on enterohepatic circulation (Figure 31-3) to sustain necessary bile acid levels. The portal vein transports bile acids from the intestine back to the liver as complexes with serum albumin.

The synthesis of bile salts is under very tight regulatory control to maintain cholesterol homeostasis and supply sufficient amounts of detergent to the intestine. This is controlled in part by a feedback mechanism on cholesterol 7α-hydroxylase, the rate-limiting enzyme in the synthetic pathway. Increased concentrations of bile acids inhibit cholesterol 7α-hydroxylase, while low levels relieve the inhibition. Elevated cholesterol levels, however, activate the enzyme, thus increasing bile acid biosynthesis. Levels of both cholesterol and bile acids affect the concentration of cholesterol 7α-hydroxylase and this regulation appears to be controlled at the transcriptional level via nuclear receptors. Therefore, binding of bile acids or cholesterol to a given nuclear receptor in turn regulates the expression of the *CYP7A1* gene, activating expression in the case of binding to cholesterol and repressing it when bile acids are bound. Thus, the proper maintenance of bile acid levels can prevent accumulation of cholesterol.

Eighty percent of gallstones in the Western world are a result of cholesterol precipitation from the bile, a condition known as cholelithiasis. The pathogenetic mechanism of gallstone formation usually involves a culmination of

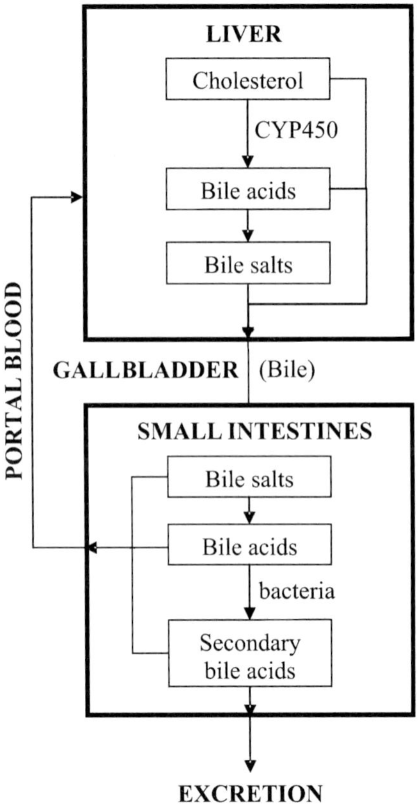

Figure 31-3. The enterohepatic circulation of bile acids.

deleterious events involving the metabolic pathways of cholesterol or the bile acids/salts. First, the cholesterol concentration in bile becomes supersaturated. Bile is a controlled mixture of cholesterol, bile acids, and phospholipids (with small amounts of bile pigments), and if cholesterol levels are elevated or bile acids/salts lowered, the ratio of the three major components changes leaving cholesterol less protected against the aqueous environment and more likely to precipitate. Elevated cholesterol levels can occur by having excess HMG-CoA reductase activity, the rate-limiting enzyme in cholesterol biosynthesis; this condition is typically seen in the obese. Alternatively, reduced levels of acyl-CoA: cholesterol acyltransferase [ACAT], the enzyme that esterifies cholesterol within cells, or reduced levels of cholesterol 7α-hydroxylase can cause elevation of cholesterol. Deoxycholate, a secondary bile acid synthesized by intestinal bacteria, inhibits *CYP7A1*. Therefore, high levels of deoxycholate resulting from prolonged exposure of bile acids to intestinal bacteria may result in high levels of cholesterol in bile. In addition to elevated cholesterol levels in bile, there must be adequate time for cholesterol crystal nucleation, which will ultimately form macroliths. Fasting, like overnight sleeping, allows for long-term storage of bile in the gallbladder and could give ample time for crystal nucleation.

COMPREHENSION QUESTIONS

[31.1] The modification to bile salts that increases the working pH range and amphipathic nature of bile salts is

A. 7α-Hydroxylation
B. Dehydroxylation by intestinal bacteria
C. Esterification
D. Conjugation to taurine or glycine

[31.2] A new drug called CT2033, a corticosteroid, has reached the clinical trial stage. This drug was designed to treat inflammation but seems to also cause an undesired side effect where there is disturbance of cholesterol and bile acid homeostasis. Which of the following is *least likely* to explain the side effects caused by CT2033?

A. It downregulates the expression of a hepatobiliary bile acid transporter gene.
B. It inhibits a hepatobiliary transporter protein decreasing bile acid secretion.
C. It competes with cholesterol for *CYP7A1* binding.
D. It binds and inhibits pancreatic lipases.

[31.3] A 53-year-old male patient with elevated levels of low-density lipoprotein (LDL) cholesterol, signs of premature cholesterol gallstone disease and substantially elevated triglycerides visited his physician for a follow-up to check his current status. The patient had received various statin, HMG-CoA-reductase inhibitors therapies for the past 2 years. However, after blood work done at this follow-up visit, complications had still not subsided. This patient has similar problems as two of his siblings. Which of the following best explains this patients dyslipidemia?

A. An influx of abnormal phospholipids in the gallbladder as a result of ileal disease
B. A loss of HMG-CoA reductase function
C. A loss of *CYP7A1* (cholesterol 7α-hydroxylase) function
D. Elevated levels of ACAT

Answers

[31.1] **D.** Conjugation of bile acids to these two amino acids through amide linkages is important for maintaining the detergent properties of bile salts over the wide pH range of the intestinal tract. Conjugation decreases the pKa of the bile salts assuring ionization and solubility in the intestines.

[31.2] **D.** Any of the situations in answers A to C could directly alter cholesterol and bile acid homeostasis. Many of the bile acid transporters are regulated by nuclear receptors. Therefore, a nuclear receptor ligand such as CT2033 could change the expression levels of the transporters, which could then result in problems with secretion of bile acids and probable accumulation of cholesterol. If the new drug were to directly bind and inhibit the transporter, the same result would occur. Since corticosteroids are derivatives of cholesterol, it could be reasonable for CT2033 to fit in the same *CYP7A1* binding pocket and compete with cholesterol. This could cause accumulation of cholesterol because *CYP7A1* is needed to convert cholesterol into bile salts. The drug could act to effectively lower active *CYP7A1* levels. However, answer D is least likely to affect cholesterol and bile acids since pancreatic lipases are involved in the breakdown of fat. Bile acids simply emulsify fats and allow pancreatic lipases to degrade the fat. Therefore, inhibiting pancreatic lipases should not have much of an effect on cholesterol and bile acid homeostasis.

[31.3] **C.** A loss in function of *CYP7A1* prevents the catabolism of cholesterol to bile salts. Elevated levels of LDL cholesterol, signs of premature cholesterol gallstone disease, and substantially elevated triglycerides are all complications that can result from blocking the enzyme that breaks down cholesterol. Therefore, high levels of cholesterol accumulate in the bile and, with decreased production of bile salts to help dissolution of cholesterol, the formation of cholesterol gallstones. Statin therapy is not as effective because it inhibits the enzyme that controls the rate of cholesterol synthesis but does nothing with respect to the degradation of cholesterol. The increase in blood triglyceride levels when *CYP7A1* is deficient is not well understood, but triglyceride levels appear to have a reciprocal relationship to bile acid synthesis. Finally, this appears to be a genetic disorder since other siblings showed the same phenotype, which would point to a possible mutated gene and likely prevent function of *CYP7A1*.

BIOCHEMISTRY PEARLS

- Bile salts are the major component of bile and are very efficient "detergents" when conjugated to amino acids because of the presence of both polar and nonpolar regions, aiding in fat digestion.
- The initial and rate-limiting step of bile acid synthesis is oxidation of cholesterol to 7α-hydroxycholesterol by a mixed function oxidase from the cytochrome P450 superfamily, cholesterol 7α-hydroxylase (*CYP7A1*).
- The synthesis of bile salts is under very tight regulatory control, mainly by a feedback mechanism on cholesterol 7α-hydroxylase, the rate-limiting enzyme in the synthetic pathway.

REFERENCES

Chiang JYL. Regulation of bile acid synthesis. Front Biosci 1998;3:D176–93.

Pullinger CR, Eng C, Salen G, et al. Human cholesterol 7α-hydroxylase (CYP7A1) deficiency has a hypercholesterolemic phenotype. J Clin Invest 2002; 110(1):109–17.

Russell DW. The enzymes, regulation, and genetics of bile acid synthesis. Annu Rev Biochem 2003;72:137–74.

Trauner M, Boyer JL. Bile salt transporters: molecular characterization, function, and regulation. Physiol Rev 2003;83(2):633–71.

CASE 32

A 63-year-old female presents to the clinic with recurrent midepigastric pain over the last 3 months. She reports some relief shortly after eating, but then the discomfort returns. She has tried various over-the-counter medications without relief. She also reports feeling tired and has had to increase the amount of ibuprofen needed for relief of her arthritis. She denies nausea, vomiting, and diarrhea. On exam she is found to have mild midepigastric tenderness and guaiac positive stool. A CBC revealed a microcytic anemia and normal white blood cell count, consistent with iron deficiency. The patient was referred to a gastroenterologist who performed an upper GI endoscopy that identified gastric ulcers. He stated that he suspected that the ibuprofen, a nonsteroidal anti-inflammatory drug (NSAID) was the causative agent and suggested switching from ibuprofen to a coxib, such as celecoxib.

- **What is the likely biochemical etiology of the disorder?**

- **Why do coxibs generally have a lower incidence of upper GI problems than other NSAIDs?**

- **What is the major difference between aspirin and other NSAIDs with regard to platelet function?**

ANSWERS TO CASE 32: NSAID-ASSOCIATED GASTRITIS

Summary: A 63-year-old female with arthritis taking an NSAID with recent onset of epigastric pain relieved with food, guaiac positive stools, and iron deficiency anemia. Endoscopic examination reveals gastric ulcers.

- **Biochemical etiology:** Primarily, NSAID inhibition of a gastric enzyme (COX-1) required for synthesis of prostaglandins that have a protective effect on the gastric mucosa. A contributory factor is direct mucosal damage due to the acidic chemistry of NSAIDs.
- **Decreased gastric side effects with coxibs:** Traditional NSAIDs, such as ibuprofen and aspirin, inhibit both COX-1 and COX-2. The coxibs are selective inhibitors of COX-2, allowing continued production of protective prostaglandins by gastric COX-1.
- **Difference between aspirin and other NSAIDs:** Aspirin covalently modifies platelet COX-1, thus irreversibly blocking thromboxane formation and reducing platelet function for the lifespan of the affected platelet (platelets cannot synthesize new proteins). The inhibitory action of other NSAIDs on platelet COX-1 is not covalent and is eventually reversed when the agents' blood levels decline.

CLINICAL CORRELATION

Nonsteroidal antiinflammatory drugs (NSAIDs), also known as prostaglandin synthesis inhibitors or cyclooxygenase (COX) inhibitors, can induce upper GI irritation or ulcers. The NSAIDs include a wide variety of medications including aspirin, ibuprofen, naproxen, and indomethacin. These medications are used for pain, inflammation, dysmenorrhea, headache, arthritis, or fever. These compounds act as antiinflammatory and antipyretic agents by inhibiting COX catalysis by prostaglandin H synthase (PGHS). PGHS has two isoenzymes: PGHS-1 (or COX-1) is generally a basal enzyme found in various tissues including platelets and gastric mucosa; PGHS-2 (or COX-2) is an inducible enzyme typically expressed in response to cytokines and mitogens at sites of inflammation or cell proliferation.

Older NSAIDs, such as aspirin and ibuprofen, inhibit both COX-1 and -2, but a newer class of NSAIDs, called coxibs, are selective for inhibition of COX-2. By sparing production of cytoprotective prostaglandins by mucosal COX-1, coxibs have fewer problems with irritation and ulceration in the upper GI tract.

Thromboxane produced by platelet COX-1 in concert with a downstream enzyme is prothrombotic, so aspirin and other NSAIDs cause platelet dysfunction and increase bleeding time. Aspirin is unusual in that it causes covalent, irreversible inhibition of the COX protein, whereas other NSAIDs have noncovalent, reversible actions. Thus, platelets, because they cannot synthesize more COX protein, are irreversibly affected by aspirin but only temporarily affected by other NSAIDs. Low-dose aspirin is often used in antithrombotic prophylaxis.

The prothrombotic and vasoconstrictive actions of COX-1-derived thromboxane in the vasculature are opposed by an antithrombotic and vasodilative prostaglandin, prostacyclin, that originates from COX-2 in vascular endothelial cells. The COX-2 selective coxibs thus tend to decrease prostacyclin levels in the vasculature without reducing the thromboxane levels. This tendency is thought to explain the small but significant increase in cardiovascular risk that recently led to withdrawal of two coxibs from the U.S. market.

APPROACH TO PROSTAGLANDIN METABOLISM

Objectives

1. Describe the biosynthetic and cell signaling pathways involving prostanoids.
2. Distinguish between the pathophysiologic roles of the two PGHS isoforms.
3. Cite the pharmacologic targets of NSAIDs and the characteristics that distinguish coxibs and aspirin from other NSAIDs.

Definitions

Eicosanoids: Oxygenated lipid signaling molecules containing 20 carbons derived from polyunsaturated fatty acids released from membrane phospholipids by the action of phospholipase A_2. These include the prostanoids produced by the cyclooxygenase pathway and the leukotrienes produced by the lipoxygenase pathway.

Prostanoids: Oxygenated lipid signaling molecules derived from polyunsaturated fatty acids released from membrane phospholipids by the action of phospholipase A_2. Prostanoids include prostaglandins, prostacyclin, and thromboxanes.

Prostaglandin: An oxygenated lipid signaling molecule that has a five-member ring system that is derived from arachidonic acid and other 20-carbon polyunsaturated fatty acids. The prostaglandins are hormone-like molecules that regulate cellular events near the area in which they are synthesized.

Thromboxane: An oxygenated lipid signaling molecule that has a six-member ring system derived from arachidonic acid and other 20-carbon polyunsaturated fatty acids. Thromboxanes are involved in platelet aggregation as well as vaso- and bronchoconstriction and lymphocyte proliferation.

PGH synthase: Prostaglandin H synthase; the enzyme that catalyzes the formation of prostaglandin H from C_{20} polyunsaturated acids. PGH synthase has two activities; the cyclooxygenase activity introduces the five-membered ring into the polyunsaturated fatty acid while also introducing an endoperoxide between carbons 9 and 10 and a hydroperoxide at carbon 15. The peroxidase activity reduces the hydroperoxide to a hydroxyl group using glutathione as the source of reducing equivalents. PGH synthase exists in two isoforms, PGHS-1 and PGHS-2. PGHS-1 is the "basal" isoform and is expressed constitutively, whereas PGHS-2 is the inducible isoform and has been implicated in cell proliferation and inflammation.

NSAIDs: Nonsteroidal antiinflammatory drugs; NSAIDs inhibit the cyclooxygenase activity of PGH synthase, thus inhibiting the production of prostaglandins and thromboxanes.

Coxibs: A class of NSAIDs that is selective for inhibition of the cyclooxygenase activity of PGHS-2, with weaker action against the PGHS-1 cyclooxygenase.

DISCUSSION

Prostanoids are oxygenated lipid-signaling molecules derived from **polyunsaturated fatty acids.** The major prostanoids synthesized from the prototypical polyunsaturated fatty acid, arachidonic acid, are prostaglandin (PG) D_2, PGE_2, PGF_{2a}, PGH_2, PGI_2 (also known as **prostacyclin**), and thromboxane (TX) A_2. The prostanoid signaling cascade begins with an external stimulus, most often the binding of a ligand to a cell surface receptor that activates one or more phospholipases A_2. The latter are enzymes that release arachidonic acid from its esterified form in membrane phospholipids such as phosphatidylethanolamine and phosphatidylinositol. Arachidonate is converted to PGH_2 by one of the isoforms of PGH synthase (PGHS-1 or -2), enzymes localized to the endoplasmic reticulum membrane and the nuclear envelope.

PGH_2 is in turn metabolized to the prostanoid lipid signals (PGD_2, PGE_2, PGF_{2a}, PGH_2, PGI_2, or TXA_2) by one of the secondary enzymes that are named for the individual prostanoid produced (Figure 32-1). The type of prostanoid produced is determined by which downstream enzyme is present; usually one downstream enzyme predominates in a given cell. For example, **the prominent secondary enzyme in platelets is thromboxane synthase,** whereas **vascular endothelial cells feature prostacyclin (PGI) synthase.** Prostanoid signaling molecules usually exit the cell that produces them to act on G-protein coupled receptors on the surface of the same cell or cells nearby (termed autocrine or paracrine actions). Some prostanoids may be further metabolized to ligands for a subset of nuclear receptors, the peroxisome proliferator-activated receptors (PPARs). The active prostanoids are rapidly converted to inactive metabolites by enzymes present in a variety of cells. As a result, prostanoid signaling molecules have very short half-lives in the circulation and are not hormones in the conventional sense.

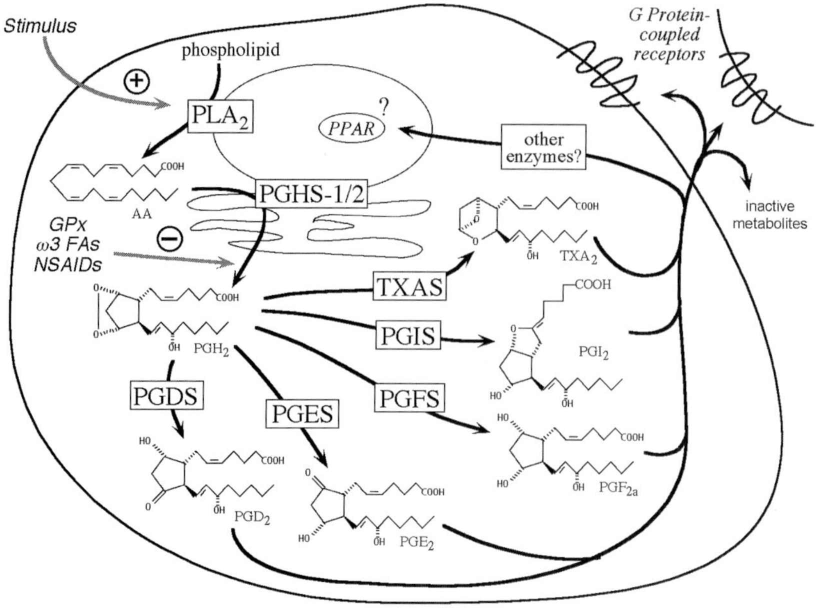

Figure 32-1. Diagram illustrating cell signaling pathways involving prostanoids.

Conversion of arachidonate to PGH_2 **is a key regulatory step in prostanoid biosynthesis.** Each PGHS isoform catalyzes two separate reactions (Figure 32-2). The first reaction (**arachidonate**→PGG_2) involves insertion of two molecules of oxygen and cyclization of the fatty acid backbone. This step is catalyzed by the **cyclooxygenase** activity of PGHS-1 or -2; it is these cyclooxygenase activities (also called COX-1 and COX-2) that are **inhibited by nonsteroidal antiinflammatory drugs** (NSAIDs). The **second step** (PGG_2→PGH_2) involves the **reduction of the hydroperoxide** on C15 to an alcohol and is **catalyzed by the peroxidase activity of PGHS-1 or -2.**

Although both PGHS isoforms have cyclooxygenase and peroxidase activities and are structurally similar proteins, they have very distinct pathophysiologic functions. Many cells, including platelets and gastric mucosal cells, have moderate levels of the "basal" isoform, PGHS-1. Functions attributed to PGHS-1 include regulating hemostasis and vascular tone, renal function, and maintaining gastric mucosal integrity. A smaller number of cells, such as macrophages, vascular endothelial cells, and fibroblasts, dramatically upregulate levels of the "inducible" isoform, PGHS-2, in response to cytokines or mitogens. PGHS-2 has been implicated in cell proliferation, inflammation, carcinogenesis, and parturition.

Many cyclooxygenase inhibitors have been developed and their structures are quite varied (Figure 32-3). All known inhibitors compete with fatty-acid

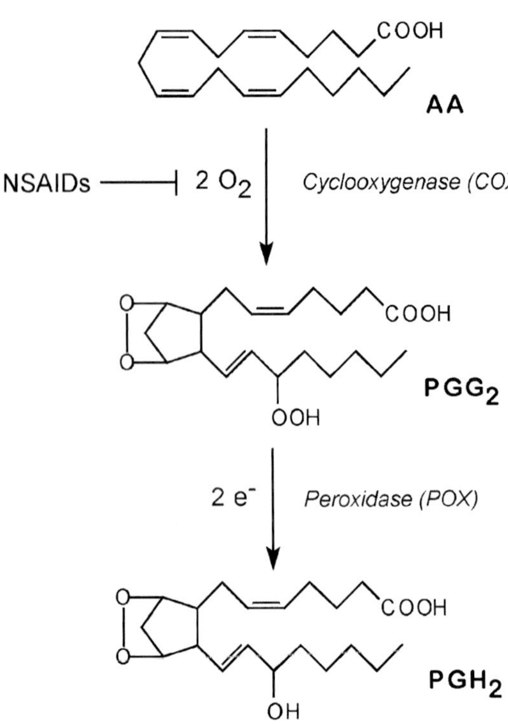

Figure 32-2. Steps in conversion of arachidonic acid (AA) to PGH2 by PGH synthase. A cosubstrate, indicated by e^-, is required to furnish two reducing equivalents for the peroxidase reaction.

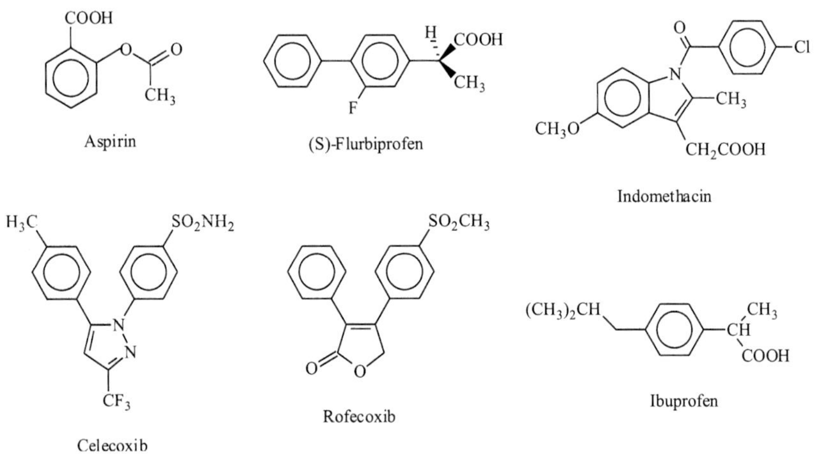

Figure 32-3. Structures of some cyclooxygenase inhibitors.

substrate for binding to the cyclooxygenase site on the enzyme. **Aspirin** was one of the earliest NSAIDs discovered and is now widely used as an analgesic and antiinflammatory agent. More recently, aspirin has emerged as a very useful **antithrombotic agent** because of its action against **platelet cyclooxygenase activity.** Aspirin shows both archetypal modes of cyclooxygenase inhibition. The first mode involves rapid reversible binding of inhibitor (I) at the cyclooxygenase site of the enzyme (E) to form an EI complex; the second mode involves a slower conversion of EI to a higher affinity complex, EI′ (Equation 1).

$$E + I \leftrightarrow EI \rightarrow EI' \qquad \text{(Equation 1)}$$

EI and EI′ cannot bind fatty acid and thus neither can catalyze the cyclooxygenase reaction. For **aspirin, conversion of EI to EI′ is accompanied by covalent modification of the protein, making the transition irreversible.** Formation of the EI′ complex produces a more powerful cyclooxygenase inhibition because the inhibitor is not readily displaced by substrate and because inhibition persists even when free inhibitor is removed. **Flurbiprofen and indomethacin** are able to form EI′ complexes with both PGHS-1 and -2, although they do not covalently modify either protein. Ibuprofen forms only EI complexes with both PGHS isoforms. The recently developed coxibs (such as celecoxib and rofecoxib) derive their exquisitely selective inhibition of PGHS-2 cyclooxygenase from their ability to form noncovalent EI′ complexes with PGHS-2 and not with PGHS-1. This selectivity has made the coxibs very useful for antiinflammatory and antiproliferative therapy with reduced gastrointestinal side effects, but it also makes them ineffective as antiplatelet agents and consequently can increase cardiovascular risks.

COMPREHENSION QUESTIONS

[32.1] The coxibs, including celecoxib (Celebrex), are a recently developed class of nonsteroidal antiinflammatory drugs (NSAIDs). The coxibs show antiinflammatory actions without affecting platelet function. These effects of the coxibs are best attributed to selective inhibition of which of the following?

A. The cytosolic isozyme of phospholipase A_2 ($cPLA_2$)
B. The cyclooxygenase activity of the "basal" prostaglandin H synthase isozyme (PGHS-1)
C. The cyclooxygenase activity of the "inducible" prostaglandin H synthase isoform (PGHS-2)
D. The microsomal isozyme of prostaglandin E synthase (mPGES-1)
E. Prostacyclin (PGI2) synthase (PGIS)

[32.2] Prostaglandins comprise a family of oxygenated lipid signaling molecules derived from polyunsaturated fatty acids such as arachidonic acid. They are involved in regulating a number of cellular processes. Some of the prostaglandins act to increase vasodilation and levels of cAMP in cells, whereas others increase vaso- and bronchoconstriction and smooth muscle contraction. In the conversion of arachidonic acid to prostaglandins, the oxygenation step is accomplished by the enzyme that synthesizes which of the following compounds?

A. Prostaglandin D_2
B. Prostaglandin E_2
C. Prostaglandin $F_{2\alpha}$
D. Prostaglandin H_2
E. Prostaglandin I_2

[32.3] Signaling via prostanoids begins by interaction of the prostanoid with its receptor. The receptor involved is usually located in which part of the cell?

A. Plasma membrane of a cell near the cell making the prostanoid
B. Nucleus of a cell in a different organ from the cell making the prostanoid
C. Endoplasmic reticulum of the cell making the prostanoid
D. Lysosomes of a cell circulating in the blood
E. Golgi of a cell circulating in the blood

Answers

[32.1] **C.** The coxibs were designed to inhibit the activity of the inducible form of PGH synthase so as not to inhibit the constitutive production of prostaglandins and thromboxanes. They inhibit the first step in the process, that catalyzed by the cyclooxygenase activity of PGH synthase-2.

[32.2] **D.** The first step in the synthesis of prostaglandins and thromboxanes is the reaction catalyzed by the cyclooxygenase activity of PGH synthase. This reaction causes the cyclization of the fatty acid while at the same time introducing an unstable endoperoxide between carbons 9 and 10 and a hydroperoxide at carbon 15 to produce PGG_2, which is rapidly reduced to PGH_2 by the peroxidase activity of PGH synthase.

[32.3] **A.** The prostanoids have a wide variety of physiologic effects, but they regulate these processes locally by binding to a receptor on the plasma membrane of a cell close to where the prostanoid was synthesized. Binding of the prostanoid to the receptor usually activates a GTP-binding protein, which acts to activate (or inhibit) adenylate cyclase or the phosphatidylinositol cascade.

BIOCHEMISTRY PEARLS

- Conversion of arachidonate to PGH_2 via cyclooxygenase catalysis is a key regulatory step in prostanoid biosynthesis.
- Aspirin has emerged as a very useful antithrombotic agent because of its action against platelet cyclooxygenase activity.
- The coxibs (such as celecoxib) derive their exquisitely selective inhibition of COX-2 cyclooxygenase from their ability to form noncovalent EI′ complexes with COX-2 and not with COX-1.
- The selectivity of coxibs for COX-2 over COX-1 gives them antiinflammatory and antiproliferative actions with reduced GI side effects, but it also makes them ineffective as antiplatelet agents and thus increases risks from myocardial infarction and stroke.

REFERENCES

Blobaum AL, Marnett LJ. Structural and functional basis of cyclooxygenase inhibition. J Med Chem 2007;50:1425–41.

Funk CD. Prostaglandins and leukotrienes: advances in eicosanoid biology. Science 2001;294:1871–5.

Grosser T, Fries S, FitzGerald GA. Biological basis for the cardiovascular consequences of COX-2 inhibition: therapeutic challenges and opportunities. J Clin Invest 2006;116:4–15.

Parfitt JR, Driman DK. Pathological effects of drugs on the gastrointestinal tract: a review. Hum Pathol 2007;38:527–36.

CASE 33

A 26-year-old female at 35 weeks gestation presents to the clinic with complaints of generalized itching. Patient reports no rash or skin changes. She denies any change in clothing detergent, soaps, or perfumes. She denies nausea and vomiting and otherwise feels well. On physical exam, there are no rashes apparent on her skin and only some excoriations from itching. Blood work reveals slightly elevated serum transaminase and bilirubin levels.

- **What is the patient's likely diagnosis?**
- **What are treatment options?**
- **What is the cause of the patient's generalized itching?**

ANSWERS TO CASE 33: CHOLESTASIS OF PREGNANCY

Summary: A 26-year-old female at 35 weeks gestation with generalized pruritus without a rash and slightly elevated liver transaminases and bilirubin.

- **Diagnosis:** Cholestasis of pregnancy.
- **Treatment options:** Oral antihistamines, cholestyramine, ursodeoxycholic acid.
- **Etiology of generalized itching:** Increased serum bile salts and accumulation of bile salts in the dermis of the skin.

CLINICAL CORRELATION

Cholestasis of pregnancy is a condition in which the normal flow of bile from the gallbladder is impeded, leading to accumulation of bile salts in the body. Generalized itching and, possibly, jaundice may result. It is speculated that the hormones such as estrogen and progesterone, which are elevated in pregnancy, cause a slowing of the gallbladder function, leading to this disorder. Uncomplicated cholestasis is usually diagnosed clinically by generalized itching in a pregnant woman, usually in the third trimester without a rash. Elevated serum levels of bile salts can help to confirm the diagnosis. Elevated bilirubin levels or liver transaminase enzymes may also be seen. The usual treatment includes antihistamine medications for the itching. Some experts recommend ursodeoxycholic acid, a naturally occurring bile acid that seems to improve liver function and may reduce the serum bile acid concentration. More severe cases may require bile salt binders such as cholestyramine or corticosteroids.

APPROACH TO BILE SALTS

Objectives

1. Be able to describe the catabolism and metabolism of bile salts.
2. Biochemical mechanism of action of cholestyramine and ursodeoxycholic acid.

Definitions

Bile acids: The major metabolites of cholesterol, which are synthesized in the liver and stored in the gallbladder for use as emulsifiers in the digestion of lipids. **Primary** bile acids are those synthesized directly from cholesterol in the liver; **Secondary** bile acids are metabolites of primary bile acids produced by the action of intestinal bacteria.

Bile salts: The ionized form of bile acids, which is the state that bile acids exist at physiologic conditions.

Cholestyramine: A synthetic, strongly basic anion exchange resin that will strongly bind bile salts when taken orally and prevent their reabsorption in the intestinal tract.

Conjugated bile salts: Bile salts whose carboxylate groups have been enzymatically condensed with the amino groups of either of the amino acids glycine or taurine to form the glyco- or tauroconjugates. This conjugation extends the pH range over which the bile salts are ionized and therefore effective emulsifying agents.

Ursodeoxycholic acid: A naturally occurring bile acid (originally isolated from the bile of bears) that is present in minor concentrations in the bile (approximately 1 to 10 percent). When administered orally, it helps to dissolve cholesterol gallstones.

DISCUSSION

Bile salt molecules secreted by the gallbladder are essential for the emulsification and absorption of fats. They are the salt forms of bile acids, which are the major product of cholesterol catabolism in the liver. Bile salts form micelles as their hydrophobic face contacts the fat (triacylglycerol), and their polar face maintains contact with the aqueous environment. This micelle formation allows water-soluble digestive enzymes to digest the entrapped triacylglycerol molecule, releasing fatty acids that are readily absorbed by the digestive system.

There **are two primary bile acids formed in the liver from cholesterol: cholic acid and chenodeoxycholic acid.** The formation of bile acids prevents cholesterol accumulation in organs; the body cannot break down the steroid ring of cholesterol. At physiologic pH, bile acids will always be in the form of bile salts. Bile salts are conjugated with glycine or taurine in the liver prior to excretion, forming glyco- or tauroconjugates. Bacterial enzymes present in the intestine produce the secondary bile salts, deoxycholate and lithocholate, by reducing the primary bile salts.

Bile salts perform an important function and are recycled by the body. The body produces 400 mg of bile salts per day from cholesterol; this represents the fate of half of the cholesterol used daily in metabolism (800 mg). However, 20 to 30 g of bile acids is maintained in the enterohepatic circulation. Less than 0.5 g per day is lost to excretion. Bile salts are produced in the liver, stored in the gallbladder, and secreted through the bile duct into the duodenum where they act on triacylglycerol molecules in the intestines. In the gut, the glycine or taurine moiety is removed from the bile salt. It is reabsorbed in the small intestine and returned to the liver for reuse via the portal vein. Bile salts are absorbed by passive diffusion along the entire small intestine, and a specialized Na^+-bile salt cotransporter is present in the lower ileum.

Intrahepatic cholestasis of pregnancy is a syndrome of unknown etiology characterized by a 100-fold increase in maternal and fetal blood bile salt levels. Bile salts are produced in both the fetal and maternal liver. The fetus transfers the bile salts across the placenta for disposal. When the function of the maternal gallbladder is slowed, bile salts can accumulate in the liver and bloodstream, ultimately resulting in the classical pruritus symptom. It is believed that pregnancy-related hormones may slow bile salt excretion from the gallbladder.

The **most successful therapy for cholestasis of pregnancy has been ursodeoxycholic acid** (Figure 33-1). Ursodeoxycholic acid is a naturally occurring bile acid, which, when administered, relieves both pruritus and liver function abnormalities. Experimental evidence suggests that it protects hepatocytes and cholangiocytes from bile acid-induced cytotoxicity and improves hepatobiliary excretion. Additionally, it decreases bile salt transfer to the fetus and improves the secretory function of placental trophoblast cells. Ursodeoxycholic acid is recycled through the enterohepatic circulation.

Cholestyramine is another treatment option for cholestasis of pregnancy. It is an oral medication that binds bile salts in the intestine and promotes their excretion in the feces. As this drug is not absorbed, it most likely has little effect on the fetus. Effects on the fetus are still under evaluation. However, cholestyramine can interfere with the absorption of **fat soluble vitamins, such as vitamins A, D, E, and K.** In rare cases, drug-induced vitamin K deficiency is believed to contribute to hemorrhaging during childbirth.

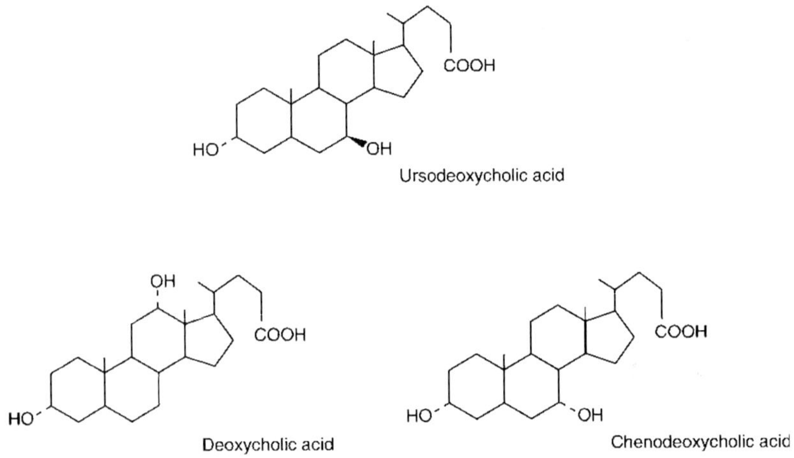

Figure 33-1. Comparison of the structures of isomers of deoxycholate.

COMPREHENSION QUESTIONS

[33.1] A 18-year-old male with sickle cell anemia develops severe right upper-abdominal pain radiating to his lower right chest and his right flank 36 hours prior to admission to the ER. Twelve hours following the onset of pain, he began to vomit intractably. In the past year he has had several episodes of mild back and lower extremity pain that he attributed to mild sickle cell crises. He reported that the present pain was not like his usual crisis pain. He also reports that his urine is the color of iced tea and his stool now has a light clay color. On examination, his temperature is slightly elevated, and heart rate is rapid. He is exquisitely tender to pressure over his right upper abdomen. The sclerae of his eyes are slightly yellowish in color.

What is the most likely cause of this patient's symptoms?

A. A cholesterol-rich gallstone
B. A defect in the synthesis of bile acids
C. A defect in heme synthesis
D. A gallstone rich in calcium bilirubinate
E. A sickle cell crisis brought on by overexertion

[33.2] A patient has been on combination statin and cholestyramine therapy to lower his serum cholesterol levels. Prior to any surgery, this patient would be well advised to be supplemented with which of the following?

A. Vitamin A
B. Vitamin B_{12}
C. Vitamin C
D. Vitamin K
E. Linolenic acid

[33.3] A 17-year-old female, whose parents were first cousins, presents to a neurologist because of recurring seizures despite being on anticonvulsive therapy. Nodules that appeared to be fatty deposits were present on her Achilles tendon and several of her joints. Plasma cholesterol concentrations were elevated, and an assay of plasma sterols indicated elevated cholestanol. Cultured skin fibroblasts did not contain any sterol 27-hydroxylase activity. A diagnosis of cerebrotendinous xanthomatosis, a genetic disease inherited in an autosomal fashion, was made. A deficiency in sterol 27-hydroxylase would lead to a decrease in the synthesis of which of the following compounds?

A. Chenodeoxycholate
B. Cortisol
C. 1,25-Dihydroxycholecalciferol
D. Estradiol
E. Testosterone

Answers

[33.1] **D.** Although a cholesterol-rich gallstone cannot be completely ruled out with the given information, because the patient has experienced several mild sickle cell crises that are accompanied by increased red blood cell destruction, his symptoms are consistent with a gallstone caused by precipitation of calcium salt of bilirubin. Large quantities of bilirubin can overwhelm the ability of the liver to convert it to the more soluble diglucuronide conjugate. As a consequence, the more insoluble unconjugated form enters the bile and is easily precipitated in the presence of calcium ion. If a large stone forms, it can obstruct the bile duct and result in the symptoms exhibited by the patient.

[33.2] **D.** Cholestyramine binds bile acids strongly so that they cannot be reabsorbed in the intestinal tract, thus increasing the flow of cholesterol to bile acid synthesis and decreasing cholesterol levels in the plasma. However, by binding bile acids, cholestyramine also decreases the absorption of fat-soluble vitamins and fatty acids, which must be taken up in micelles formed with bile acids. Although absorption of vitamin A and linolenic acid may be compromised, the patient needs to be concerned about vitamin K, which is required for blood clot formation.

[33.3] **A.** The enzyme sterol 27-hydroxylase catalyzes the hydroxylation of carbon 27 of the steroid side chain in the conversion of cholesterol to the primary bile acids. It is a mitochondrial, cytochrome P450 enzyme that has a broad specificity and can act on cholesterol as well as its reduced and hydroxylated metabolites. A deficiency in this enzyme leads to decreased bile acid synthesis and increased conversion of cholesterol to cholestanol.

BIOCHEMISTRY PEARLS

- ❖ There are two primary bile acids formed in the liver from cholesterol: cholic acid and chenodeoxycholic acid.
- ❖ Ursodeoxycholic acid has been reported to have good efficacy in treating the symptoms of pregnancy-related cholestasis, relieving both pruritus and liver function abnormalities.
- ❖ Cholestyramine can interfere with the absorption of fat-soluble vitamins, such as vitamins A, D, E, and K.

REFERENCES

Brites D. Intrahepatic cholestasis of pregnancy: changes in maternal-fetal bile acid balance and improvement by ursodeoxycholic acid. Ann Hepatol 2002;1(1):20–8.

Devlin TM, ed. Textbook of Biochemistry with Clinical Correlations, 5th ed. New York: Wiley-Liss, 2002.

Germain AM, Carvajal JA, Glasinovic JC, et al. Intrahepatic cholestasis of pregnancy: an intriguing pregnancy specific disorder. J Soc Gynecol Investig 2002;9(1):10–14.

CASE 34

A 49-year-old female presents to your clinic for follow-up after initiating a new medication (lovastatin) for her elevated cholesterol. She is currently without complaints and is feeling well. On repeat serum cholesterol screening, there is noted to be a decrease in the cholesterol level. The patient asks if she needs to continue the medication and what the potential side effects and benefits might be. Her physician explains that this medication inhibits the rate-limiting step and key enzyme in cholesterol biosynthesis.

What is the mechanism of action of this medication?

ANSWER TO CASE 34: STATIN MEDICATIONS

Summary: A 49-year-old female with history of elevated cholesterol on lovastatin, which appears to be improving serum cholesterol levels.

- **Mechanism action of lovastatin:** β-Hydroxy-β-methylglutaryl-coenzyme A (HMG-CoA) reductase inhibitor.

CLINICAL CORRELATION

Hyperlipidemia is one of the most treatable risk factors of coronary heart disease. Initially, when the fasting low-density lipoprotein (LDL) cholesterol is measured and found to be elevated, life style modification is recommended such as dietary adjustments, exercise, and weight loss. The lipids are measured again after a 3- to 6-month interval. If the LDL cholesterol level is again found to be above threshold, then pharmacologic therapy is entertained. One of the most common medications is a statin agent, acting to inhibit HMG-CoA reductase. The potential side effects include elevated liver function tests, increased muscle creatine phosphokinase (CPK) secondary to myopathy and rarely rhabdomyolysis. Other agents that may be considered include bile acid sequestrants, nicotinic acid, fibric acid, and fish oils.

APPROACH TO CHOLESTEROL SYNTHESIS

Objectives

1. Know about the role of HMG-CoA reductase in cholesterol synthesis.
2. Know the mechanism of action of statin medications.
3. Understand the role of cholesterol on steroid synthesis.
4. Be aware that niacin decreases lipolysis in adipose tissue and very-low-density lipoprotein (VLDL) synthesis in liver.

Definitions

Cytochrome P450 enzyme system: The cytochromes P450 are mixed-function oxidases that require both NADPH and O_2. They are involved in a number of reactions in the conversion of lanosterol to cholesterol, as well as important steps in the synthesis of steroid hormones. Cytochromes P450 are very important in the detoxification of xenobiotics and in the metabolism of drugs.

HMG-CoA reductase: β-Hydroxy-β-methylglutaryl-CoA reductase; the enzyme that catalyzes the rate-limiting and committed step in the synthesis of cholesterol. It converts β-hydroxy-β-methylglutaryl-CoA (HMG-CoA) to mevalonate.

Isoprenoid: Any of a number of hydrophobic compounds derived from the polymerization of isopentenyl pyrophosphate and its isomer, dimethylallyl pyrophosphate. The isoprene unit is a five-carbon branched hydrocarbon (2-methyl-1,3-butadiene).

Statin: Any of a number of drugs that competitively inhibit the rate-limiting enzyme in cholesterol biosynthesis, HMG-CoA reductase.

DISCUSSION

Cholesterol is synthesized mainly in the liver by a three-stage process. All 27 carbon atoms in the cholesterol molecule are derived from acetyl-CoA. The first stage is the synthesis of the activated five-carbon isoprene unit, **isopentenyl pyrophosphate.** Six molecules of isopentenyl pyrophosphate then condense to form **squalene** in a sequence of reactions that also synthesize isoprenoid intermediates that are important in protein isoprenylation modifications. The characteristic four-ring structure of cholesterol is then formed by cyclizing of the linear squalene molecule. Several demethylations, the reduction of a double bond, and the migration of another double bond result in the formation of cholesterol. Figure 34-1 provides an overview of cholesterol biosynthesis.

The key enzyme in the synthesis of cholesterol is β-hydroxy-β-methylglutaryl-CoA reductase **(HMG-CoA reductase), which catalyzes the synthesis of mevalonate from HMG-CoA in an irreversible, rate-limiting reaction.** Mevalonate is the immediate six-carbon precursor to isopentenyl pyrophosphate. HMG-CoA reductase is localized on the membrane of the endoplasmic reticulum and spans the membrane. The active site for this enzyme is found on the cytosolic side of the membrane. HMG-CoA reductase is inhibited by cholesterol in a feedback mechanism and the levels of mRNA for the enzyme are also regulated by the levels of cholesterol. Low concentrations of cholesterol increase the level of mRNA for HMG-CoA reductase, whereas high concentrations of cholesterol decrease the mRNA level. Because the enzyme HMG-CoA reductase is the **rate-limiting step** of cholesterol biosynthesis, this enzyme is the target for many cholesterol lowering drugs.

The five major classes of **steroid hormones** are derived from cholesterol by the pathway illustrated in Figure 34-2. Hydroxylation is important in these conversions. The hydroxylation reactions require NADPH and O_2 and are carried out by the **cytochrome P450 enzyme system. The enzyme 21-hydroxylase is required for the synthesis of mineralocorticoids and glucocorticoids.**

Another important hormone derived from cholesterol is **vitamin D.** This steroid-like hormone is involved in regulating calcium and phosphorus metabolism. The complete synthesis of vitamin D requires **ultraviolet light** to convert 7-dehydrocholesterol to previtamin D_3. The reaction scheme is shown in Figure 34-3. The active hormone 1,25-dihydroxycholecalciferol (calcitriol)

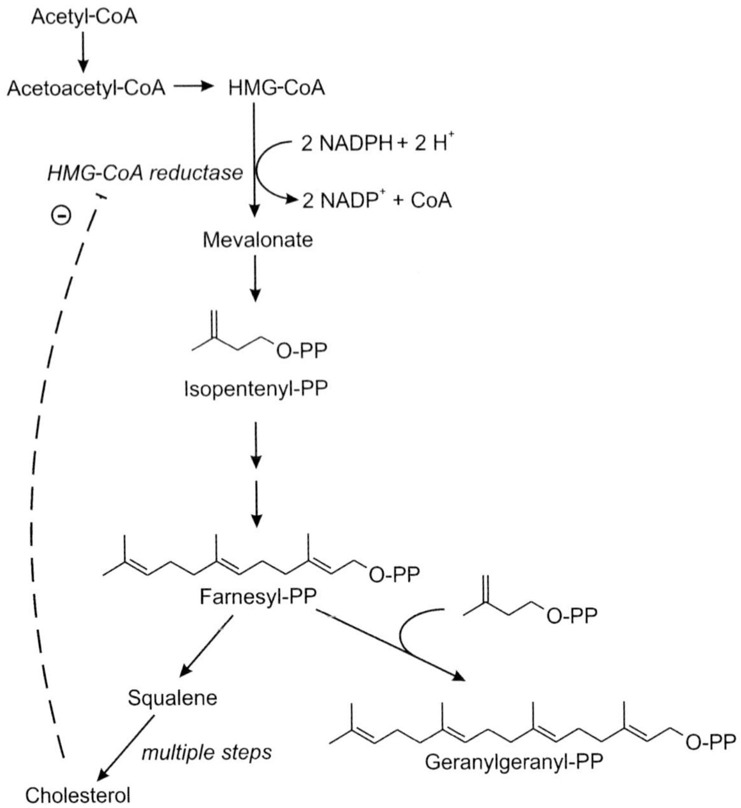

Figure 34-1. Overview of cholesterol synthesis.

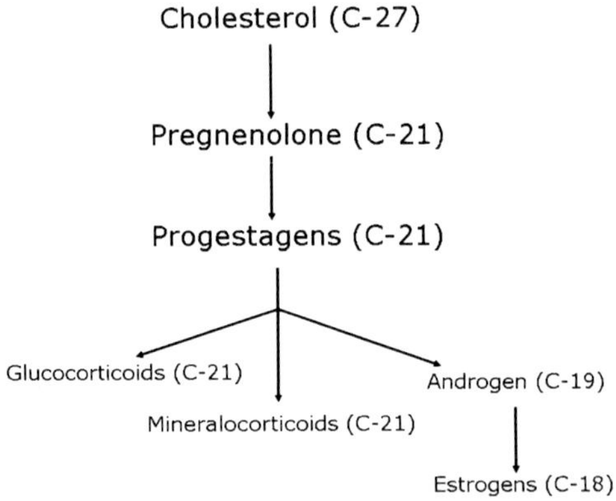

Figure 34-2. General biosynthetic pathway of the steroid hormones.

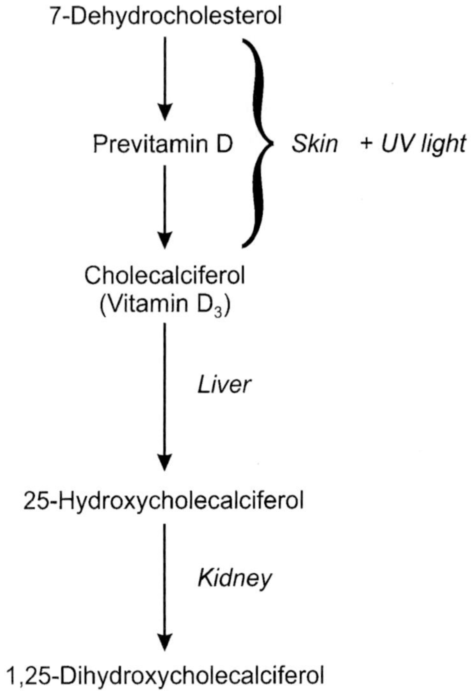

Figure 34-3. Biosynthesis of active vitamin D_3.

requires reactions that occur in the liver and kidneys and acts in a manner similar to steroid hormones to activate transcription, thereby regulating gene expression.

Lovastatin is a member of a class of drugs (atorvastatin and simvastatin are others in this class) called statins that are used to treat hypercholesterolemia. **The statins act as competitive inhibitors of the enzyme HMG-CoA reductase.** These molecules mimic the structure of the normal substrate of the enzyme (HMG-CoA) and act as transition state analogues. While the statins are bound to the enzyme, HMG-CoA cannot be converted to mevalonic acid, thus inhibiting the whole cholesterol biosynthetic process. Recent studies indicate that there may be important secondary effects of statin therapy because some of the medical benefits of statins are too rapid to be a result of decreasing atherosclerotic lesions. Statin therapy has been associated with reduced risks of dementia, Alzheimer disease, ischemic cerebral stroke, and other diseases that are not correlated with high cholesterol levels. Although this is still an active area of research, it appears that the pleiotropic effects of statins may be a result of a reduction in the synthesis of isoprenoid intermediates that are formed in the pathway of cholesterol biosynthesis.

Niacin is a vitamin that is used in high doses to treat hypercholesterolemia. Niacin acts to decrease VLDL and LDL plasma levels. Its mechanism of action is not clearly understood but probably involves inhibition of VLDL secretion, which in turn decreases the production of LDL. Niacin inhibits the release of free fatty acids from adipose tissue which leads to a decrease of free fatty acids entering the liver and decreased VLDL synthesis in the liver. This decreases the availability of VLDL for conversion to LDL (containing cholesterol esters). Niacin also increases high-density lipoprotein (HDL) (the "good cholesterol") by an unknown mechanism.

COMPREHENSION QUESTIONS

[34.1] Which of the following compounds directly inhibits the expression of the HMG-CoA reductase gene?

A. Squalene
B. HMG-CoA
C. Lanosterol
D. Isopentenyl pyrophosphate
E. Cholesterol

[34.2] You decide to treat a patient who has very high levels of serum cholesterol with the statin drug Lipitor (atorvastatin). You know that this drug acts in the metabolic pathway leading to the synthesis of cholesterol. The substrate for the enzyme inhibited by the statin drugs is which of the following?

A. Acetoacetyl-CoA
B. HMG-CoA
C. Farnesol pyrophosphate
D. Isopentenyl pyrophosphate
E. Mevalonate

[34.3] Which of the following vitamins can be used in high doses to treat hypercholesterolemia?

A. Niacin
B. Riboflavin
C. Pyridoxine
D. Folic acid
E. Thiamine

Answers

[34.1] **E.** The major regulatory enzyme of cholesterol metabolism, HMG-CoA reductase, is regulated by three distinct mechanisms. The first is phosphorylation by a cAMP dependent protein kinase. Phosphorylation of HMG-CoA reductase inactivates the enzyme. The other two mechanisms involve the levels of cholesterol. The degradation of the enzyme is controlled by cholesterol levels. The half-life of HMG-CoA reductase is regulated by cholesterol levels with high concentrations of cholesterol leading to a shorter half-life. The final regulatory mechanism involves control of the expression of the HMG-CoA reductase gene. High levels of cholesterol lead to a decrease in the mRNA levels coding for HMG-CoA reductase.

[34.2] **B.** The first step in the biosynthesis of cholesterol is the condensation of two molecules of acetyl-CoA to form acetoacetyl-CoA. The addition of a third acetyl-CoA molecule gives rise to HMG-CoA. HMG-CoA reductase is converted to mevalonate. HMG-CoA reductase is the target of the statin drugs and the substrate used by this enzyme is HMG-CoA.

[34.3] **A.** Niacin is the vitamin that can be used, in high doses, to treat hypercholesterolemia. Niacin acts to decrease VLDL and LDL plasma levels. Its mechanism of action is not clearly understood but probably involves inhibition of VLDL secretion, which in turn decreases the production of LDL. Niacin inhibits the release of free fatty acids from adipose tissue, which leads to a decrease of free fatty acids entering the liver and decreased VLDL synthesis in the liver. This decreases the availability of VLDL for conversion to LDL (containing cholesterol esters). Niacin also increases HDL (the "good cholesterol") by an unknown mechanism.

BIOCHEMISTRY PEARLS

- **HMG-CoA reductase,** which catalyzes the synthesis of mevalonate from HMG-CoA in an irreversible, rate-limiting reaction.
- The enzyme 21-hydroxylase is required for the synthesis of mineralocorticoids and glucocorticoids.
- The statins act as competitive inhibitors of the enzyme HMG-CoA reductase.

REFERENCES

Berg JM, Tymoczko JL, Stryer L. Biochemistry, 5th ed. New York: Freeman, 2002:722–31.

Devlin TM, ed. Textbook of Biochemistry with Clinical Correlations, 5th ed. New York: Wiley-Liss, 2002:742–52.

Granner DK. The diversity of the endocrine system. In: Murray RK, Granner DK, Mayes PA, et al. Harper's Illustrated Biochemistry, 26th ed. New York: Lange Medical Books/McGraw-Hill, 2003.

Liao JK. Isoprenoids as mediators of the biological effects of statins. J Clin Invest 2002;110:285–288.

Mayes PA, Botham KM. Cholesterol synthesis, transport & excretion. In: Murray RK, Granner DK, Mayes PA, et al. Harper's Illustrated Biochemistry, 26th ed. New York: Lange Medical Books/McGraw-Hill, 2003.

Type II Hyperlipidemia, The Merck Manual of Diagnosis and Therapy. http://www.merck.com/mrkshared/mmanual/section2/chapter15/15c.jsp, 2004.

CASE 35

A 9-year-old boy is brought to the ER by his parents after 2 days of worsening nausea/vomiting and abdominal pain. The abdominal pain is located in the epigastric region and radiates to his back. He has had several episodes of similar pain in the past but none quite as severe. His parents deny fever/chills and change in bowel habits. In the ER, the patient is afebrile and in moderate distress. Both the liver and spleen appear to be enlarged and he has epigastric tenderness. Several small yellow-white papules were noted on his back and buttocks. Laboratory tests reveal elevated amylase and lipase levels. On further questioning, the father reports having high triglyceride levels and several members of the mother's family have had early heart disease. Laboratory tests performed after hospitalization revealed elevated triglyceride levels and reduced lipoprotein lipase activity.

- **What is etiology of the boy's abdominal pain?**

- **What is the likely underlying biochemical disorder?**

- **What is the role of lipoprotein lipase?**

ANSWERS TO CASE 35: HYPERTRIGLYCERIDEMIA (LIPOPROTEIN LIPASE DEFICIENCY)

Summary: A 9-year-old boy with acute abdominal pain consistent with pancreatitis, hepatosplenomegaly, eruptive xanthomas, and a family history of hypertriglyceridemia and heart disease.

- **Etiology of abdominal pain:** Acute pancreatitis
- **Underlying biochemical disorder:** Disorder of lipoprotein metabolism
- **Role of lipoprotein lipase:** Hydrolysis of triglycerides from VLDL and chylomicrons

CLINICAL CORRELATION

Lipoprotein lipase (LPL) is an enzyme found on the capillary endothelial surface of adipose tissue, heart, and skeletal muscle and is required, along with apoC-II, for the hydrolysis of triglycerides. ApoC-II, found on the surface of chylomicrons and VLDL, serves as an activator of LPL. A deficiency in LPL results in elevated levels of triglycerides (VLDL and chylomicrons). The cholesterol level may be normal or slightly elevated. LPL deficiency is inherited in an autosomal recessive pattern. Patients with LPL deficiency often present with recurrent episodes of pancreatitis in their childhood and may have other clinical signs of hypertriglyceridemia such as: xanthomas, hepatosplenomegaly, and lipemia retinalis. Reduced serum LPL activity, after an injection of intravenous heparin, confirms the diagnosis of either LPL or apoC-II deficiency. The initial therapeutic intervention consists primarily of dietary modification (reduction of fat intake).

APPROACH TO LIPID TRANSPORT

Objectives

1. Describe the metabolism and transport of lipoproteins.
2. Understand the rationale for serum blood test results with the different hypertriglyceridemias.

Definitions

Apolipoprotein C-II (apoC-II): The apolipoprotein on the surface of chylomicrons and very-low-density lipoproteins (VLDLs) that binds to and activates lipoprotein lipase.

Apolipoprotein E (apoE): The apolipoprotein on the surface of several lipoproteins including chylomicrons, chylomicron remnants, VLDL, VLDL remnants, and IDL. It mediates the binding of apoE-containing lipoproteins with the LDL receptor and the chylomicron remnant (apoE) receptor.

Lipoprotein lipase (LPL): An enzyme bound to the surface of the capillary endothelium by heparan sulfate proteoglycans. LPL catalyzes the hydrolysis of triglyceride in chylomicrons and VLDL to free fatty acids and glycerol.

Hepatic lipase (HL): An enzyme bound to the surface of sinusoidal endothelial cells of liver. It catalyzes the hydrolysis of mono-, di-, and triglycerides as well as the phospholipids phosphatidyl choline and phosphatidyl ethanolamine.

DISCUSSION

Triglycerides (triacylglycerols, TG) are safely transported in the bloodstream packaged into lipoproteins called chylomicrons or very-low-density lipoproteins (VLDLs). Chylomicrons are formed in the epithelial cells of the intestine and are responsible for the transport of dietary lipids. VLDL is synthesized in the liver and transport endogenously synthesized lipids from the liver to peripheral tissues. Both of these lipoproteins are composed of a core composed primarily of TG enveloped by a monolayer composed of phospholipids, free cholesterol, and apolipoproteins.

Dietary triglycerides are hydrolyzed by pancreatic lipase in the lumen of the small intestine. Colipase, a protein secreted along with pancreatic lipase, binds to the TG and the pancreatic lipase and improves the hydrolytic process. TG is broken down to free fatty acids and 2-monoacylglycerols, which form micelles along with bile salts and other lipid soluble compounds such as cholesterol and fat-soluble vitamins. The free fatty acids and monoglycerides are absorbed by the microvilli of the intestinal epithelial cells. In the epithelial cell, the fatty acids and monoglycerides are reformed into triglycerides, which are packaged with phospholipids, cholesterol, and apolipoprotein B-48 into chylomicrons.

The newly synthesized chylomicrons are secreted into the lymph and enter the bloodstream via the thoracic duct. In the bloodstream, the chylomicron particles obtain proteins from high-density lipoproteins (HDL), including apoC-II and apoE, which are important for the function of the chylomicron (Figure 35-1).

In the capillary beds in adipose tissue, muscle tissue (especially cardiac muscle) and in lactating mammary glands, apoC-II binds to and activates lipoprotein lipase (LPL), which is bound to the endothelial surface of the capillaries by heparan sulfate. LPL hydrolyzes the TG in the core of the chylomicron to free fatty acids and glycerol. The fatty acids are taken up by the adipose or muscle cells; glycerol is recycled back to the liver. In muscle, the fatty acids are oxidized to produce ATP and in adipose they are reformed into TG for storage. The TG-depleted chylomicron remnant remains in the blood stream until it binds to the chylomicron remnant receptor located on hepatocytes, a process

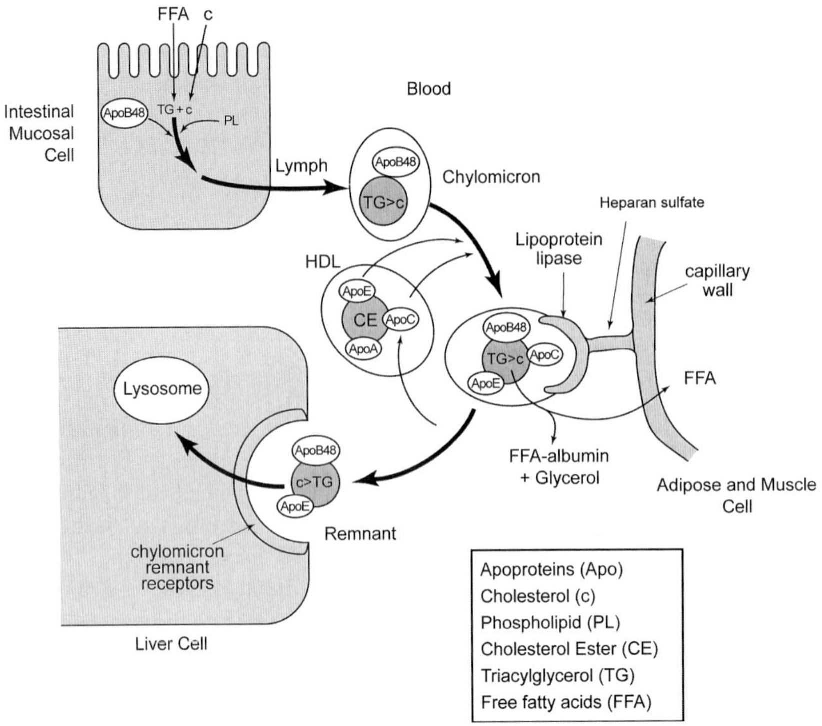

Figure 35-1. Formation and metabolism of chylomicrons. (Abbreviations used are defined in the enclosed box.)

mediated by apoE. The remnants are taken into the hepatocytes by endocytosis and degraded in the lysosome to fatty acids, amino acids, cholesterol, glycerol, and phosphate.

Chylomicrons appear in the blood stream shortly after consumption of a meal containing fat. However, the clearance rate for chylomicrons is fast and blood is usually free of chylomicrons following an overnight fast.

VLDLs are lipoproteins synthesized in the liver from endogenous TG, cholesterol, and phospholipids, along with several apolipoproteins including, apoB-100, apoE, and apoC-II. Following synthesis, the VLDL is secreted into the bloodstream, where it obtains more apoE and apoC-II from HDL (Figure 35-2). In a process similar to that of chylomicron degradation, the VLDL apoC-II binds to and activates LPL in the capillary beds of adipose, muscle and mammary tissue. LPL degrades the VLDL TG core, releasing free fatty acids and glycerol. About half of the resulting VLDL remnants bind to receptors in liver cells that recognized apoE and are taken up by endocytosis. The remaining VLDL receptors are further degraded to intermediate-density lipoproteins (IDL), which have their residual TG removed by the action of hepatic lipase to yield low-density lipoproteins (LDL). Hepatic lipase (HL) is synthesized and

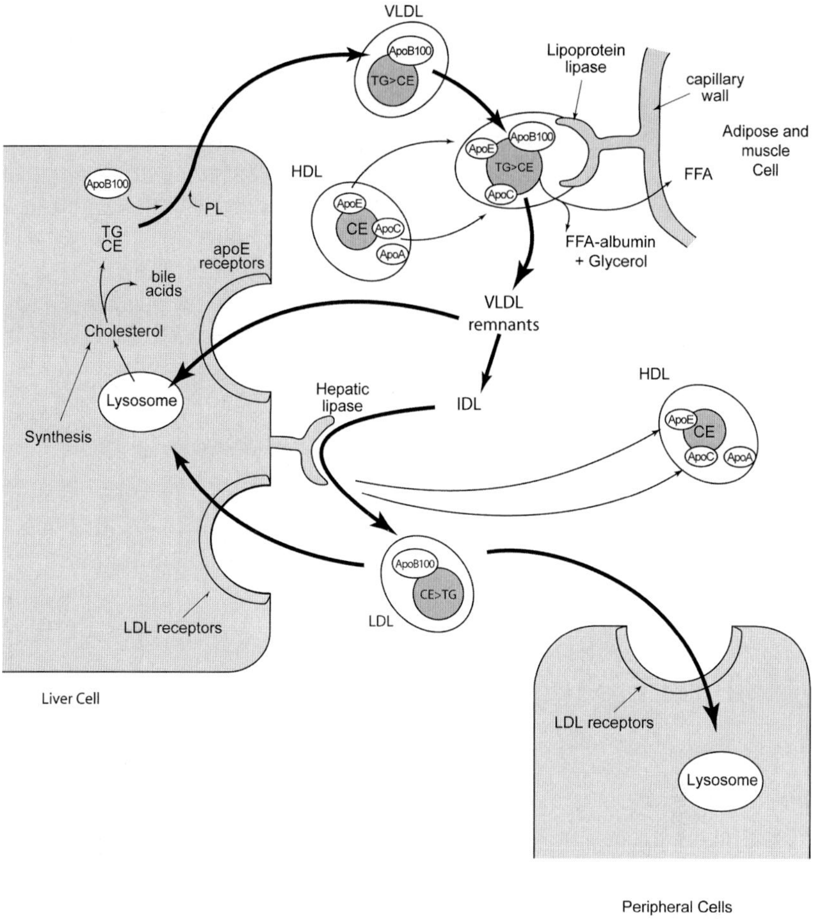

Figure 35-2. Formation of VLDL and metabolism into LDL. (Abbreviations are as used in Figure 35-1.)

secreted from liver and is tethered to the external surfaces of liver cells by heparan sulfate. LDL is taken up by receptor-mediated endocytosis upon binding to the LDL receptor in liver and peripheral tissues.

Elevated serum TG levels can occur due to a number of factors. Hypertriglyceridemia can be the result of a genetic disorder in one of the proteins involved in lipoprotein metabolism, or it can arise secondarily to a number of other disorders, including diabetes mellitus, obesity, and alcohol abuse, and as a side effect of some medications such as β-blockers, oral estrogens, and some diuretics. Genetic deficiencies in LPL, apoC-II, or HL can give rise to elevations in circulating TG, as can over production of apoB-100 or increased $apoE_2$ levels. $ApoE_2$ has a decreased affinity for hepatic receptors than $apoE_3$ and chylomicron remnants and VLDL remnants containing $apoE_2$ are cleared more slowly from the circulation.

The most common genetic defect leading to hypertriglyceridemia is a deficiency in LPL, which results in increased levels of both chylomicrons and VLDL. Individuals who are homozygous for the defective gene usually present with symptoms of chylomicronemia (TG levels >2000 mg/dL, abdominal pain, pancreatitis, xanthomas, *lipemia retinalis*) in childhood. Clinical diagnosis of LPL deficiency requires measurement of LPL activity in plasma following intravenous injection of heparin, which displaces the LPL from its heparan sulfate tether. However, heparin also releases HL into the plasma and the post-heparin plasma must be treated with antibodies specific to HL to remove it. Alternatively, LPL can also be measured in adipose tissue, which has no HL activity. A deficiency in apoC-II will also show evidence of decreased postheparin plasma LPL activity. An increase in LPL activity when normal apoC-II is added to the assay indicates that a defect in apoC-II is the culprit.

A deficiency in HL can also lead to elevated plasma TG levels, however, these increases in TG are usually found in VLDL remnants, LDL, and HDL, which all become more buoyant in their densities. Definitive diagnosis of HL deficiency is made by demonstrating the absence of HL activity in postheparin plasma.

COMPREHENSION QUESTIONS

[35.1] A teenage boy presents with moderate to severe epigastric pain. Physical examination reveals extensive eruptive xanthomas and hepatosplenomegaly. A blood sample reveals milky plasma. Which of the following is the most likely lipoprotein to be elevated in this patient's plasma?

A. Chylomicrons
B. Chylomicron remnants
C. HDL
D. IDL
E. LDL

[35.2] Laboratory results for a patient with uncontrolled Type I diabetes mellitus reveal hyperglycemia (634 mg/dL) and hypertriglyceridemia (498 mg/dL). The most likely cause of the hypertriglyceridemia in this patient is which of the following?

A. Deficiency in apoprotein C-II
B. Increased hepatic triglyceride synthesis
C. Decreased lipoprotein lipase activity
D. Deficiency in LDL receptors
E. Absence of hormone-sensitive lipase

[35.3] A 25-year-old female was referred to a lipid research center for investigation of moderate hypertriglyceridemia because the plasma lipid and lipoprotein profiles showed abnormalities. Both HDL and LDL were more buoyant and showed elevations in TG content with the mass of TG approximately the same as that of cholesterol. A deficiency in which of the following is the most likely cause of this patient's lipid abnormality?

A. Lecithin-cholesterol acyltransferase (LCAT)
B. Lipoprotein lipase
C. Apoprotein C-II
D. Hepatic lipase
E. Apoprotein B-100

Answers

[35.1] **A.** This patient presents with classical symptoms of hypertriglyceridemia: milky plasma, eruptive xanthomas, enlarged liver and spleen, and symptoms of pancreatitis. The milky plasma is due to abnormally high chylomicron levels, indicating that TG is not being removed from the transporter of dietary TG and it is accumulating in the plasma. The other lipoproteins listed have low levels of TG, are much smaller, and would not be expected to lead to milky plasma if elevated.

[35.2] **C.** Decreased lipoprotein lipase activity is the result of the failure of the pancreatic β-cells to produce and secrete insulin. Insulin stimulates the synthesis of lipoprotein lipase; in the absence of insulin, lipoprotein lipase activity in the capillary beds is low. While a deficiency in apoC-II would also lead to hypertriglyceridemia, this genetic defect is rare. If a Type I diabetes mellitus patient is not given insulin, the increased glucagon/insulin ratio would stimulate gluconeogenesis and β-oxidation in the liver rather than synthesis of triglycerides. Neither a deficiency of LDL receptors nor hormone-sensitive lipase would be expected to increase circulating TG levels.

[35.3] **D.** Hepatic lipase. The abnormal buoyancies of the LDL and HDL fractions due to their increased TG content indicate that IDL is not being processed to LDL and HDL is not being remodeled. Both of these processes are accomplished by hepatic lipase. A deficiency in LCAT would lead to elevated serum cholesterol levels, almost all as free cholesterol. Deficiencies in lipoprotein lipase or apoC-II would lead to hypertriglyceridemia as increased chylomicrons and/or VLDL. An apoB-100 deficiency would lead to increased levels of LDL, but they would have the normal TG/cholesterol ratio.

BIOCHEMISTRY PEARLS

- Triglycerides (triacylglycerols, TG) are safely transported in the bloodstream packaged into lipoproteins called chylomicrons or very-low-density lipoproteins (VLDLs).
- The most common genetic defect leading to hypertriglyceridemia is a deficiency in lipoprotein lipase, which results in increased levels of both chylomicrons and VLDL.
- Individuals who are homozygous for the defective gene for lipoprotein lipase usually present with symptoms of chylomicronemia (TG levels >2000 mg/dL, abdominal pain, pancreatitis, xanthomas, *lipemia retinalis*) in childhood.

REFERENCES

Brunzell JD, Deeb SS. Familial lipoprotein lipase deficiency, apo C-II deficiency, and hepatic lipase deficiency. In: Scriver CR, Beaudet AL, Sly WS, et al., eds. The Metabolic and Molecular Bases of Inherited Disease, 8th ed. New York: McGraw-Hill, 2001.

Haymore BR, Parks JR, Oliver TG, et al. Hypertriglyceridemia. Hosp Physician 2005;41(3):17–24.

CASE 36

During your medical school training, you spend some time in a pediatric clinic in a third-world country. One of the first patients you encounter is an 8-month-old girl brought to the clinic because of excessive exhaustion and fatigue. On further questioning of the mother, she reports that she was previously breast-feeding but had to stop to return to work. To feed all of her other children, she has had to dilute her formula with water to make the formula last longer for the entire family. After your physical exam is performed, you diagnose the infant with severe malnutrition and aid the mother with resources to increase food intake for their household.

- **What is this syndrome called (with deficiencies in both calories and protein)?**

- **What is the difference between this syndrome and kwashiorkor syndrome?**

- **What physical findings might differentiate the two syndromes?**

ANSWERS TO CASE 36: STARVATION

Summary: An 8-month-old girl presents with exhaustion and excessive starvation secondary to deficient intake of calories and protein.

- **Diagnosis:** Marasmus
- **Physical findings of kwashiorkor and not marasmus:** Subcutaneous fat, distended abdomen, hepatomegaly, and fatty liver

CLINICAL CORRELATION

Protein-energy malnutrition is caused by inadequate food intake or diseases that interfere with food absorption or digestion. The two major types of malnutrition are marasmus and kwashiorkor. In marasmus, a child usually between the ages of 1 to 3 years has inadequate caloric intake leading to loss of subcutaneous fat, loose wrinkled skin, and either flat or distended abdomen resulting from atropic abdominal wall muscles. Often, children are susceptible as they go from breast milk to solid food. The affected child usually has the appearance of an "old person's face." In kwashiorkor, the main issue is lack of protein, leading to edema, sparse hair, enlarged liver, and a distended abdomen. The edema of the face and legs is different from that of marasmus. The therapy for both of these diseases is caloric replacement.

APPROACH TO STARVATION RELATED DISEASES

Objectives

1. Understand the metabolic changes in starvation.
2. Be familiar with the formation of ketone bodies in starvation.
3. Know about the oxidation of fatty acids.
4. Be familiar with the metabolic change in fasting states as compared to starvation.

Definitions

Marasmus: Malnutrition resulting from inadequate intake of protein and calories.

Kwashiorkor: Malnutrition resulting from inadequate intake of protein though the intake of total calories is adequate.

Ketone bodies: The short chain fatty acid metabolites acetoacetate and β-hydroxybutyrate and acetone.

Triglyceride: A glycerol molecule with each hydroxyl group esterified with a fatty acid moiety.

Diglyceride: A glycerol molecule with two hydroxyl groups esterified with a fatty acid moiety.

Monoglyceride: A glycerol molecule with one hydroxyl group esterified with a fatty acid moiety.

DISCUSSION

Malnutrition and its ultimate form starvation arise from many different causes and are present even in affluent societies. The case description reveals that the child lives in a third-world country, and the physical findings reveal that the child suffers from protein-calorie-deficient starvation, or marasmus.

Fasting and starvation represent changes from the baseline metabolic interactions between tissues that exist in the fed state. Each of three states—fed, fasting, and starvation—must be considered from the standpoint of the whole body primarily because the constituent tissues have different requirements for their nutritional sources. For example, **red blood cells have an absolute requirement for glucose as the exclusive food source from which energy is derived.** Although other tissues use fatty acids and amino acids as well, the red blood cell cannot because it lacks mitochondria and therefore the enzymes required for most of the metabolic steps required in β-oxidation of fatty acids and metabolism of the carbon skeletons of amino acids. **Brain tissue normally has an exclusive preference for glucose,** the exception being in **advanced starvation** when the **brain** can use **ketone bodies** for energy production.

The metabolic interactions of tissues in the fed state are shown in Figure 36-1. Glucose, fatty acids from triglycerides, and amino acids are provided by the diet and used differentially by the tissues. In the **liver, glucose is used for storage as glycogen or converted to fatty acids for formation into triglycerides for storage in adipose tissue.** Amino acid carbon skeletons are used for metabolic intermediates for energy production or fatty acid synthesis. Resting muscle takes up glucose and stores it as glycogen and uses amino acids for protein synthesis. Resting muscle prefers fatty acids and ketone bodies over glucose to satisfy its energy demands. In adipose tissue, glucose and fatty acids are taken up. Glucose metabolism provides energy and glycerol 3-phosphate for triglyceride formation and storage using fatty acids transported to adipose cells as triglycerides in lipoprotein particles. The brain and red blood cells take up glucose from blood to meet energy demands.

During circumstances of no food intake for 16 to 20 hours (the postfeeding state), substantial changes occur in the interactions between tissues, as shown in Figure 36-2. The liver shifts from consumption of glucose for glycogen storage to mobilization of its glycogen stores to release glucose to the bloodstream to supply the glucose requirements of the brain and red blood cell. Because the hepatic glycogen supply is depleted fairly quickly, metabolic signals increase liver gluconeogenesis, depleting tricarboxylic acid cycle intermediates and

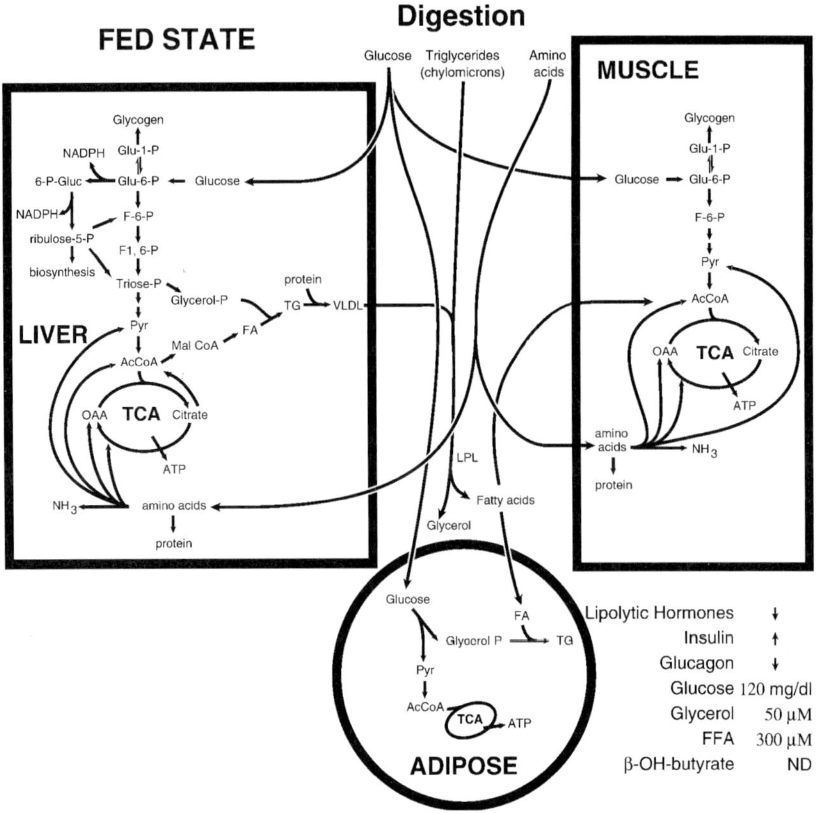

Figure 36-1. The metabolic flow during the fed state. Increased blood glucose triggers release of insulin and a decrease in the release of glucagon and lipolytic hormones.

prompting the use of **amino acid carbon skeletons from protein breakdown for new glucose formation.** The energy required for gluconeogenesis is derived by increasing β-oxidation of fatty acids mobilized from adipose storage sites. Fatty acid synthesis is simultaneously inhibited to prevent a futile cycle. As the tricarboxylic acid cycle's supply of four-carbon intermediates in liver mitochondria is drained for gluconeogenesis, rapid β-oxidation of fatty acids produces acetyl-CoA faster than the tricarboxylic acid cycle can metabolize the carbon atoms of acetyl-CoA to CO_2 and free CoA. The result is a high ratio of acetyl-CoA to free CoA and thus a slowing of β-oxidation and compromise of liver mitochondrial ATP formation. The conversion of the free CoA pool to acetyl-CoA is reversed by the formation of the **ketone body acetoacetate** (and later its reduced product, β-hydroxybutyrate) regenerating free

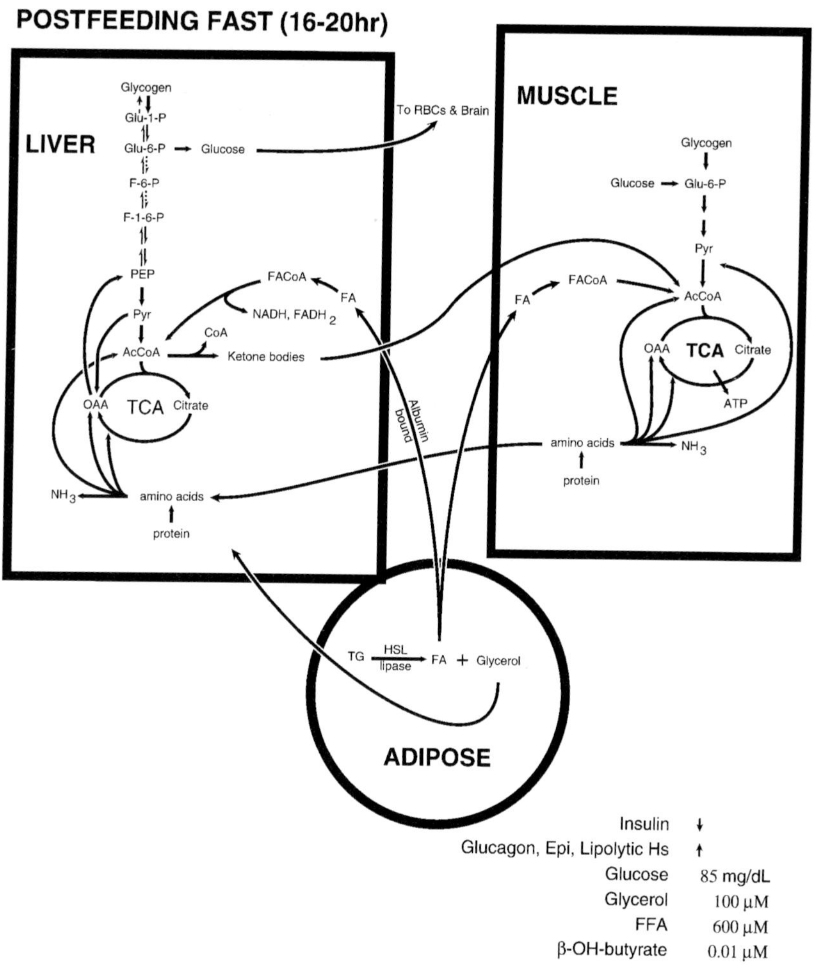

Figure 36-2. The metabolic flow following a postfeeding fast. Blood glucose levels begin to decrease, triggering homeostatic mechanisms to prevent it from decreasing dramatically.

CoA (Figure 36-3). This occurs only in liver mitochondria because of its critical role in gluconeogenesis. The ketone bodies are transported out of the liver mitochondria and the liver into the bloodstream for transport to other tissues where they reenter metabolism by being converted to acetoacetyl-CoA at the expense of succinyl-CoA and then cleaved by β-ketothiolase to produce two molecules of acetyl-CoA for metabolism in the TCA cycle.

The **β-oxidation of fatty acids that occurs in the mitochondrial matrix provides the energy for gluconeogenesis in the liver.** Fatty acids transported from adipose tissues by blood albumin cross the hepatic plasma membrane and

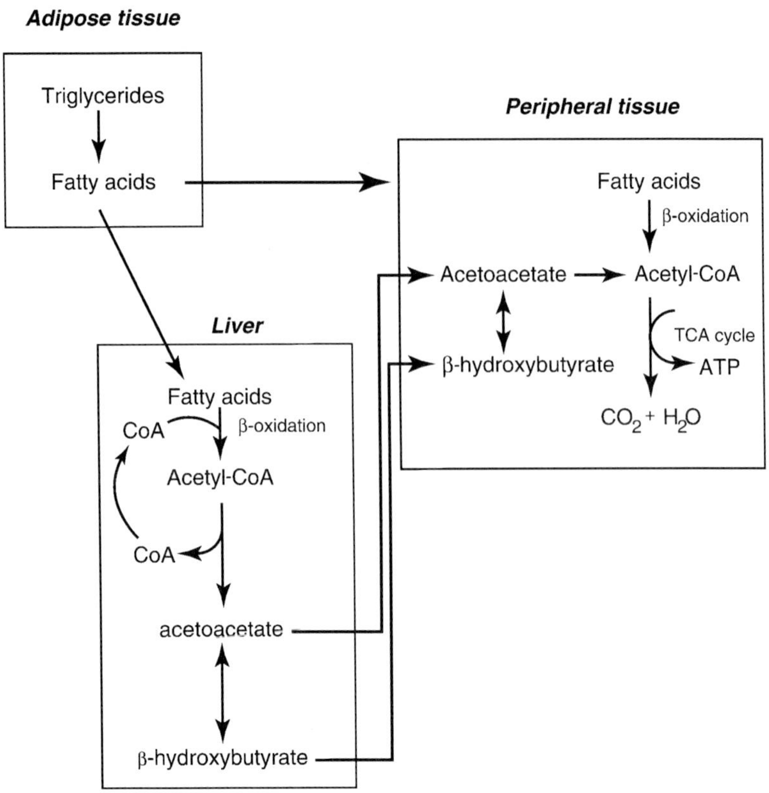

Figure 36-3. Mobilization of fatty acids during times in which the liver is synthesizing glucose via the gluconeogenic pathway. The β-oxidation of fatty acids by the liver produces the energy needed for gluconeogenesis, but because the TCA cycle is slowed because of depletion of C_4 acids (used for glucose synthesis), ketone bodies (acetoacetate and β-hydroxybutyrate) are formed from acetyl-CoA to regenerate CoA for continued β-oxidation. The ketone bodies are exported to extrahepatic tissues, where they are used as an energy source.

are activated by fatty acid thiokinases producing fatty acid-CoA and requiring ATP, as shown in Figure 36-4. The fatty acids are transported across the mitochondrial inner membrane as carnitine derivatives utilizing the carnitine shuttle. All the rest of the reactions occur in the mitochondrial matrix beginning with the oxidation of the fatty acid by flavin adenine dinucleotide (FAD)-linked fatty acyl-CoA dehydrogenase producing *trans*-Δ^2-enoyl-CoA. This product is hydrated by enoylhydratase producing L-3-hydroxyacyl-CoA. This product undergoes a second oxidation catalyzed by NAD-linked L-3-hydroxy fatty acyl-CoA dehydrogenase producing 3-ketoacyl-CoA. This product is

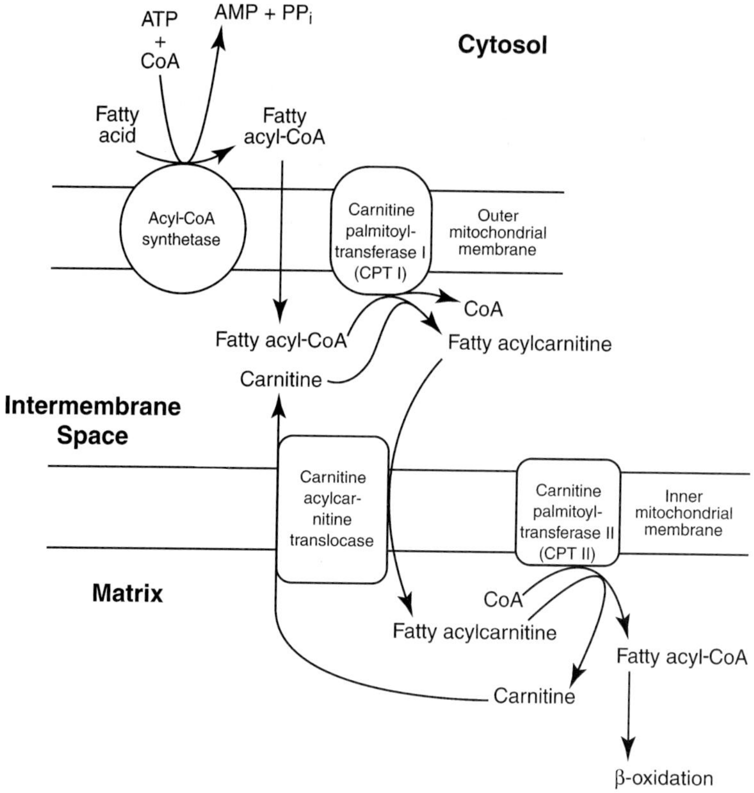

Figure 36-4. The activation of fatty acids and transport into the mitochondrion via the carnitine shuttle. *(Reproduced, with permission, from: D.B. Marks, et al.,* Basic Medical Biochemistry: A Clinical Approach, *Philadelphia: Lippincott Williams & Wilkins, 1996:361.)*

cleaved by β-ketothiolase producing one molecule of acetyl-CoA and a new activated fatty acyl-CoA, two carbon atoms shorter than at the outset in a reaction that requires another molecule of free CoA. The newly produced fatty acyl-CoA repeats the cycle of steps in β-oxidation releasing another acetyl-CoA and onward until the last cleavage step that hydrolyzes acetoacetyl-CoA to two molecules of acetyl-CoA.

While the postfeeding fast represents a normal state reflective of the alternation of feeding and not feeding, the starvation state shown in Figure 36-5 represents an abnormal state and reflects a dramatic increase in the metabolic changes observed in the postfeeding state, illustrated in Figure 36-3. Thus starvation represents an intensification of the metabolic adjustments of the fasting state with some significant differences seen only in prolonged starvation. Two marked changes in plasma concentrations occur, a decrease in glucose

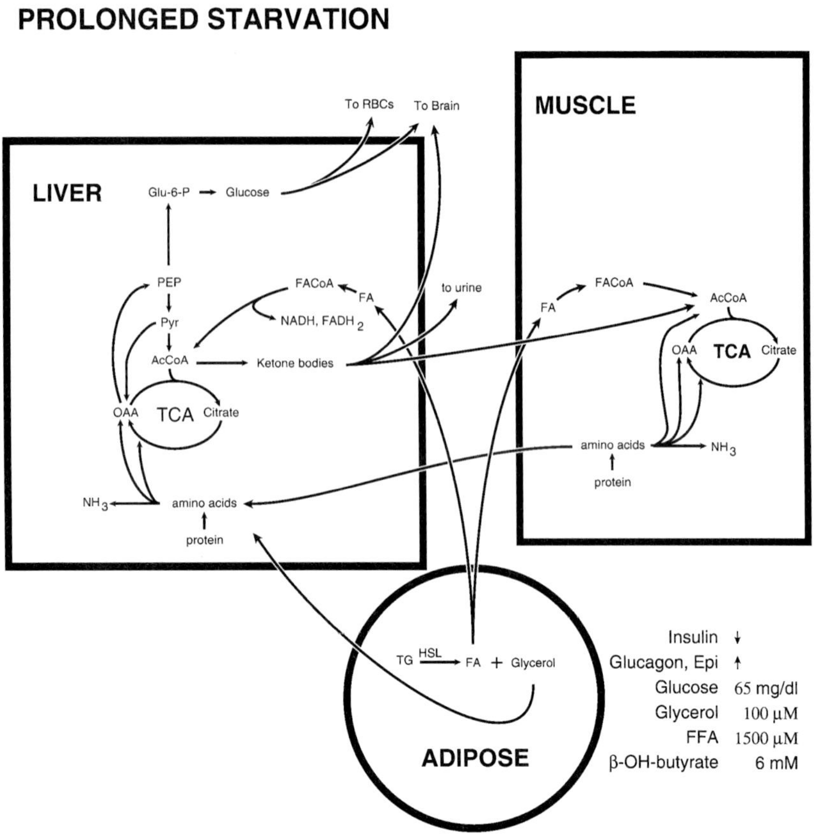

Figure 36-5. The metabolic flow during prolonged starvation. The brain adapts to use ketone bodies as a source of energy, thus decreasing its utilization of glucose.

concentration and a dramatic increase in the concentration of ketone bodies, reflecting altered metabolic poise. In the liver the tricarboxylic acid (TCA) cycle is slowed by the drainage of four-carbon intermediates to gluconeogenesis, fatty acid breakdown continues apace, and body proteins continue to be broken down to replenish the tricarboxylic acid cycle intermediates. In muscle the fuels used for energy generation are fatty acids and ketone bodies. Muscle activity decreases as result of the mobilization of muscle protein, which itself slows as the period of starvation increases. In adipose tissue the breakdown of triglycerides to fatty acids is accelerated. In brain and the central nervous system an adaptive change occurs allowing this tissue to use ketone bodies as an energy source relieving both the total body demand for glucose and the use of muscle protein as a carbon source for gluconeogenesis in the liver. Ketone body concentration increases in the blood reflecting the accelerated breakdown of triglycerides in adipose tissue and the slower hepatic tricarboxylic acid cycle. In spite of the utilization of ketone bodies by peripheral tissues and the brain and central nervous system (CNS) tissues after 5 to 6 weeks of starvation, ketone body blood levels rise, spill over into the urine and are excreted in significant quantity wasting material that could be used for energy generation because the tissues capable of using ketone bodies as fuel are already using the maximum possible amount. Thus the major differences between starvation and the postfeeding fast are the adaptive ability of the brain and central nervous system to use ketone bodies to satisfy some of their energy demand and in the levels of circulating ketone bodies that are high enough to spill over into the urine in significant quantities.

COMPREHENSION QUESTIONS

An 8-month-old child presents with exhaustion, irritability, and malnutrition. The family history reveals poverty and inadequate nutrition in all members. The 8-month-old was fed diluted formula, and the tentative diagnosis of marasmus was made.

[36.1] In cases of starvation many metabolic changes take place to meet the body's metabolic demands. Which of the following illustrates starvation-triggered changes in intermediary metabolism?

A. Increased dependence of liver on glucose for its energy supply
B. Increased synthesis of proteins in muscle tissue
C. Increased use of ketone bodies for energy source in brain
D. Decreased mobilization of triglycerides by adipose tissue
E. Adaptation of red blood cells to use ketone bodies for energy

Use the reactions below for Question 36.2:

1. Acetoacetyl-CoA + acetyl-CoA → β-hydroxy-β-methylglutaryl-CoA + CoA
2. Acetoacetate + NADH → β-hydroxybutyrate + NAD^+
3. β-hydroxy-β-methylglutaryl-CoA + H_2O → acetoacetate + acetyl-CoA
4. Acetyl-CoA + acetyl-CoA → acetoacetyl-CoA + CoA
5. Acetoacetate + succinyl-CoA → acetoacetyl-CoA + succinate

[36.2] Using the above reactions, which of the following correctly describes the pathway of ketone body formation?

A. 3 → 2 → 1 → 4
B. 4 → 1 → 3 → 2
C. 4 → 2 → 3 → 1
D. 5 → 1 → 2 → 3
E. 5 → 2 → 3 → 1

[36.3] During starvation muscle activity decreases, and muscle protein is broken down to provide a carbon source for the liver production of glucose via gluconeogenesis. Which of the following amino acids remains in the muscle cell to provide a source of energy for the muscle?

A. Alanine
B. Aspartate
C. Leucine
D. Glutamate
E. Threonine

Answers

[36.1] **C.** In starvation a major metabolic adjustment is that the brain activates the ketone body metabolic pathway and uses ketone bodies for energy, thus sparing somewhat the breakdown of body proteins to generate amino acid carbon skeletons for gluconeogenesis in the liver. In starvation liver derives most of its energy from β-oxidation of fatty acids. Muscle proteins are broken down to generate carbon skeletons during starvation. Triglyceride stores in adipose tissue are being used to provide fatty acids for β-oxidation in the liver. Red blood cells are not able to use ketone bodies because they have no mitochondria.

[36.2] **B.** The synthesis of ketone bodies begins with the combination of two molecules of acetyl-CoA to generate one molecule of free CoA and a molecule of acetoacetyl-CoA, which combines with another molecule of acetyl-CoA to yield another free CoA molecule and β-hydroxy-β-methylglutaryl-CoA (HMG-CoA). HMG-CoA undergoes hydrolysis to produce one molecule acetyl-CoA and one molecule of acetoacetate, which can be reduced to β-hydroxybutyrate. The reaction of succinyl-CoA and acetoacetate is a reaction in the pathway of ketone body utilization but not in the pathway of ketone body formation.

[36.3] **C.** Leucine but none of the other amino acids listed is a branched-chain amino acid. The muscle has a very active branched-chain amino acid metabolic pathway and uses that pathway to provide energy for its own use. The products of leucine metabolism are acetyl-CoA and acetoacetate, which are used in the tricarboxylic acid cycle. Acetoacetate is activated by succinyl-CoA and cleaved to two molecules of acetyl-CoA in the β-ketothiolase reaction. The other branched-chain amino acids, valine, and isoleucine, yield succinyl-CoA and acetyl-CoA as products of their catabolism.

BIOCHEMISTRY PEARLS

- Red blood cells have an absolute requirement for glucose as the exclusive food source from which energy is derived.
- **Brain tissue normally has an exclusive preference for glucose,** the exception being in **advanced starvation** when the brain can use **ketone bodies** for energy production.

REFERENCE

Devlin TM, ed. Textbook of Biochemistry with Clinical Correlations, 5th ed. New York: Wiley-Liss, 2002:629–42, 713–23, 862–85.

CASE 37

A 45-year-old male with history of hepatitis C and now cirrhosis of the liver is brought to the emergency center by family members for acute mental status changes. The family reports that the patient has been very disoriented and confused over the last few days and has been nauseated and vomiting blood. The family first noticed disturbances in his sleep pattern followed by alterations in his personality and mood. On examination, he is disoriented with evidence of icteric sclera. His abdomen is distended with a fluid wave appreciated. He has asterixis and hyperreflexia on neurologic exam. His urine drug screen and ethyl alcohol (EtOH) screen are both negative. A blood ammonia level was noted to be elevated, and all other tests have been normal.

- **What is the most likely cause of the patient's symptoms?**
- **What is asterixis?**
- **What was the likely precipitating factor of the patient's symptoms?**

ANSWERS TO CASE 37: CIRRHOSIS

Summary: A 45-year-old male presents with cirrhosis, likely secondary to hepatitis C, with acute mental status change coinciding with recent onset of hematemesis. Patient has an elevated serum ammonia level and otherwise negative workup.

- **Diagnosis:** Hepatic encephalopathy likely secondary to elevated ammonia levels.
- **Asterixis:** Nonspecific to hepatic encephalopathy. Nonrhythmic asymmetric tremor with loss of voluntary control of extremities while in a sustained position. It is also known as "liver flap."
- **Precipitating factor:** Increased nitrogen load from upper gastrointestinal bleed.

CLINICAL CORRELATION

Cirrhosis is a chronic condition of the liver with diffuse parenchymal injury and regeneration leading to distortion of the liver architecture and increased resistance of blood flow through the liver. The patient usually manifests malaise, lethargy, palmar erythema, ascites, jaundice, and hepatic encephalopathy in the late stages. Toxins accumulating in the blood stream affect the patient's mental status. The most common etiologies of cirrhosis are toxins such as alcohol, viral infections such as hepatitis B or C infection, or metabolic diseases in children (Wilson disease, hemochromatosis, or α_1-antitrypsin deficiency). Treatment depends on the exact etiology, although the common therapy includes avoidance of liver toxins, salt restriction, and possibly procedures to reduce the portal pressure.

APPROACH TO AMINO ACID METABOLISM AND AMMONIA

Objectives

1. Be familiar with the urea cycle.
2. Know about amino acid metabolism.
3. Be aware of the biochemical means of removing excess ammonia.

Definitions

Glutamate dehydrogenase: A mitochondrial enzyme present in all tissues that metabolizes amino acids. It catalyzes the oxidative deamination of glutamate to α-ketoglutarate using NAD^+ as the electron acceptor to also produce nicotinamide adenine dinucleotide (NADH) and ammonia. The enzyme uses the reducing equivalents of nicotinamide adenine dinucleotide phosphate (NADPH) to perform the reverse reaction.

Ornithine: An α-amino acid similar in structure to lysine but having one methylene group less in the side chain. It is carbamoylated to form citrulline to begin the urea cycle and is regenerated in the final step that releases urea.

Transaminase: An aminotransferase; a pyridoxal phosphate-requiring enzyme that transfers an amino group from an α-amino acid to an α-keto acid.

Urea cycle: The series of reactions that occur in the liver to synthesize urea for the excretion of nitrogen. The two nitrogen atoms present in urea arise from ammonium ion and the α-amino group of aspartate. The cycle also requires CO_2 (HCO_3^-) and the expenditure of four high-energy phosphate bonds and produces fumarate.

DISCUSSION

Amino acids differ from carbohydrates and fats in that they contain nitrogen as part of their molecular structure. For the carbons in amino acids to enter into the energy generating metabolic pathways, the amino groups must first be removed so that they can be detoxified and excreted. The amino acid nitrogen is excreted predominantly as urea, but some is also excreted as free ammonia in order to buffer the urine.

The **first step** in the **catabolism of most amino acids** is the **transfer of the α-amino group from the amino acid to α-ketoglutarate (α-KG).** This process is catalyzed by transaminase (aminotransferase) enzymes that require pyridoxal phosphate as a cofactor. The products of this reaction are glutamate (Glu) and the α-ketoacid analog of the amino acid destined for catabolic breakdown. For example, aspartate is converted to its α-keto analog, oxaloacetate, by the action of aspartate transaminase (AST), which also produces Glu from α-KG. The transamination process is freely reversible, and the direction in which the reaction proceeds is dependent on the concentrations of the reactants and products. These reactions do not effect a net removal of amino nitrogen; the amino group is only transferred from one amino acid to another.

For net removal of amino nitrogen, a second enzymatic reaction must take place that removes the amino group from Glu for disposal. The net removal of the amino nitrogen is accomplished by the mitochondrial enzyme **glutamate dehydrogenase (GDH),** which catalyzes the oxidative deamination of Glu to α-KG in a reaction that uses NAD^+ as the electron acceptor. The enzyme can also catalyze the reverse reaction to produce Glu, but in this case it uses the reducing equivalents from NADPH instead of NADH. The oxidative deamination reaction is allosterically activated by adenosine diphosphate (ADP) and guanosine diphosphate (GDP), whereas the reductive amination is activated by GTP and adenosine triphosphate (ATP). The overall process for net removal of the amino group from α-amino acids is summarized in Figure 37-1.

Ammonia is produced by almost all cells in the body; however, only the liver has the enzymatic machinery to convert it to urea. Therefore, extrahepatic ammonia must be transported to the liver. However, ammonia in the blood is toxic to cells, and therefore the nitrogen from amino acid catabolism is transported in blood either as glutamine or alanine. Glutamine is synthesized from Glu and ammonia in an ATP-requiring reaction that is catalyzed by glutamine synthetase. Alanine is formed from pyruvate in a transamination reaction catalyzed by alanine transaminase (ALT).

Glutamine and alanine are transported to the liver in the blood, where they are taken up by cells in the periportal region. Ammonia is released by the combined action of ALT (in the case of alanine), glutaminase (in the case of glutamine) and GDH. The α-amino group of alanine is transferred to α-KG to form Glu and pyruvate. Glutaminase catalyzes the hydrolysis of the side-chain amide group releasing ammonia and Glu. Ammonia and Glu enter the mitochondria, where Glu is oxidatively deaminated by GDH. The ammonia that is released by glutaminase and GDH then enters the urea cycle (Figure 37-2), which includes enzymes that are located both in the mitochondria and the cytosol.

Ammonia is condensed with bicarbonate and ATP in the mitochondrion to form carbamoyl phosphate in a reaction catalyzed by carbamoyl phosphate synthetase I. Two molecules of ATP are used in this reaction; one provides the phosphate, and the other is hydrolyzed to ADP and inorganic phosphate (P_i) to provide the energy that drives the reaction to products. The activated carbamoyl group is then transferred to the amino acid ornithine by the mitochondrial enzyme ornithine transcarbamoylase to form citrulline. Citrulline then is transported out of the mitochondrion to the cytosol, where the rest of the reactions

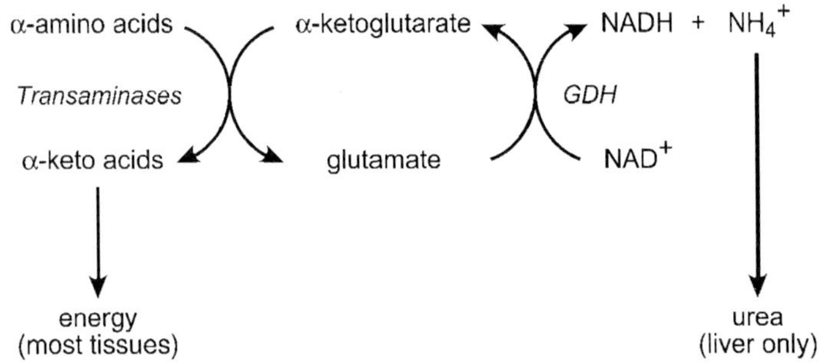

Figure 37-1. Summary of amino acid catabolism.

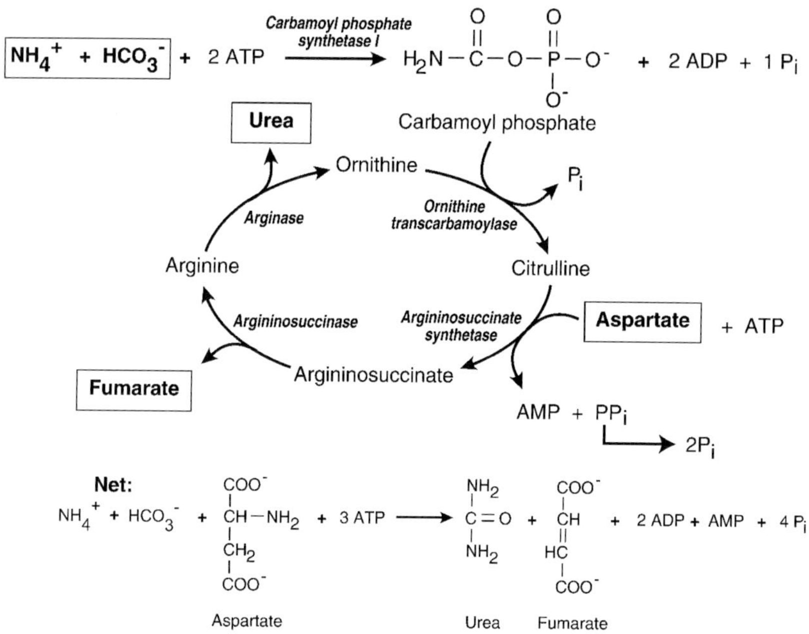

Figure 37-2. Urea cycle.

that are required to synthesize urea take place. The second nitrogen of urea comes directly from the amino acid aspartate. The side chain of citrulline condenses with the α-amino group of aspartate to form argininosuccinate in a reaction that is thermodynamically driven by the conversion of ATP to AMP and inorganic pyrophosphate (PP_i). The rapid hydrolysis of PP_i by pyrophosphatase releases energy and removes the PP_i, thus making the formation of argininosuccinate thermodynamically irreversible. Argininosuccinate is then cleaved to arginine and fumarate by argininosuccinase (argininosuccinate lyase). Arginase then hydrolyzes the guanidino group of arginine, releasing urea and regenerating ornithine, which can then reenter the mitochondrion and accept another carbamoyl group from carbamoyl phosphate. The urea is transported to the kidney for excretion.

Glutamine is also used by the kidney as a source of ammonia that is used to buffer the urine. Ammonia is released from glutamine by the same enzymes that are active in the liver. The free ammonia accepts a proton to form ammonium ion, thus decreasing the acidity of the urine.

Although most of the ammonia detoxified by the liver arises from the breakdown of amino acids in dietary protein or endogenous protein that is being turned over, ammonia is also produced by bacteria in the gut. This is absorbed into the portal venous blood and taken directly to the liver for conversion into urea.

When liver function is severely impaired or when collateral links between the portal and venous blood vessels arise as occurs in **cirrhosis,** the **ability of the liver to detoxify ammonia to urea is compromised** resulting in **hyperammonemia.** This can be **exacerbated** by an **increase in the ammonia load,** such as can occur from **gastrointestinal bleeding.** When blood ammonia levels rise, ammonia builds up inside the cells and drives the GDH reaction to form Glu, thus depleting α-KG stores and slowing the tricarboxylic acid (TCA) cycle. This is particularly devastating to the brain, which must have an active TCA cycle to produce the energy needed for brain function.

COMPREHENSION QUESTIONS

[37.1] An $8^1/_2$-month-old infant was admitted to the hospital in a coma and a temperature of 39.4°C (102.9°F). His pulse was elevated, his liver was enlarged, and an electroencephalogram was grossly abnormal. Since the infant could not retain milk given by gavage feeding, intravenous glucose was administered. He improved rapidly and came out of the coma in 24 hours. Analysis of his urine showed abnormally high amounts of glutamine and uracil, which suggested a high blood ammonium ion concentration. The laboratory confirmed this.

Considering the data, which enzyme may be defective in this patient?

A. Arginase
B. Carbamoyl phosphate synthetase I
C. Glutamate dehydrogenase
D. Glutaminase
E. Ornithine transcarbamoylase

[37.2] A newborn male infant was diagnosed as having phenylketonuria (PKU), and immediately placed on diet low in phenylalanine (Phe); careful compliance with the diet and frequent monitoring of the patient's plasma Phe level resulted in the level being maintained at the lower limit of the normal range. The patient appeared to be developing normally until 4 months of age, when he developed truncal hypotonia and spasticity of the limbs. Despite being on a low-phenylalanine diet, at 5 months the patient had several grand mal (epileptic) seizures. After an abnormal Phe-loading test, the patient's urine was found to have a markedly elevated urinary biopterin concentration.

Which of the following enzymes is most likely deficient in this patient?

A. Dihydropteridine reductase
B. GTP cyclohydrolase I
C. Phenylalanine hydroxylase
D. Tryptophan hydroxylase
E. Tyrosine hydroxylase

[37.3] Which treatment regimen would be most beneficial to this patient in Question 37.2?

A. A low-Phe diet with biopterin supplementation
B. A low-Phe diet with cobalamin (vitamin B_{12}) supplementation
C. A low-Phe diet + L-dopa (3,4-dihydroxyphenylalanine)
D. A low-Phe diet + L-dopa and 5-hydroxytryptophan
E. A diet completely free of Phe

Answers

[37.1] **E.** The patient exhibits signs of a defect in the urea cycle. The presence of elevated uracil in addition to ammonia and glutamine points to an accumulation of carbamoyl phosphate. If ornithine transcarbamoylase is deficient, carbamoyl phosphate will accumulate in the mitochondria and leak into the cytosol, providing the starting compound for the synthesis of uracil.

[37.2] **A.** The patient, despite being put on a low-Phe diet, exhibits neurologic problems resulting from an inability to synthesize catecholamine and indoleamine neurotransmitters. This is caused by a deficiency in dihydropteridine reductase (DHPR). DHPR regenerates tetrahydrobiopterin (BH_4), which is oxidized to dihydrobiopterin by phenylalanine hydroxylase, as well as tyrosine hydroxylase and tryptophan hydroxylase (tryptophan 5-monooxygenase). If phenylalanine hydroxylase were deficient, a diet low in Phe would alleviate the effects. Since the urinary biopterin concentration is elevated, a deficiency in GTP cyclohydrolase I is eliminated because that is an enzyme in the biosynthetic pathway of BH_4. Phe hydroxylase, Tyr hydroxylase, and Trp hydroxylase activities are low because of a lack of BH_4.

[37.3] **D.** Because of the DHPR deficiency, the activities of Phe hydroxylase and Tyr hydroxylase are low, hence the synthesis of catecholamine neurotransmitters are depressed. The synthesis of the indoleamine neurotransmitter serotonin is also depressed because BH_4 is required for the hydroxylation of tryptophan. The best treatment is to decrease the Phe load by a low-Phe diet and provide the precursors for the catecholamine and indoleamine neurotransmitters that occur after the enzymes affected by the deficiency of BH_4, which would be L-dopa and 5-hydroxytryptophan.

BIOCHEMISTRY PEARLS

- ❖ Ammonia is produced by almost all cells in the body; however, only the liver has the enzymatic machinery to convert it to urea.
- ❖ Liver disease such as cirrhosis affects the ability of the liver to detoxify ammonia to urea resulting in hyperammonemia.
- ❖ Hyperammonemia can be exacerbated by an increase in the ammonia load, such as can occur from gastrointestinal bleeding.

REFERENCES

Coomes MW. Amino acid metabolism. In: Devlin TM, ed. Textbook of Biochemistry with Clinical Correlations, 5th ed. New York: Wiley-Liss, 2002.

Rodwell VW. Catabolism of proteins & of amino acid nitrogen. In: Murray RK, Granner DK, Mayes PA, et al. Harper's Illustrated Biochemistry, 26th ed. New York: Lange Medical Books/McGraw-Hill, 2003.

CASE 38

A 1-year-old girl is brought to her pediatrician's office with concerns about her development. She had an uncomplicated birth outside the United States at term. The mother reports that the baby is not achieving the normal milestones for a baby of her age. She also reports an unusual odor to her urine and some areas of hypopigmentation on her skin and hair. On exam, the girl is noted to have some muscle hypotonia and microcephaly. The urine collected is found to have a "mousy" odor.

- **What is the most likely diagnosis?**

- **What is the biochemical basis of the hypopigmented skin and hair?**

ANSWERS TO CASE 38: PHENYLKETONURIA (PKU)

Summary: A 1-year-old girl born outside the United States with developmental delays, hypotonia, hypopigmentation, and foul smelling urine.

- **Likely Diagnosis:** Phenylketonuria (PKU)
- **Biochemical basis of hypopigmentation:** Phenylalanine is competitive inhibitor of tyrosinase (key enzyme in melanin synthesis)

CLINICAL CORRELATION

Elevated phenylalanine can be caused by a variety of different enzyme deficiencies resulting in impaired conversion of phenylalanine to tyrosine. The most common deficiency is in phenylalanine hydroxylase (autosomal recessive) resulting in the classic picture of PKU. Two other enzyme deficiencies leading to PKU include dihydropteridine reductase and 6-pyruvoyl-tetrahydropterin synthase, an enzyme in the biosynthetic pathway of tetrahydrobiopterin. With PKU, the baby appears normal at birth but then fails to reach normal developmental milestones. If unrecognized, the child will develop profound mental retardation and impairment of cerebral function. A mousy odor of the skin, hair, and urine can often be detected clinically. Areas of hypopigmentation develop secondary to the disruption of melanin synthesis. In the United States, all children are screened for PKU in hopes to prevent the serious life-long complications. Treatment consists of dietary modifications with limitation of phenylalanine intake and supplementation of tyrosine. The diagnosis of PKU and initiation of diet modification needs to be implemented prior to 3 weeks of age to prevent mental retardation and the other classic signs of PKU.

APPROACH TO PHENYLKETONURIA

Objectives

1. Describe the biochemical conversion of phenylalanine to tyrosine.
2. Describe the biochemical events that occur when conversion of phenylalanine to tyrosine is inhibited.

Definitions

Hypopigmentation: Lack of color in skin or hair due to the absence or low quantity of the skin and hair pigment melanin, a product of tyrosine (and phenylalanine) metabolism.

Tetrahydrobiopterin: A four-electron-reduced form of the reducing agent biopterin required to supply electrons to phenylalanine hydroxylase for conversion of phenylalanine to its hydroxylated product, the amino acid tyrosine.

Phenylketonuria: The presence of elevated amounts of phenylketones, primarily phenylpyruvate, in urine; a primary indication of disturbance of phenylalanine metabolism resulting from elevated transamination of phenylalanine due to reduction in phenylalanine hydroxylation to tyrosine.

DISCUSSION

Phenylketonuria is a disease readily diagnosable in childhood and it is important for an optimal clinical outcome for it to be diagnosed as early as possible. Laboratory tests can be performed in the neonatal period and now genetic testing can identify the trait before birth.

Phenylketonuria obtains from an elevated level of phenylpyruvate in the urine of the patient. As shown in Figure 38-1, phenylpyruvate is the cognate α-ketoacid of the amino acid phenylalanine. It is formed by transaminating phenylalanine and α-ketoglutarate to yield glutamate and phenylpyruvate. This reaction is freely reversible and therefore driven by elevated concentrations of reactants or products. For large scale conversion of phenylalanine to phenylpyruvate an increase in the concentration of phenylalanine must occur to drive the transamination reaction toward the formation of phenylpyruvate. Such an elevated phenylalanine level will be reflected in the blood as hyperphenylalaninemia, which is defined as a plasma phenylalanine level greater than 120 μmol/L. The most likely cause is a disturbance in the phenylalanine hydroxylase reaction. Disease states resulting from disturbances of the phenylalanine hydroxylase reaction can be found (1) at the level of the enzymes in the reaction reflected in absent or altered proteins, (2) at the metabolic level reflected in cognate effects on other metabolic processes, and (3) at the cognitive level reflected in changes in brain function and mental retardation.

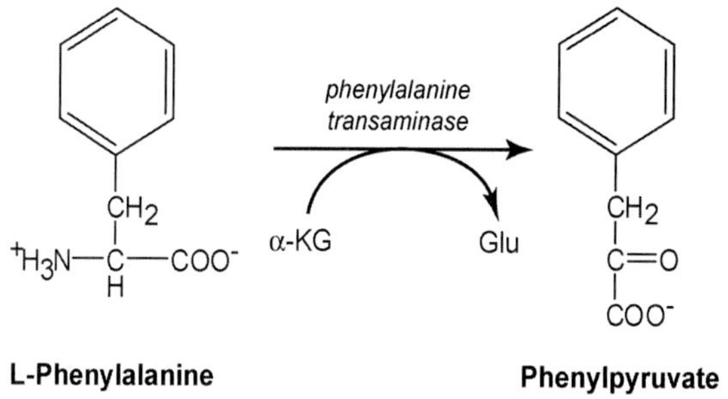

Figure 38-1. Transamination of phenylalanine to yield phenylpyruvate, a phenylketone.

Phenylalanine hydroxylation reaction and its components. As shown in Figure 38-2 the phenylalanine hydroxylation reaction is the pathway whereby dietary phenylalanine can be converted into tyrosine relieving some of the dietary requirement for tyrosine. It is catalyzed by phenylalanine hydroxylase, a monooxygenase requiring molecular oxygen (O_2) and a specific two electron donor tetrahydrobiopterin. One atom of molecular oxygen appears in the product tyrosine as a hydroxyl group in the *para* position, the remaining oxygen atom appearing in the product water.

The provision of reducing equivalents to phenylalanine hydroxylase is dependent on reduction of dihydrobiopterin by NADH catalyzed by the enzyme dihydropteridine reductase, as shown in Figure 38-2. This reduction is dependent on the availability of biopterin and therefore on the biopterin synthetic pathway. Thus any genetic or protein folding defect in either dihydropteridine reductase or the biopterin biosynthetic enzymes would compromise the efficacy of phenylalanine hydroxylation to tyrosine resulting in hyperphenylalaninemia and also phenylketonuria resulting from increase transamination of phenylalanine to phenylpyruvate.

The principal of enzyme in this pathway is phenylalanine hydroxylase. The gene for phenylalanine hydroxylase is located on chromosome 12 at band region q 23.2 and comprises 100 kb of genomic DNA. Several hundred alleles causing disease states have been recognized for this gene, more than 60 percent of which are classified as missense alleles. European and Chinese populations show an order of magnitude higher incidence than persons of African descent. This population specific expression of disease causing alleles may explain the wide range of incidence reported for this disease entity (5 to 350 cases/million live births).

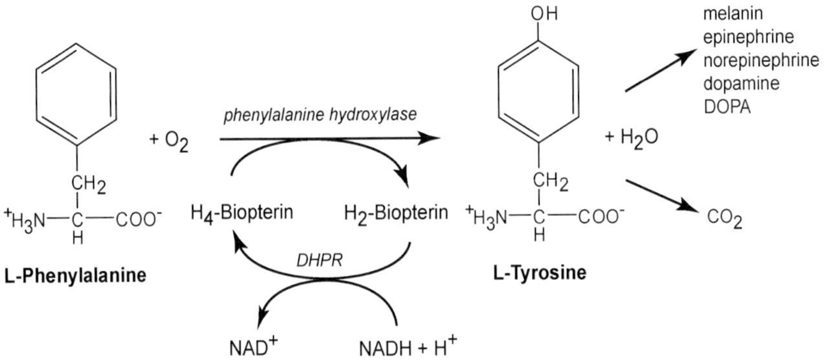

Figure 38-2. Normal conversion of phenylalanine into the amino acid tyrosine catalyzed by phenylalanine hydroxylase.

Fates of tyrosine. Tyrosine can be degraded by oxidative processes to acetoacetate and fumarate which enter the energy generating pathways of the citric acid cycle to produce ATP as indicated in Figure 38-2. Tyrosine can be further metabolized to produce various neurotransmitters such as dopamine, epinephrine, and norepinephrine. Hydroxylation of tyrosine by tyrosine hydroxylase produces dihydroxyphenylalanine (DOPA). This enzyme, like phenylalanine hydroxylase, requires molecular oxygen and tetrahydrobiopterin. As is the case for phenylalanine hydroxylase, the tyrosine hydroxylase reaction is sensitive to perturbations in dihydropteridine reductase or the biopterin synthesis pathway, anyone of which could lead to interruption of tyrosine hydroxylation, an increase in tyrosine levels, and an increase in transamination of tyrosine to form its cognate α-keto acid, *para*-hydroxyphenylpyruvate, which also would appear in urine as a contributor to phenylketonuria.

Tyrosine is also the precursor to melanin formation in melanocytes, the first step of which is catalyzed by tyrosinase as shown in Figure 38-3. This reaction is a two step reaction in which dihydroxyphenylalanine (DOPA) is an intermediate in the formation of dopaquinone. Ring closure of the alanine portion of dopaquinone forms a pyrrole ring and subsequent reactions give rise to the melanins, the primary dark pigment associated with skin color being eumelanin. The absence of tyrosinase gives rise to classic albinism. Phenylalanine is a competitive inhibitor with tyrosine for tyrosinase. Thus in a situation wherein phenylalanine hydroxylase activity is deficient, not only does the α-keto cognate transamination product phenylpyruvate increase but so also does the level of phenylalanine. Thus, excess phenylalanine inhibits tyrosinase and melanin formation resulting in hypopigmentation of skin and hair in affected persons.

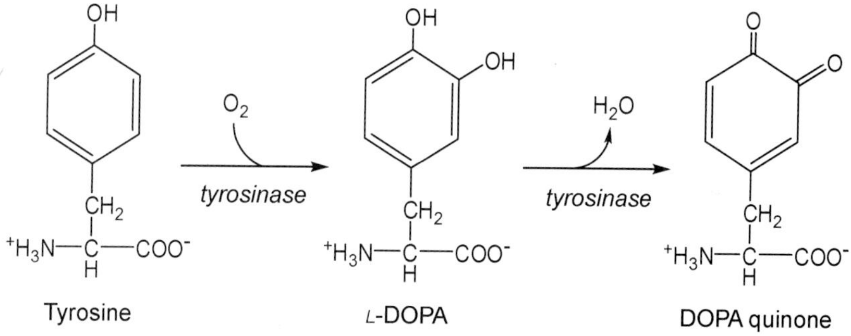

Figure 38-3. Conversion of tyrosine to dopaquinone by the enzyme tyrosinase, a Cu^{+2}-dependent enzyme in melanocytes.

COMPREHENSION QUESTIONS

[38.1] A 3-month-old boy presents with elevated levels of phenylalanine, *para*-hydroxyphenylpyruvate and phenylpyruvate in the serum. His skin color is pale. On your differential diagnosis is phenylketonuria. Which of the following would be consistent in such a case?

A. Elevated levels of homogentisic acid in the serum
B. A deficiency in vitamin B_{12} (cobalamin)
C. Elevated levels of pyridoxal phosphate in the serum
D. The urine in the boy's diaper smells like fresh maple syrup
E. Phenylalanine hydroxylase activity is only 2 percent of normal

[38.2] A 1-year-old girl presents at your clinic the day after you saw the 3-month-old boy. The symptoms are the same so you order a test on phenylalanine hydroxylase to confirm your diagnosis of phenylketonuria. To your surprise the phenylalanine hydroxylase activity is well within the normal range. Which of the following might you check next to support your diagnosis?

A. Tyrosine: α-ketoglutarate transaminase
B. Tyrosinase
C. Homogentisic acid oxidase
D. Dihydropteridine reductase
E. Dopamine hydroxylase

[38.3] Skin color is the aggregate result of the expression of a number of genes modified by ethnic origin and genetic inheritance. Hypopigmentation may be caused by which of the following?

A. Excess formation of melanin
B. Excess phenylalanine in the serum and tissues
C. Hyposecretion of melatonin
D. Excessive stimulation of tyrosinase
E. Low levels of *para*-hydroxyphenylpyruvate

Answers

[38.1] **E.** The correct response is very low levels of phenylalanine hydroxylase, a key enzyme in the metabolic sequelae of phenylketonuria, that is, elevated phenylalanine, phenylpyruvate, and *para*-hydroxyphenylpyruvate in blood. Homogentisic acid is an intermediate in the breakdown of tyrosine to fumarate and acetoacetate. Vitamin B_{12} is required in the metabolism of branched-chain amino acids not phenylalanine. The α-keto acids of the branched chain amino acids produce the maple-syrup odor.

[38.2] **A.** The correct response is dihydropteridine reductase. This enzyme reduces dihydrobiopterin to tetrahydrobiopterin the obligate electron donor for phenylalanine hydroxylase. Tyrosinase is the first enzyme on the pathway to melanin. Dopamine hydroxylase and tyrosine transaminase are enzymes on other tyrosine metabolic tracts. Homogentisic acid oxidase is an enzyme on the pathway of tyrosine to fumarate and acetoacetate.

[38.3] **B**. Excess phenylalanine inhibits tyrosinase the first step toward melanin production, thus resulting in hypopigmentation. Excess melanin leads to hyperpigmentation. Melatonin is a hormone involved in the sleep cycle. Excessive stimulation of tyrosinase would lead to more melanin and therefore hyperpigmentation. *Para*-hydroxyphenylpyruvate means less transamination and perhaps more tyrosine converted to melanin and hyperpigmentation.

BIOCHEMISTRY PEARLS

- ❖ Phenylketonuria (PKU) is an autosomal recessive disorder of amino acid metabolism affecting approximately 1/10,000 of infants in the North America.
- ❖ It is most often due to deficiency of the enzyme phenylalanine hydroxylase which causes the accumulation of harmful metabolites, including phenylketones.
- ❖ The gene for phenylalanine hydroxylase is located on chromosome 12 at band region q 23.2.
- ❖ If untreated, PKU leads to mental retardation, seizures, psychoses, eczema, and a distinctive "mousy" odor.
- ❖ Phenylketonuria is a disease readily diagnosable in childhood and it is important for an optimal clinical outcome for it to be diagnosed as early as possible.

REFERENCES

Scriver CR, Beaudet AL, Sly WS, et al. The Metabolic and Molecular Basis of Inherited Disease, 8th ed, New York: McGraw Hill, 2001:1667–776.

Devlin TM, ed. Text Book of Biochemistry with Clinical Correlations, 5th edition, New York: Wiley-Liss, 2002:797–9, 881–2.

CASE 39

The mother of a 16-year-old female calls the clinic because of concerns about her daughter's eating habits. The mother states the she will not eat anything and is obsessed with exercise and losing weight. She also states that her daughter has been more withdrawn from friends and family. After discussion with the mother, the patient comes in for a physical examination. The patient is 5 ft 1 in tall and weighs 85 lb. She is in no acute distress but appears to have a depressed affect. The patient states she is worried that her friends will think she is fat if she eats more. She denies any binge eating. Her physical examination is normal, other than dry skin and thin fine hair on extremities. Laboratory tests reveal that she is anemic and has a low albumin and magnesium level. She has normal liver and thyroid tests.

- **What is the most likely diagnosis?**

- **What potential medical problems may develop in a patient with this disorder?**

- **How can this disorder affect her menstrual cycles?**

ANSWERS TO CASE 39: ANOREXIA NERVOSA

Summary: A thin 16-year-old girl who is obsessed with her body appearance and weight to the point of not wanting to eat and exercising excessively.

- **Diagnosis:** Anorexia nervosa. This is differentiated from bulimia because she denies binge eating with associated guilty feelings.
- **Medical complications:** Dry skin, lanugo, bradycardia, hypotension, dependent edema, hypothermia, anemia, osteoporosis, infertility, cardiac failure, and even death.
- **Menstrual complications:** Amenorrhea secondary to depression of the hypothalamic-pituitary axis. Infertility will result secondary to anovulation.

CLINICAL CORRELATION

Anorexia nervosa is a disease affecting primarily young women who have distorted body images. Although their weight is less than 30 percent under ideal body weight, they see themselves as overweight. Anorectics often use diuretic and laxative agents to accomplish their weight loss. Patients with bulimia, who usually induce emesis, may be at normal weight or even above ideal body weight; in contrast, anorectics are almost always under ideal body weight. Often, affected individuals become amenorrheic, have fine lanugo hair, and become hypothermic. Therapy must be multifaceted and include family and individual counseling, behavioral modification, and possibly medication. Severe cases may be fatal.

APPROACH TO AMINO ACID AND NEGATIVE PROTEIN BALANCE

Objectives

1. Know about protein digestion and amino acid absorption.
2. Be familiar with nitrogen addition and removal from amino acids.
3. Understand amino acid metabolism in various tissues (muscle, gastrointestinal [GI], kidney).
4. Know the special products derived from amino acids.

Definitions

Anorexia nervosa: A mental disorder in which the patient has an extreme fear of becoming obese and therefore an aversion to food. The disorder usually occurs in young women and can result in death if the condition is not treated successfully.

Nitrogen balance: The condition wherein the amount of nitrogen ingested (via protein) is equal to that excreted. **Negative nitrogen balance** is a condition wherein more nitrogen is excreted than ingested, usually during prolonged periods of calorie restriction. Positive nitrogen balance results when more nitrogen is ingested than excreted, as in growing children.

Pyridoxal phosphate: The coenzyme that is required for transaminase (aminotransferase) reactions, as well as other enzymes. It is the active form of pyridoxine (vitamin B_6).

DISCUSSION

Proteins are polymers of α-amino acids covalently linked via peptide bonds. α-Amino acids consist of a central (α) carbon atom to which an amino group, a carboxylic acid group, a hydrogen atom, and a side-chain group are covalently linked. **Twenty different amino acids are used for protein synthesis, each of which is encoded by at least one codon (the three nucleotide genetic code)** and **differ only in their side-chain group.** Amino acids can be divided into two classes, essential and nonessential amino acids. **Essential amino acids cannot be synthesized by humans** (unlike nonessential amino acids) and therefore must be ingested to meet the requirements of the organism. Certain nonessential amino acids can become pseudoessential if the starting material from which they are synthesized becomes limiting (e.g., methionine-derived cysteine). Amino acids are also used for the synthesis of nonprotein biomolecules (e.g., nucleotides, neurotransmitters, and antioxidants), as well as participating in critical whole body processes, such as interorgan nitrogen transfer and acid–base balance. Unlike carbohydrate and fatty acids, **no storage form of excess amino acids exists** per se. Instead, dietary amino acids in excess to the body's synthetic needs are used as an energy source and/or **converted to glycogen and lipid.** During periods of **insufficient nutrient ingestion** (e.g., starvation, anorexia nervosa), **noncritical skeletal muscle and liver proteins are preferentially degraded** to release utilizable amino acids, to meet both the biosynthetic and energy needs of the body (Figure 39-1).

The major sites of ingested protein digestion are the stomach and the small intestine. Gastric, pancreatic, and intestinal peptidases hydrolytically cleave peptide bonds. The released amino acids, dipeptides, and tripeptides are transported into small intestinal epithelial cells, wherein dipeptides and tripeptides are further degraded to free amino acids. The latter are subsequently released into the circulation. A healthy, well-fed adult is generally in nitrogen balance. This means that the amount of nitrogen ingested (as protein) is equal to that excreted (primarily as urea). When rates of dietary amino acid incorporation into new protein exceed rates of amino acid degradation and nitrogen excretion, the individual is said to be in positive nitrogen balance. Growing children are normally in positive nitrogen balance. In contrast, an individual is said to be in **negative nitrogen balance when more nitrogen is excreted**

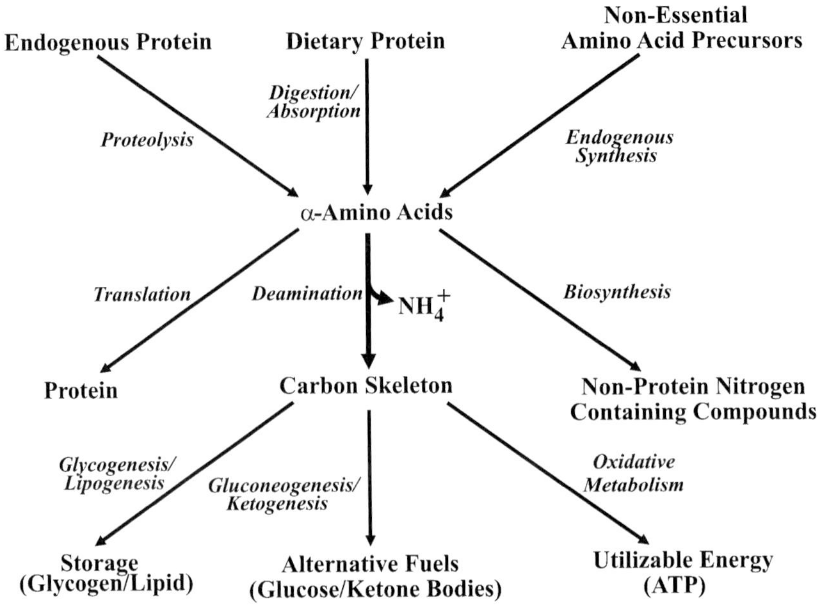

Figure 39-1. Schematic diagram showing the source and utilization of α-amino acids in metabolic processes.

than ingested. This occurs during prolonged periods of calorie restriction (e.g., starvation, anorexia nervosa), when protein is degraded to release amino acids as a utilizable energy source.

A general tactic employed during catabolism of amino acids is the removal of the α-amino group, followed by conversion of the remaining carbon skeleton into a major metabolic intermediate. The **main way in which α-amino groups are removed** is through **transamination,** the transfer of an amino group from an α-amino acid to an α-keto acid. This transamination reaction is catalyzed by a class of enzymes called transaminases (or aminotransferases). These enzymes use pyridoxal phosphate, a vitamin B_6 derivative, during the catalytic process. Pyridoxal phosphate acts as an initial acceptor of the α-amino group, forming a Schiff base (–CH = N–) intermediate. The pyridoxamine phosphate intermediate subsequently donates the amino group to an acceptor α-keto acid, forming a new α-amino acid. The acceptor α-keto acid is most often α-ketoglutarate, therefore resulting in glutamate generation. Not all amino acids are substrates for transaminases. Certain amino acids are initially converted into an intermediate that is subsequently transaminated (e.g., asparagine is hydrolyzed to asparate, the latter of which is transaminated by aspartate aminotransferase, forming oxaloacetate). Given that the majority of amino acids use transaminases in their pathways of degradation (although a select few amino acids can also be directly deaminated, e.g., threonine), glutamate formation increases markedly during periods of increased amino acid catabolism. Ultimately, the majority of this glutamate is oxidatively deaminated in the liver, where the released ammonium is incorporated into urea, via the urea cycle, and subsequently excreted in urine (Figure 39-2).

The carbon skeleton released during amino acid catabolism can undergo multiple fates, depending on the metabolic situation during which it was formed, the cell type within which it was generated, and the amino acid from which it is derived. For example, during periods of **excess amino acid ingestion, the amino acid–derived carbon skeleton is either used as a metabolic fuel or converted to glycogen or lipid.** In contrast, when rates of amino acid catabolism are increased as a result of **prolonged caloric insufficiency,** a large portion of the carbon skeleton is used by the liver for the **synthesis of either glucose or ketone bodies,** depending on the specific amino acid. Indeed, glucogenic amino acids (those amino acids whose carbon skeleton can generate net glucose through gluconeogenesis) are essential for maintenance of blood glucose levels during prolonged caloric insufficiency. Maintenance of blood glucose is also made possible by increased reliance of skeletal muscle on amino acids as a fuel source, thereby decreasing glucose utilization.

Certain amino acids are preferentially used in a tissue-specific manner. For example, **skeletal muscle has a relatively high capacity for branched-chain amino acid (leucine, isoleucine, and valine) utilization.** Following

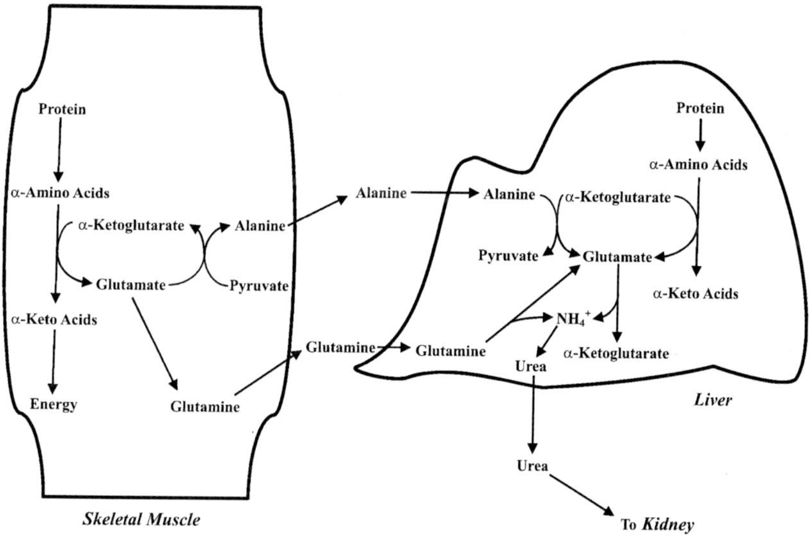

Figure 39-2. General pathway for the degradation of amino acids showing the relationship between extrahepatic tissues and the liver, which is the site of the formation of urea.

transamination, the carbon skeleton is oxidatively metabolized as an energy source during situations such as sustained exercise and caloric insufficiency. The amino group is transported in the circulation to the liver as either glutamine (formed by the enzymatic addition of an amino group to the side-chain group of glutamate) or alanine (formed by enzymatic transfer of the α-amino group from glutamate to pyruvate). Once at the liver, the transferred amino groups are ultimately used in urea synthesis, as described above (see Figure 39-2).

In contrast to skeletal muscle, which is the primary site of endogenous glutamine synthesis, **rapidly dividing cells (e.g., lymphocytes, enterocytes) preferentially use glutamine.** The reason for this is that rapidly dividing cells require both energy as well as precursors for **biosynthetic reactions.** The carbon skeleton of **glutamine** enters **intermediary metabolism via α-ketoglutarate, a Krebs cycle intermediate,** providing the required energy for cellular processes. In addition, the **amino groups of glutamine** are used in **purine and pyrimidine biosynthesis,** which in turn are required for the synthesis of both ribonucleic acid (RNA) and deoxyribonucleic acid (DNA). Indeed, lymphocyte proliferation is greatly accelerated when glutamine is used as a metabolic substrate (as opposed to glucose), leading to suggestions that glutamine deficiency may result in immunosuppression and therefore increased susceptibility to infection. An additional reason for high rates of glutamine utilization in enterocytes appears to be for the synthesis of citrulline. The latter is transported to the kidney, where it is converted to arginine. Arginine is not only important in protein synthesis but also essential in cellular signaling (via nitric oxide production) and in maintaining adequate levels of urea cycle intermediates (Figure 39-3).

Glutamine also plays an important role in maintenance of whole body acid–base balance. High rates of catabolism of positively charged and sulfur-containing amino acids result in the net formation of hydrogen ions. To maintain an acid–base balance, the kidney uses glutamine as a gluconeogenic precursor, resulting in glucose, bicarbonate (HCO_3^-) ion, and ammonium (NH_4^+) ion formation. The bicarbonate is released into the circulation, where it associates with a proton (forming CO_2 and H_2O), thereby increasing blood pH. In contrast, the ammonium ion is excreted (see Figure 39-3).

As noted above, amino acids are not only used for protein synthesis but are also essential for the biosynthesis of other biomolecules. These include carnitine (a derivative of lysine); creatine (a derivative of glycine and arginine); glutathione (a derivative of glutamate, cysteine, and glycine); serotonin and melatonin (derivatives of tryptophan); dopamine, norepinephrine, and epinephrine (derivatives of tyrosine); as well as purines and pyrimidines (requiring aspartate and glutamine for biosynthesis). Alterations in dietary intake of amino acids can influence the rates at which these macromolecules are synthesized. For example, ingestion of a tryptophan-rich meal elevates neuronal serotonin synthesis, resulting in lethargy. Tryptophan crosses the blood–brain barrier via a transporter that is also specific for the branched-chain amino

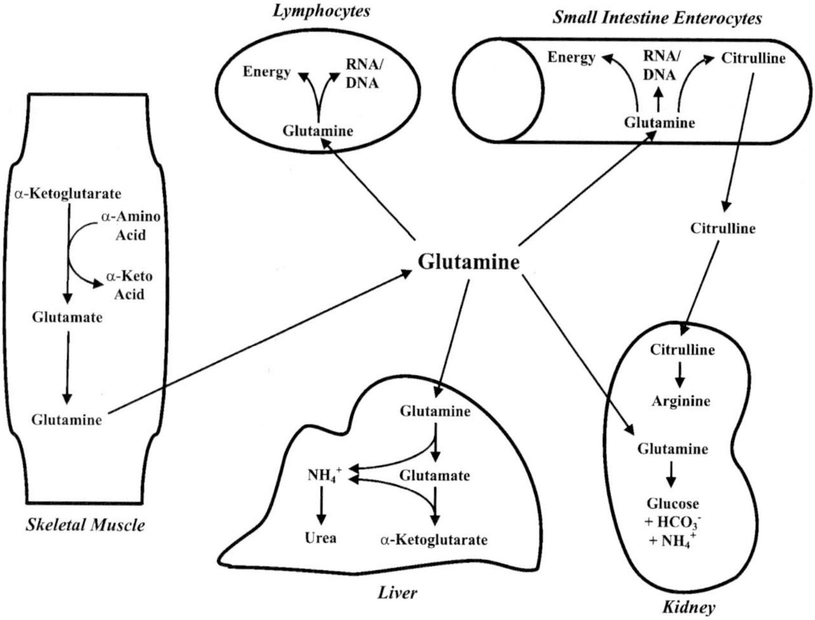

Figure 39-3. Schematic diagram showing the central role of glutamine as a transporter of amino acid nitrogen to various tissues.

acids. Conditions that influence the tryptophan-to-branched-chain amino acid ratio in the circulation have been shown to affect serotonin levels in the brain, because of competition for this transporter. For example, increased skeletal muscle branched-chain amino acid catabolism during starvation is associated with an increase in the blood tryptophan-to-branched-chain amino acid ratio, increased tryptophan uptake in the brain, and increased serotonin synthesis. The latter then influences wakefulness.

Amino acids also play an important role in cellular signaling. The guanidino group from arginine is used in the synthesis of nitric oxide, a ubiquitous, highly reactive signaling molecule that regulates multiple physiologic processes, including blood pressure, immune response, learning, and metabolism. Much less understood is the mechanism by which leucine affects cellular signaling. This amino acid has been termed a "pseudohormone" following the observations that one or more of its early metabolites (e.g., α-ketoisocaproate) is/are among the strongest regulators of protein turnover identified to date.

COMPREHENSION QUESTIONS

For Questions 39.1 to 39.3 refer to the following case.

A 12-year-old girl is presented at the clinic with reports of frequent fainting and lethargy. The girl is 5 ft tall and weighs 80 lb. Skin fold tests show an abnormally low percentage of body fat mass.

[39.1] Which of the following is *least likely* to be consistent with the patient's symptoms?

A. Anorexia nervosa
B. Bulimia
C. Type I diabetes
D. Type II diabetes

[39.2] Which of the following hormones is likely to be severely diminished in the patient described above?

A. Cortisol
B. Epinephrine
C. Glucagon
D. Insulin

[39.3] Which of the following metabolic fluxes would be most consistent in this patient?

A. Protein→amino acids
B. Glucose→fatty acids
C. Glucose→glycogen
D. Fatty acids→triacylglycerol

Answers

[39.1] **D.** Individuals with type II diabetes tend to be either overweight (a body mass index [BMI] between 25 and 30) or obese (BMI greater than 30). This may be due in part to chronic hyperinsulinemia. In contrast, decreased insulin levels associated with type I diabetes, anorexia, and bulimia, will promote lipolysis and therefore decreased adipose mass.

[39.2] **D.** The low body weight and fat mass observed in the patient are consistent with a metabolically "fasted" state. During such a condition, circulating insulin levels will be low, whereas counterregulatory hormones (e.g., glucagon, epinephrine, and cortisol) will be elevated.

[39.3] **A.** Consistent with a metabolically fasted state, decreased circulating insulin levels will signal attenuation of biosynthetic reaction and augmentation of catabolic reactions. The latter will provide necessary fuels to sustain the energetic needs of the body. Elevation of counter-regulator hormones (e.g., glucagon, epinephrine, and cortisol) will synergistically stimulate catabolic processes. Increased proteolysis (protein→amino acids) in tissues such as skeletal muscle and the liver provide amino acids as a direct fuel source, as well as ketogenic and gluconeogenic precursors (depending on the individual amino acid).

BIOCHEMISTRY PEARLS

- Twenty different amino acids are used for protein synthesis, each of which is encoded by at least one codon (the three-nucleotide genetic code) and differ only in their side-chain group.
- Amino acids can be divided into one of two classes, essential and nonessential amino acids. Essential amino acids cannot be synthesized by humans.
- Nitrogen balance means whether there is more or less nitrogen (protein) ingested than excreted.
- With prolonged caloric insufficiency, a large portion of the carbon skeleton is used by the liver for the synthesis of either glucose or ketone bodies, depending on the specific amino acid.
- Rapidly dividing cells (e.g., lymphocytes, enterocytes) preferentially use glutamine for energy and biosynthetic purposes.

REFERENCE

Newsholme EA, Leech AR. Biochemistry for the Medical Sciences. New York: Wiley, 1983.

❖ CASE 40

A 20-year-old female was brought to the emergency department after being found on the dormitory room floor nauseated, vomiting, and complaining of abdominal pain. Her friends were concerned when she did not show up for a biochemistry final at the local university. The patient had been under a lot of stress with finals, a recent breakup with a boyfriend, and trying to find a job. In the dormitory room, one of her friends noticed an empty bottle of Tylenol (acetaminophen) near the bed with numerous pills lying on the ground near their friend. On arrival to the emergency department, the patient was found to be in moderate distress and vomiting. The patient was quickly assessed, and laboratory work was obtained. Patient had a hypokalemia noted on electrolytes and elevated liver enzymes. Her white blood cell count was normal. Her urine drug screen was negative, and her acetaminophen blood level was above 200 μg/mL. The emergency department physician prescribes oral *N*-acetylcysteine to help prevent toxicity from the acetaminophen.

◆ **What is the pathophysiology of the liver toxicity?**

◆ **What is the biochemical mechanism whereby the *N*-acetylcysteine helps in this condition?**

ANSWERS TO CASE 40: ACETAMINOPHEN OVERDOSE

Summary: A 20-year-old college student under increasing stress was found in moderate distress with nausea, vomiting, and abdominal pain with empty bottle of Tylenol (acetaminophen) at her bedside. Oral *N*-acetylcysteine is prescribed.

- **Pathophysiology:** Acetaminophen is metabolized via the cytochrome P450 enzymes into a deleterious product *N*-acetyl benzoquinoneimine, an unstable intermediate, which causes arylated derivatives of protein, lipid, ribonucleic acid (RNA), and deoxyribonucleic acid (DNA), causing destruction of these compounds. Because the liver has high levels of cytochrome P450 enzymes, it is the major organ affected by acetaminophen overdose.
- **Biochemical mechanism of *N*-acetylcysteine:** As glutathione is used to conjugate the acetaminophen toxic metabolite, the antidote *N*-acetylcysteine helps to facilitate glutathione synthesis by increasing the concentrations of one of the reactants of the first synthetic step.

CLINICAL CORRELATION

The patient described has all the initial signs of a deliberate overdose of acetaminophen. Normally acetaminophen is cleared by conjugation with either glucuronic acid or sulfate followed by excretion. Metabolism also takes place, producing an active intermediate capable of binding tissue macromolecules. These conjugative and metabolic pathways involve a number of enzymes that may themselves be compromised to such an extent that the threshold for the concentration that constitutes an overdose is substantially lowered. More typically, overdose concentrations are the result of deliberate ingestion, as in this clinical case, or accidental ingestion, often involving either a child who finds a bottle of acetaminophen and consumes its contents or a disoriented elderly person who loses track of how many tablets have been consumed. Usually, the acetaminophen serum level is drawn and plotted on a nomogram to determine the possibility of hepatic damage. Hepatocyte necrosis with clinical manifestations of nausea and vomiting, diarrhea, abdominal pain, and shock may ensue. Few survivors of an overdose have long-term hepatic disease. The initial therapy is gastric lavage, activated charcoal, supportive care, and administration of *N*-acetylcysteine.

APPROACH TO GLUTHATHIONE AND ACETAMINOPHEN

Objectives

1. Know about the role of glutathione in protection of acetaminophen overdose.
2. Understand that acetaminophen overdose can lead to liver toxicity.
3. Know about the effect of glutathione.
4. Be aware of the mechanism of action of *N*-acetylcysteine in treating acetaminophen toxicity.

Definitions

Phase I drug metabolism: Oxidative metabolism of drugs usually mediated by cytochrome P450 leading to hydroxylation or epoxidation of substrate compounds.

Phase II drug metabolism: Conjugative metabolism of oxidized drugs usually involving hydration of epoxides by epoxide hydrase producing phenolic derivatives, conjugation by uridine diphosphate (UDP)-glucuronyl transferase to produce glucuronide adducts or S-alkylated adducts, or sulfation by sulfotransferase producing sulfated derivatives.

Drug toxicity: Aberrant reaction to a therapeutic agent often depending on individual variations in either the quantity or activity level of specific drug metabolizing enzymes toward the drugs or on individual genetic polymorphisms of drug metabolism enzymes giving higher or lower activities or product profile produced by the genetic variant versus the normally expressed enzyme.

DISCUSSION

The **major pathway of removal of acetaminophen is by formation of a glucuronide conjugate.** The reactions required for formation of acetaminophen glucuronide are shown in Figure 40-1 and depend on the generation of activated glucuronic acid. The first phase is the formation of activated glucuronic acid from glucose. Glucose is phosphorylated to glucose 6-phosphate by hexokinase in an adenosine triphosphate (ATP)-requiring reaction, which constitutes the first step of glycolysis, the basal pathway for cellular energy generation. Phosphoglucomutase, which plays a critical role in glycogen formation, converts glucose 6-phosphate to glucose 1-phosphate. Glucose 1-phosphate is activated to uridine diphosphate (UDP)-glucose by UDP-glucose pyrophosphorylase using UTP and producing pyrophosphate as an additional product. Pyrophosphate is rapidly hydrolyzed to 2 mol of phosphate by pyrophosphatase; this pulls the reaction toward the formation of UDP-glucose. This reaction is also on the pathway to the formation of glycogen. The last step in this activation phase

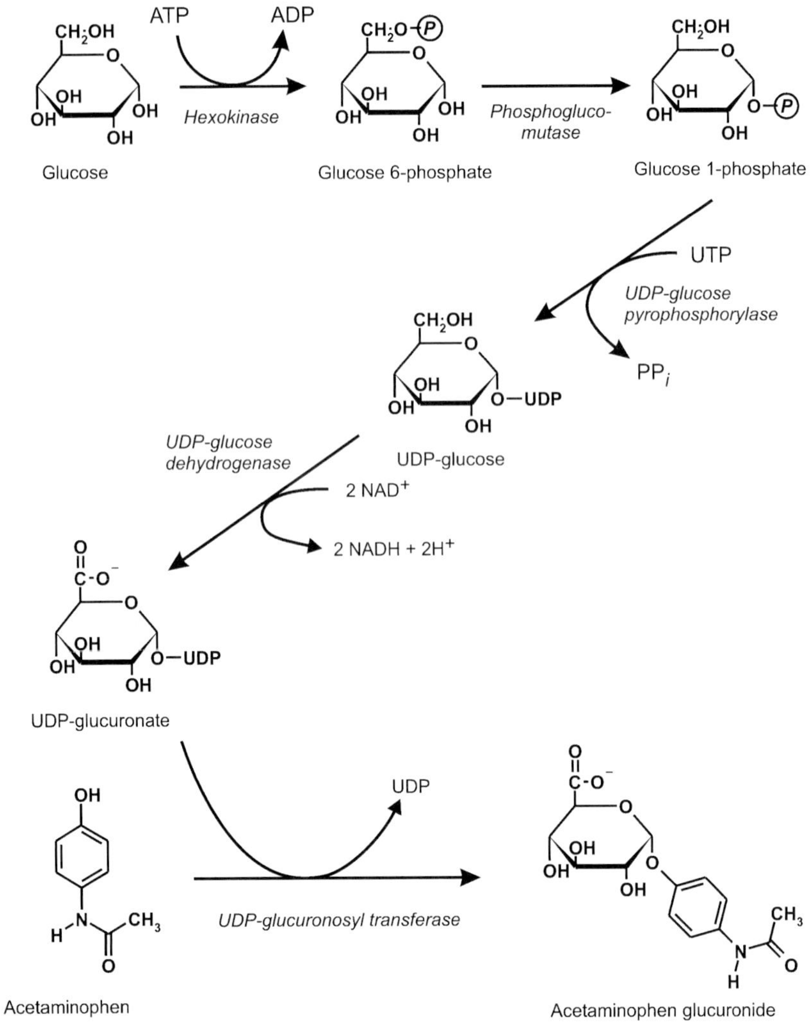

Figure 40-1. Formation of UDP-glucuronate and acetaminophen glucuronide.

is the oxidation of the sixth carbon of glucose in UDP-glucose to the acid level forming UDP-glucuronic acid. This reaction is catalyzed by UDP-glucose dehydrogenase, which also produces 2 mol of nicotinamide adenine dinucleotide (NADH) as an additional product. UDP-glucuronyl transferase, in the next phase of the reaction, catalyzes the transfer of the glucuronide group to the hydroxyl group of acetaminophen forming acetaminophen glucuronide, which is eliminated without toxic effects to the organism.

Alternatively, acetaminophen can be conjugated with organic sulfate for elimination. This alternative pathway shown in Figure 40-2 also consists of

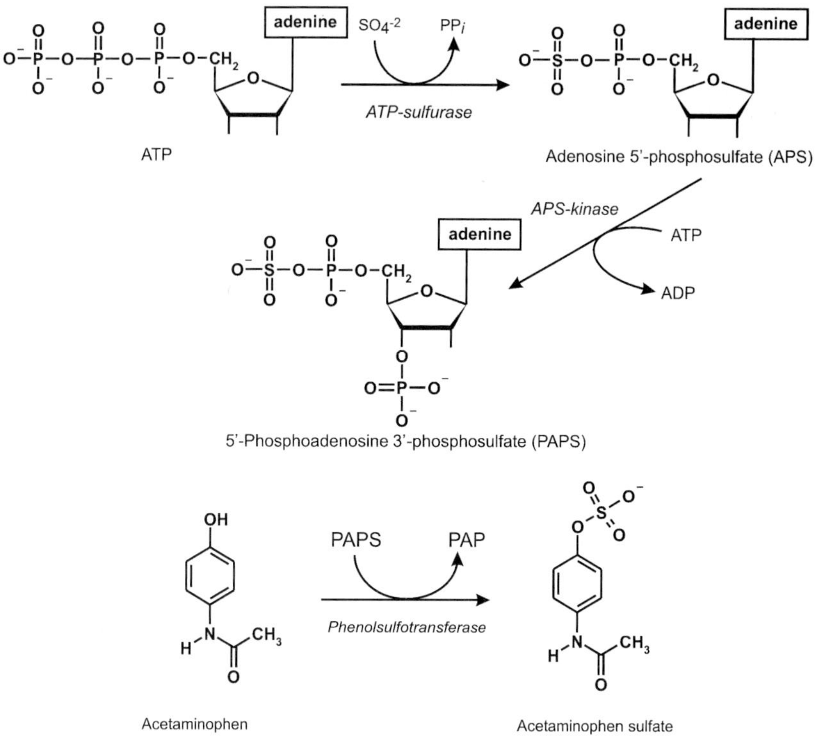

Figure 40-2. Formation of 5′-phosphoadenosine 3′-phosphosulfate (PAPS) and sulfated acetaminophen.

two phases. The first phase consists of preparing activated sulfate for transfer as 5′-phosphoadenosine 3′-phosphosulfate (PAPS), and the second consists of transfer of the sulfate moiety to acetaminophen. In the first step of the first phase, ATP-sulfurylase catalyzes the formation of pyrophosphate and adenosine-5′-phosphosulfate (APS) from ATP and sulfate. In the second step, APS-kinase catalyzes the formation of PAPS and adenosine diphosphate (ADP) from ATP and APS. (The structure for PAPS is shown in Figure 40-2.) In the second phase, phenolsulfotransferase, one of a group of sulfotransferases, transfers the sulfate group from PAPS to acetaminophen to yield adenosine-3′,5′-bisphosphate and acetaminophen sulfate, which is eliminated. In the event that this pathway and the glucuronide forming pathway are both overwhelmed by overdose, more acetaminophen is metabolized by the cytochrome P450 pathway.

Oxidative metabolism of acetaminophen by the cytochrome P450 system, shown in Figure 40-3, is **catalyzed by various cytochromes P450 such as *CYP2E1*** and others. The **deleterious product is *N*-acetyl benzoquinoneimine, an unstable intermediate** shown in Figure 40-3, which can

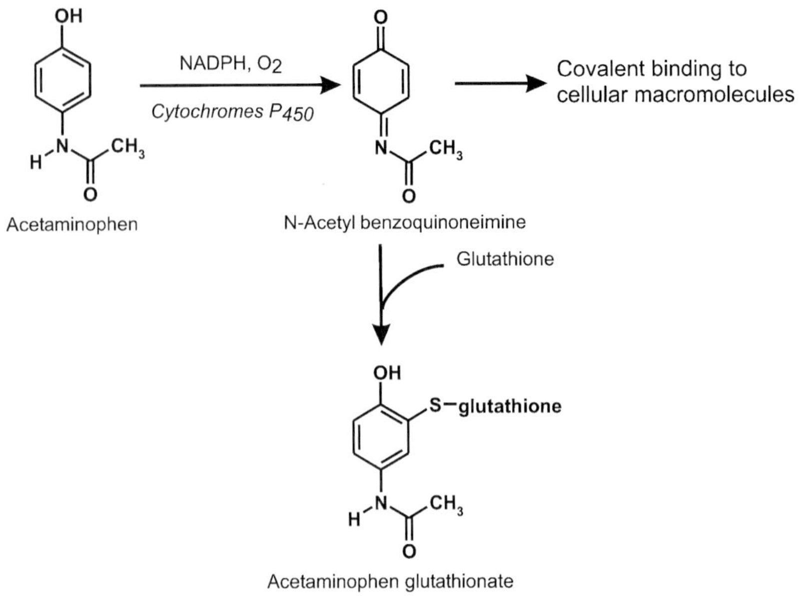

Figure 40-3. Metabolism of acetaminophen.

react with cellular macromolecules damaging them and the integrity of the cells wherein the metabolic alteration occurs. The tissue with the **highest concentration of cytochromes P450 is the liver,** and it is there (along with kidney) that acetaminophen causes the most damage. In the liver, *N*-acetyl benzoquinoneimine can form **arylated derivatives of protein, lipid, RNA, and DNA, causing destruction of these compounds as well as any larger structure** with which they are associated, for example, cellular and subcellular membranes, leading to hepatocyte lysis and loss of cellular contents, such as enzymes, to the circulation. However, *N*-acetyl benzoquinoneimine may also disrupt Ca^{2+} balance, leading to dramatically increased intracellular Ca^{2+} concentrations, that is, 20 μM versus 0.1 μM, which is the normal concentration. Ca^{2+} is a potent and therefore well-regulated signal. Thus dramatic increases in Ca^{2+} concentration would have deleterious effects on the balance of many cellular processes, especially energy generation. The low serum calcium and elevated liver enzyme serum levels seen in the patient reflect hepatocyte lysis.

What mechanisms protect against this metabolite-induced destruction of cellular integrity? The primary defense against radical metabolite intermediate-mediated damage is the glutathione (GSH) system. Glutathione is a tripeptide with an active sulfhydryl group that plays a role in protection of cellular macromolecules from attack by radicals such as organic hydroperoxides or active metabolic intermediates such as *N*-acetyl benzoquinoneimine. The formation of glutathione is summarized in Figure 40-4. γ-Glutamylcysteine synthetase catalyzes the formation of the dipeptide γ-glutamylcysteine from the amino acids glutamate and cysteine using energy provided by the hydrolysis of ATP to ADP and phosphate. Glutathione synthetase catalyzes the addition of glycine to γ-glutamylcysteine to form the tripeptide glutathione, again using 1 mol of ATP.

The role of the glutathione in detoxifying *N*-acetyl benzoquinoneimine is shown in Figure 40-3. The adduct acetaminophen glutathionate is no longer toxic to cells and can be excreted without further damage. **Any process depleting glutathione levels would compromise the ability of the cell to protect itself against *N*-acetyl benzoquinoneimine.** This includes deficiencies of any of the enzymes involved in the synthesis of glutathione or that keep glutathione in its reduced state as well as any other radical generating process that consumes glutathione. In the case of acetaminophen overdose, in addition to giving activated charcoal to absorb excess stomach and intestinal acetaminophen or its conjugates, a strategy to replenish glutathione concentration is important. The most frequently administered compound is ***N*-acetylcysteine.** Easily obtainable and readily soluble, it serves to **facilitate glutathione synthesis by increasing the concentrations of one of the reactants of the first synthetic step.** The committed dipeptide product being made, glutathione is synthesized to replenish depleted supplies. *N*-acetylcysteine seems most effective when given less than 10 hours after ingestion of acetaminophen but is recommended within the first 35 hours after ingestion.

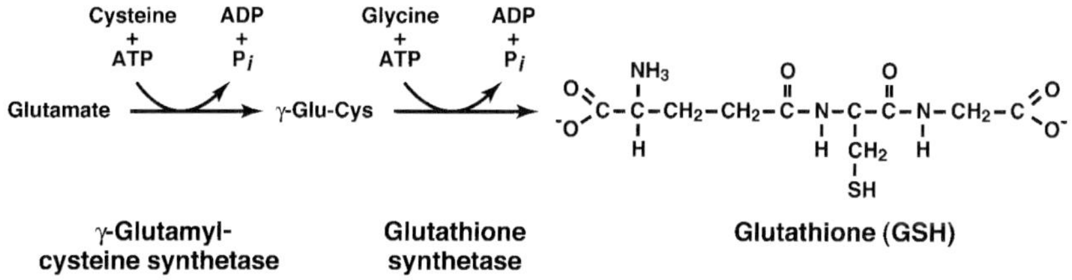

Figure 40-4. Glutathione formation.

COMPREHENSION QUESTIONS

[40.1] A patient admitted to the emergency room with nausea and vomiting showed low serum potassium, elevated blood enzymes, and acetaminophen blood level above 200 μg/mL. The patient was diagnosed with acetaminophen overdose. Acetaminophen is a widely used analgesic. What is the most probable explanation of how a high dose of acetaminophen might have led to a toxic condition?

A. Acetaminophen itself is toxic.
B. Acetaminophen is metabolized to a potential toxic product.
C. Acetaminophen is metabolized to a potential toxic product that is fully conjugated.
D. Acetaminophen is metabolized to a potential toxic product that is partially conjugated.
E. None of the above.

[40.2] In acetaminophen toxicity, which of the following compounds is a potential toxic intermediate?

A. Acetaminophen glucuronide
B. Acetaminophen sulfate
C. *N*-Acetyl benzoquinoneimine
D. Acetaminophen glutathionate
E. *N*-Acetyl-p-aminophenol

[40.3] Glutathione is a critical tripeptide involved in conjugation reactions and in reactions that protect cells from reactive oxygen species. Which of the following components compose glutathione?

A. Glutamic acid, alanine, methionine
B. Glutamine, alanine, cysteine
C. Glutamate, glycine, cysteine
D. Alanine, glycine, cysteine
E. Methionine, glycine, cysteine

Answers

[40.1] **D.** Acetaminophen is not itself toxic, but its intermediate metabolite *N*-acetyl benzoquinoneimine can be toxic unless it is adequately conjugated. In the case presented, a high dose of acetaminophen overwhelmed the conjugative processes, allowing the toxic intermediate to interact with body components and thus causing the nausea, vomiting, and elevated blood enzymes observed.

[40.2] **C.** *N*-acetyl benzoquinoneimine is toxic, whereas acetaminophen glucuronide, acetaminophen sulfate, and acetaminophen glutathionate are nontoxic acetaminophen conjugates. *N*-acetyl-p-aminophenol is another name for acetaminophen.

[40.3] **C.** Glutathione (γ-glutamylcysteinylglycine) is a tripeptide of glutamic acid, cysteine, and glycine in which the amino terminal glutamate residue is in a peptide linkage through its side-chain carboxyl group to the cysteine residue.

BIOCHEMISTRY PEARLS

- ❖ The major pathway of removal of acetaminophen is by formation of a glucuronide conjugate.
- ❖ Acetaminophen is oxidized by the cytochrome P450 system, yielding the deleterious product *N*-acetyl benzoquinoneimine, an unstable intermediate that can react with cellular macromolecules and thus damage them.
- ❖ The liver has a high concentration of cytochrome P450 and is particularly susceptible to acetaminophen toxicity.
- ❖ *N*-acetylcysteine is the antidote to acetaminophen toxicity and facilitates glutathione synthesis by increasing the concentrations of one of the reactants of the first synthetic step.

REFERENCE

Goodman AG, Gilman LS, eds. The Pharmological Basis of Therapeutics, 10th ed. New York: Mc-graw-Hill, 2001.

❖ CASE 41

A 37-year-old female presents to your clinic to discuss her plans for a new vegetarian diet. The patient heard from a friend about a new vegetarian diet that promised rapid weight loss. The diet consists of many leafy vegetables with no pork, chicken, beef, eggs, or milk. She is also planning on working out regularly with the goal of running a marathon within the year. After listening to the patient, you refer her to a nutritionist for further assistance and guidance.

◆ **What is an essential amino acid, and how many are there?**

◆ **List the essential amino acids.**

ANSWERS TO CASE 41: VEGETARIAN DIET—ESSENTIAL AMINO ACIDS

Summary: A 37-year-old female who is planning to undertake a radical new vegetarian diet is in your office for counseling.

- **Essential amino acids:** The amino acids that cannot be synthesized by the body. There are a total of nine essential amino acids.
- **List of essential amino acids:** Histidine, isoleucine, leucine, lysine, methionine, phenylalanine, threonine, tryptophan, and valine.

CLINICAL CORRELATION

The vegetarian should be careful to ensure a balanced ingestion of proteins, fats, carbohydrates, and vitamins. Most animal proteins contain all the essential amino acids; however, vegetable proteins often lack one or more of them. Often, plant amino acids are of low biologic value and incompletely digested. Vegans often benefit from nutritional consultation, which would allow the patient to make sure that she ate foods that complemented each other in providing the essential amino acids.

APPROACH TO ESSENTIAL AMINO ACIDS

Objectives

1. Be aware of the importance of essential amino acids.
2. Know about the synthesis of other amino acids.
3. Be aware of some of the problems with inadequate essential amino acid intake.

Definitions

Essential amino acids: Amino acids that the human body cannot synthesize (or cannot synthesize in sufficient quantities to meet cellular needs) and must be taken in the diet. If essential amino acids are deficient, the result is a condition of negative nitrogen balance.

Nonessential amino acids: Those amino acids that are synthesized by the human body in sufficient quantities to meet cellular needs.

Phenylalanine hydroxylase: The enzyme that converts the essential amino acid phenylalanine to the amino acid tyrosine using tetrahydrobiopterin and molecular oxygen. A genetic deficiency in this enzyme gives rise to the disease phenylketonuria.

Phenylketones: Normally minor metabolites of phenylalanine that result from the transamination of phenylalanine and further reduction. These include phenylpyruvate and phenyllactate. These metabolites are elevated when the conversion of phenylalanine to tyrosine is impaired.

PKU: Phenylketonuria; the pathologic condition of increased excretion of phenylketones in the urine because of impaired conversion of phenylalanine to tyrosine. Classical PKU is because of a genetic deficiency of phenylalanine hydroxylase; however, other causes are deficiencies in dihydropteridine reductase or in the biosynthesis of tetrahydrobiopterin.

DISCUSSION

All 20 amino acids are needed for normal cellular growth and function. Amino acids are the basic building blocks for all proteins synthesized in the cell. In addition, the metabolism of amino acids provides carbon and nitrogen units that are used in the synthesis of numerous important biomolecules including neurotransmitters, heme, purines, pyrimidines, polyamines, and various cellular-signaling molecules. The carbon skeletons of amino acids can also be used as an energy source. After removal of the amino group, amino acids can either be directly oxidized or converted to glucose in the liver, providing carbon units to other tissues for the production of adenosine triphosphate (ATP) through glycolysis and the tricarboxylic acid (TCA) cycle. Certain amino acids can be directly interconverted to intermediates of the TCA cycle to provide a rapid source of carbon units. Finally, catabolism of amino acids provides a means of nitrogen and carbon removal from the body through their metabolism to urea and CO_2. Thus, all amino acids are necessary for life.

Apart from their biologic importance, amino acids are classified as essential or nonessential based on their ability to be synthesized in the body. **The term *essential* amino acid is used to identify those amino acids that must be taken in through the diet** (Table 41-1). There are 10 such amino acids for which biosynthetic pathways do not exist in cells of the human body. In contrast, there are 11 amino acids that are termed *nonessential,* for which the human body has biosynthetic pathways for their generation. One of these amino acids, **tyrosine,** is **synthesized from the essential amino acid phenylalanine** (Figure 38-2). In addition, it should be noted that arginine is listed as both an essential and nonessential amino acid. Arginine is considered nonessential because biosynthetic pathways for its generation do exist in certain cells of the body. Arginine can be synthesized from the amino acid glutamate. **Glutamate** is first converted to **ornithine,** which is then converted to **arginine** by enzymes of the urea cycle. The urea cycle is found only in the liver, and thus the production of arginine through this pathway is limited. The production of arginine through this pathway is likely sufficient for healthy adults but may not be sufficient in times of growth when increased protein synthesis augments the need for amino acids. Thus, in growing children and in adults following surgery or trauma, arginine becomes an essential amino acid.

Table 41-1
ESSENTIAL AND NONESSENTIAL AMINO ACIDS

	NONESSENTIAL AMINO ACIDS (can be synthesized in the body)	
ESSENTIAL AMINO ACIDS (must be obtained from the diet)	Synthesized from Glucose	Synthesized from an Essential Amino Acids
Histidine	Alanine	Tyrosine
Isoleucine	Arginine	
Leucine	Asparagine	
Lysine	Aspartate	
Methionine	Cysteine	
Phenylalanine	Glutamate	
Threonine	Glutamine	
Tryptophan	Glycine	
Valine	Proline	
Arginine†	Serine	

†Required for growth

There is a **continuous need for amino acids for protein synthesis, energy utilization, and the production of biologic mediators.** The most readily available source of all amino acids, but particularly the essential amino acids, is the diet. Protein is taken in through the diet and digested to smaller peptides and amino acids in the stomach and small intestine by specific proteolytic enzymes known as proteases. Because of their different specificities, enzymes work to cleave specific peptide bonds within proteins. Individual digestive enzymes are not capable of completely digesting proteins themselves, but in concert with many different enzymes, most proteins can be efficiently digested. Once released by the digestive enzymes, amino acids are absorbed by epithelial cells of the small intestine for distribution and utilization throughout the body.

Composition of the diet is an important consideration when trying to understand and plan for the uptake of essential amino acids. Not all dietary constituents are equal with regard to the type and amounts of proteins present or the amino acids that can be derived from these proteins by the human digestive tract. Proteins derived from vegetable matter may not contain all the essential amino acids needed and digestion of certain plant proteins can be insufficient to produce certain individual amino acids. In contrast, proteins found in animal products are readily digestible and contain all essential amino acids. Therefore, careful consideration should be given to the dietary intake of individuals who may have or will be undergoing an increased level of exertion.

Regulation of the dietary intake of amino acids can also be important when considering the treatment of certain defects in amino acid biosynthesis. **Phenylalanine** is an essential amino acid that is also used to generate the nonessential amino acid tyrosine. The enzyme that carries out this reaction is the mixed function oxidase **phenylalanine hydroxylase** (PAH). **Inherited deficiencies in PAH are associated with a condition known as phenylketonuria (PKU;** see Case 38). The absence of PAH results in elevations of phenylalanine and various phenylketones, the accumulation of which is associated with the neurologic defects seen in this disorder. PKU can be treated by controlling the dietary intake of phenylalanine. Diets low in phenylalanine will help prevent excessive elevations in phenylalanine. Phenylalanine can not be completely eliminated from the diet because it is an essential amino acid needed for protein synthesis. In the absence of PAH activity, tyrosine becomes an essential amino acid because it cannot be generated from phenylalanine.

COMPREHENSION QUESTIONS

[41.1] All amino acids are needed for the production of proteins in cells and for the synthesis of important biomolecules. Which of the following amino acids, all of which can be synthesized by the human body, must be taken in the diet because it is not synthesized in sufficient quantities to meet the body's needs?

A. Asparagine
B. Glutamine
C. Methionine
D. Proline
E. Tyrosine

[41.2] A young child is in an automobile accident that requires surgical intervention and substantial recovery time in the hospital. A consultation with a nutritionist results in a specific dietary plan. The plan included the supplementation of an amino acid that is typically considered a nonessential amino acid. Which of the following amino acids is an essential amino acid under conditions of enhanced growth or surgical recovery?

A. Alanine
B. Arginine
C. Glycine
D. Serine
E. Tyrosine

[41.3] As part of a standard neonatal screen, an infant is diagnosed with a loss of function genetic defect in the enzyme phenylalanine hydroxylase. Defects in this enzyme can result in a condition known as phenylketonuria (PKU), which results from the toxic effects of phenylalanine derived phenylketones. Fortunately, this condition can be managed by regulating the amount of phenylalanine provided in the diet. Which of the following nonessential amino acids will need to be supplied in the diet of this infant?

A. Alanine
B. Aspartate
C. Glycine
D. Serine
E. Tyrosine

Answers

[41.1] **C.** Methionine can be synthesized by the methylation of homocysteine by the enzyme methionine synthase, which requires the participation of vitamin B_{12} and 5-methyltetrahydrofolate. It is actually the homocysteine component of methionine that is required, since this reaction has the capacity to synthesize enough methionine. However, there is no good dietary source of homocysteine. This conversion of homocysteine to methionine also serves the purpose of making THF available for other biosynthetic reactions.

[41.2] **B.** Arginine is considered nonessential because biosynthetic pathways for its generation do exist in certain cells of the body. Arginine is synthesized from the amino acid glutamate. Glutamate is first converted to ornithine, which is then converted to arginine by enzymes of the urea cycle. The urea cycle is found only in the liver, and thus the production of arginine through this pathway is limiting. The production of arginine through this pathway is likely sufficient for healthy adults, but it may not be sufficient in times of growth where increased protein synthesis augments the need for amino acids. Thus, in growing children and in adults following surgery, arginine becomes an essential amino acid.

[41.3] **E.** Phenylalanine is an essential amino acid that is also used to generate the nonessential amino acid tyrosine. Because tyrosine is made from phenylalanine, it becomes an essential amino acid when phenylalanine levels are limited because of the absence of phenylalanine hydroxylase activity.

BIOCHEMISTRY PEARLS

- The term *essential* amino acid is used to identify those amino acids that must be taken in through the diet and cannot be manufactured.
- There are ten such essential amino acids for which biosynthetic pathways do not exist in cells of the human body.
- Phenylalanine is an essential amino acid that is also used to generate the nonessential amino acid tyrosine.

REFERENCE

Marks DB, Marks AD, Smith CM. Basic Medical Biochemistry: A Clinical Approach. Baltimore, MD: Lippincott Williams & Wilkins, 1996:569–646.

CASE 42

A 38-year-old vegetarian (vegan) Caucasian female presents to her primary care doctor with fatigue and tingling/numbness in her extremities (bilateral). The symptoms have been gradually getting worse over the last year. Upon further questioning she reports frequent episodes of diarrhea and weight loss. On exam, she is pale and tachycardic. Her tongue is beefy red and a neurologic exam reveals numbness in all extremities with decreased vibration senses. A CBC demonstrates megaloblastic anemia.

- **What is the most likely diagnosis?**

- **What is the most likely underlying problem for this patient?**

- **What are the two most common causes of megaloblastic anemia and how would this patient's history and examination differentiate the two?**

ANSWERS TO CASE 42: COBALAMIN DEFICIENCY (VITAMIN B_{12})

Summary: A 38-year-old vegetarian female with gradually worsening fatigue, neurologic and GI symptoms, and megaloblastic anemia.

- **Diagnosis:** Cobalamin (vitamin B_{12}) deficiency.
- **Underlying problem:** Lack of cobalamin intake with complete vegetarian diet (vegan).
- **Causes of megaloblastic anemia:** Folate and cobalamin deficiency. Patients with folate deficiency have similar hematologic and GI findings but do *not* have the neurologic symptoms as with cobalamin deficiency.

CLINICAL CORRELATION

The two most common etiologies of megaloblastic anemia are deficiencies in folate or cobalamin. Cobalamin deficiency can occur from a lack of intake (such as with complete vegetarians), absence of intrinsic factor (either inherited or from removal/damage to gastric mucosa), intestinal organisms, or ileal abnormalities (tropical sprue). Patients present with anemic symptoms such as fatigue, weakness, palpitations, vertigo, and tachycardia. GI symptoms include sore, beefy-red tongue, weight loss, and diarrhea. Both folate and cobalamin deficiencies have similar anemic and GI symptoms. However, cobalamin deficiency also can present with numerous neurologic manifestations including: numbness, parathesia, weakness, ataxia, abnormal reflexes, diminished vibratory sensation, and disturbances in mentation (from irritability to psychosis). Treatment consists of identifying/treating the underlying cause of deficiency and replacement of cobalamin or folate.

Objectives

1. Describe the role of cobalamin in red blood cell formation.
2. Explain why cobalamin deficiency leads to megaloblastic anemia.
3. Develop a comprehension of the role of cobalamin in metabolism.

Definitions

Cobalamin (vitamin B_{12}) deficiency: Inadequate uptake of cobalamin from the diet; often due to lack of intrinsic factor an intestinal transport protein or less often due to unaugmented vegetarian diet that strictly avoids meat or meat products, the source of dietary cobalamin.

Megaloblastic anemia: A disturbance in erythroid cell synthesis due to an impaired DNA synthesis. This results in cells with small nuclei and normal cytoplasm, and a high RNA to DNA ratio. Impaired DNA synthesis is due to decreased thymidylate synthetase conversion (C_1 transfer) of dUMP to dTMP as a result of insufficient cobalamin or folate pool impairment.

One-carbon pool: Folate derivatives that carry a single carbon in various oxidation states (formyl, methenyl, methylene, and methyl) for transfer to acceptor molecules; that is, transfer to deoxyuridine monophosphate to form deoxythymidine monophosphate; transfer to homocysteine to form methionine.

DISCUSSION

The structure of cobalamin (vitamin B_{12}) is shown in Figure 42-1. Cobalamin is somewhat analogous to heme in its structure having as its base a tetrapyrrole ring. Instead of iron as a metal cofactor for heme, cobalamin has cobalt in a coordination state of six with a benzimidazole group nitrogen coordinated to one axial position, the four equatorial positions coordinated by the nitrogens of the four pyrrole groups and the sixth position occupied by either a

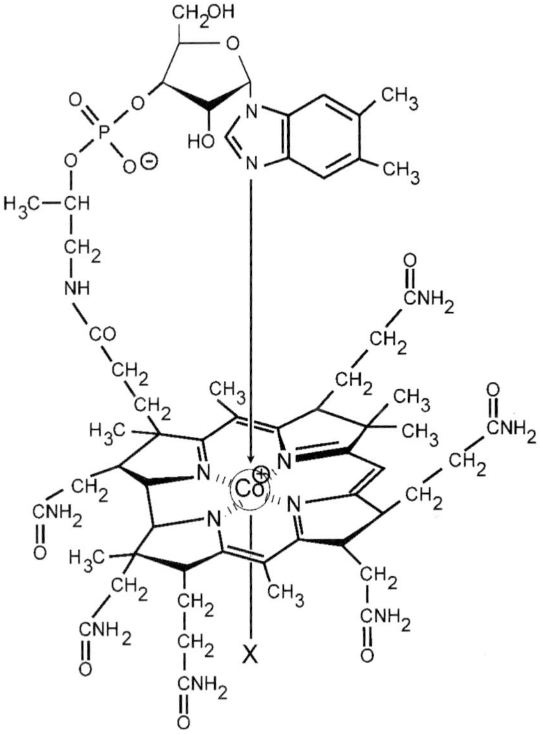

Figure 42-1. The structure of cobalamin, vitamin B_{12}. X = deoxyadenosine in deoxyadenyosylcobalamin; X = CH_3 in methylcobalamin; X = CN^- in cyanocobalamin, the commercial form found in vitamin tablets.

deoxyadenosine group, a methyl group or a CN^- group in the commercially available form in vitamin tablets. Dietary cobalamin is absorbed in the Co^{3+} oxidation state and must be reduced by intracellular reductases to the Co^+ form for use.

Dietary cobalamin is absorbed from animal food sources by a multistage process shown in Figure 42-2. Cobalamin absorption requires the presence of a protein (the intrinsic factor, IF) secreted from the parietal cells of the stomach to bind cobalamin and aid in its absorption in the ileum. The protein is released into the ileum while the cobalamin is transported to the blood stream where it binds specialized serum proteins, the transcobalamins (TC), which transport it to other tissues such as liver where cobalamin can be stored (usually several milligrams are present in liver). In the absence of the intrinsic factor

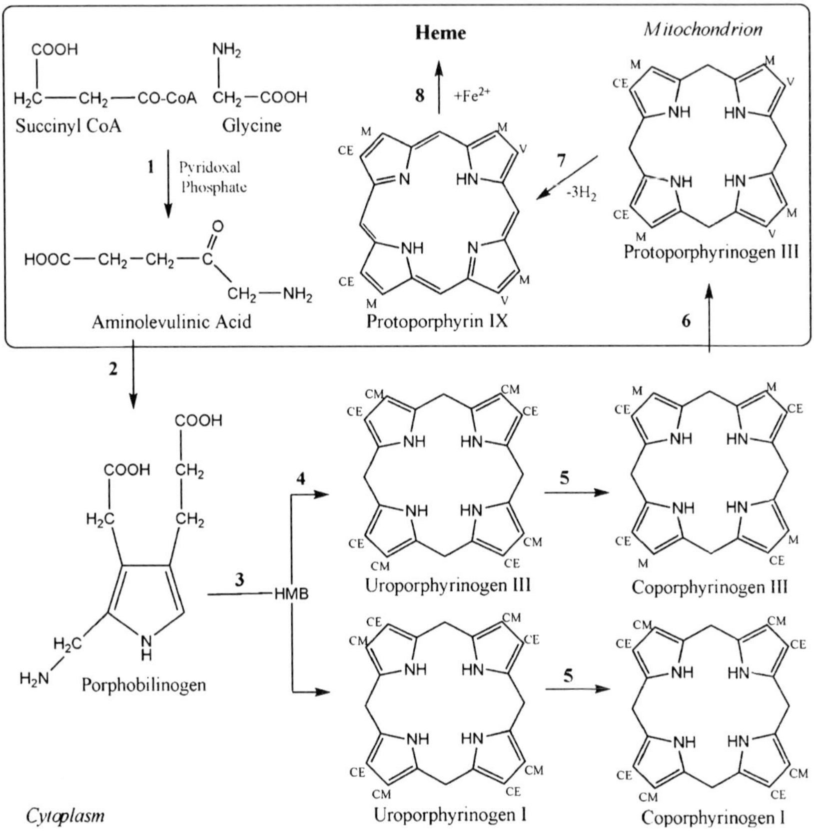

Figure 42-2. Absorption, transport and storage of vitamin B_{12}. IF = intrinsic factor, a glycoprotein secreted by gastric parietal cells; TC = transcobalamins, blood proteins that carry cobalamin to the liver. *(Reproduced, with permission, from D.B. Marks, et al.* Basic Medical Biochemistry: A Clinical Approach. *Philadelphia: Lippincott Williams & Wilkins, 1996:619.)*

inadequate amounts of cobalamin are absorbed (the dietary requirement is approximately 200 ng/day) resulting in megaloblastic anemia. When the root cause of the resultant megaloblastic anemia is absence of or inadequate amounts of intrinsic factor the condition is called pernicious anemia. Other conditions or choices may also eventuate in cobalamin deficiency–induced megaloblastic anemia. This condition is also observed in vegetarians who strictly avoid meat and meat products.

The cause of megaloblastic anemia seen in strict unsupplemented vegetarians is attributed to the effects of cobalamin deficiency on DNA synthesis, specifically the thymidylate synthetase reaction which converts dUMP→dTMP. Inadequate dTMP restricts DNA but not RNA synthesis leading to the appearance of large erythroid cells with small nuclei containing a high ratio of RNA to DNA. These cells are removed from the circulation, thus stimulating erythrogenesis and giving rise to anemia with an elevated presence of megaloblasts.

This process focuses on the role of cobalamin in folate metabolism. As shown in Figure 42-3 cobalamin is required for the conversion of homocysteine into methionine. Cobalamin must first undergo methyl transfer to form methyl cobalamin. It receives the methyl group from N^5-methyltetrahydrofolate thus regenerating tetrahydrofolate to participate in other one-carbon transfers in purine metabolism or pyrimidine remodeling. If there is a cobalamin deficiency then the methionine synthase reaction cannot occur, N^5-methyltetrahydrofolate accumulates and the other C-1 donor forms of tetrahydrofolate cannot be formed. If N^5,N^{10}-methylenetetrahydrofolate, which is required for the methylation of dUMP to dTMP, cannot be formed the thymidylate synthase reaction will be slowed and dTMP levels will drop. An added complication is that the thymidylate synthase reaction produces dihydrofolate unlike all the other C-1 pool reactions, which produce tetrahydrofolate. The dihydro-form must be reduced to the tetrahydro-form by dihydrofolate reductase, which can be inhibited by many drugs. Thus, in the absence of cobalamin methionine synthesis from homocysteine ceases allowing the "trapping" of the folate pool as N^5-methyltetrahydrofolate, diminishing levels of N^5,N^{10}-methylenetetrahydrofolate, and impairing dTMP formation, and therefore DNA synthesis. Cells requiring regeneration due to turnover feel the brunt of this situation early and thus megaloblastic anemia is a result.

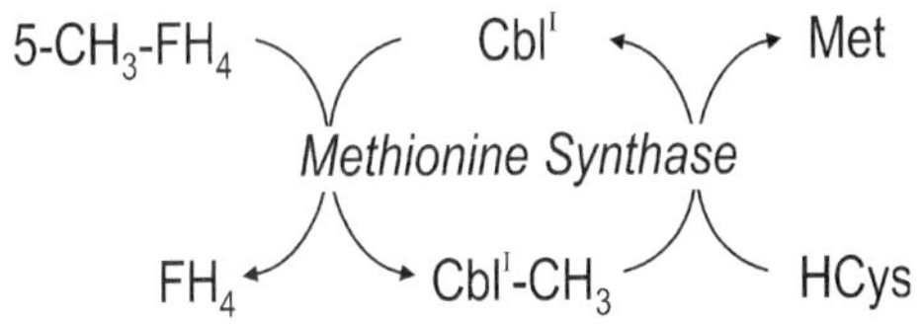

Figure 42-3. The role of cobalamin as a cofactor in the methylation of homocysteine to methionine.

Role of cobalamin in metabolism: Cobalamin plays a vital role in the catabolism of odd-chain fatty acids, threonine, methionine, and the branched-chain amino acids (leucine, isoleucine, and valine) as shown in Figure 42-4. The degradation of each of the compounds named above produces the same metabolite, propionyl-CoA. This activated three-carbon fatty acid enters the energy generating metabolic pathway at the level of the citric acid cycle as succinyl-CoA. The process requires three specialized enzymes. Propionyl-CoA carboxylase adds another carbon to propionyl-CoA to form D-methylmalonyl-CoA in a reaction which requires ATP and the CO_2-binding cofactor biotin. A racemase converts the D isomer of methylmalonyl-CoA to the L isomer. The last step is catalyzed by methylmalonyl-CoA mutase, a deoxyadenosylcobalamin requiring enzyme that moves the acyl-CoA group from the methylene carbon to the methyl carbon to form succinyl-CoA. In cases of cobalamin deficiency this reaction is compromised and leads to an accumulation of methylmalonyl-CoA in serum, which has been suggested as a possible source of neurologic defects seen in cobalamin deficiency by decreasing lipid synthesis. On the other hand, impaired biosynthesis of phosphatidyl choline due to decreased levels of methionine and *S*-adenosylmethionine (SAM) may play a role for the neurologic symptoms of cobalamin deficiency by compromising the repair of demyelination.

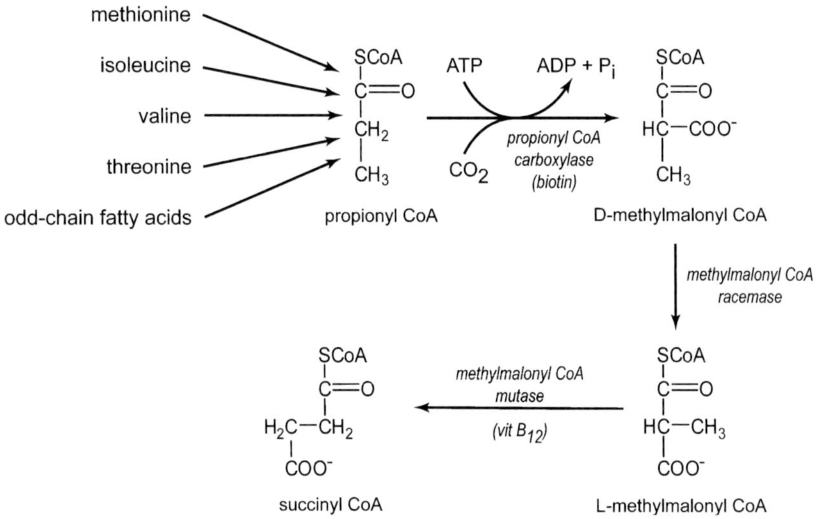

Figure 42-4. The role of vitamin B_{12} in the conversion of propionyl-CoA to succinyl-CoA.

COMPREHENSION QUESTIONS

[42.1] A patient with a tentative diagnosis of cobalamin deficiency is awaiting a comprehensive blood analysis. Which of the following perturbations would NOT fit with the putative diagnosis?

A. Elevated levels of methylmalonic acid
B. Elevated levels of propionic acid
C. Elevated levels of *para*-hydroxyphenylpyruvate
D. Decreased levels of erythrocytes
E. Elevated levels of megaloblasts

[42.2] In evaluating the case presented above your resident asks questions of the role of cobalamin in metabolism. Which of the following statements is true?

A. Cyanocobalamin is the principal form of cobalamin used physiologically
B. Cobalamin is equally active with an iron cofactor
C. Enterocytes produce an intrinsic factor required for uptake of cobalamin in the gut
D. Cobalamin is transported in the blood to tissues by proteins named transcobalamins
E. Cobalamin is active in its 3+ oxidation state

[42.3] Megaloblastic anemia has two most likely causes, deficiency of folate and deficiency of cobalamin. Often treatment of patients with cobalamin deficiency improves in terms of their hematologic features with treatment with folate but not in their neurologic symptoms. What is the most likely explanation for this explanation?

A. Cobalamin deficiencies are not serious
B. Excess folate blunts the trapping of folate as N^5-methyltetrahydrofolate
C. Folate in high concentrations can serve as cofactor for the conversion of homocysteine to methionine
D. Excess folate directly inhibits the destruction of red blood cells
E. Excess folate stimulates erythropoietic tissues to synthesize cobalamin in situ

Answers

[42.1] **C.** *para*-Hydroxyphenylpyruvate is the α-keto acid cognate of tyrosine and is not affected by cobalamin levels. All others result from cobalamin deficiency.

[42.2] **D.** Cobalamin is transported in the blood by transcobalamins. Cyanocobalamin, the pharmacologic preparation of cobalamin available in vitamin pills, is an active as is. Cobalamin is not active with an iron cofactor. The intrinsic factor is produced by parietal cells in the stomach. Cobalamin must be reduced to the Co^{+} state for activity.

[42.3] **B.** Excess folate, by overwhelming the folate pool trapped as N^5-methyltetrahydrofolate, can allow for formation of N^5,N^{10}-methylenetetrahydrofolate which is required for the thymidylate synthase reaction for DNA synthesis and red blood cell formation. Folate is not recognized as a methyl donor by methionine synthase. Folate does not inhibit destruction of erythrocytes. Cobalamin is an important critical vitamin not synthesized by humans.

BIOCHEMISTRY PEARLS

- ❖ Vitamin B_{12} (cobalamin) plays a critical role in DNA synthesis and neurologic function.
- ❖ Cobalmin deficiency can lead to a wide spectrum of hematologic, neuropsychiatric, and cardiovascular disorders that can often be reversed by early diagnosis and prompt treatment.
- ❖ Cobalamin absorption from the gastrointestinal tract requires the presence of a protein (the intrinsic factor, IF) secreted from the parietal cells of the stomach to bind cobalamin and aid in its absorption in the ileum.

REFERENCES

Devlin TM, ed. Text book of Biochemistry with Clinical Correlations, 5th ed, New York: Wiley-Liss, 2002:795–7, 1154–7.

Issalbacher K, et al. Harrison's Principles of Internal Medicine, 13th ed, New York: McGraw-Hill, 1994:1726–32.

Scriver CR, Beaudet AL, Sly WS, et al. The Metabolic and Molecular Basis of Inherited Disease, 8th ed, New York: McGraw Hill, 2001:2165–93.

CASE 43

A 46-year-old male presents to the emergency department with severe right toe pain. The patient was in usual state of health until early in the morning when he woke up with severe pain in his right big toe. The patient denies any trauma to the toe and no previous history of such pain in other joints. He did say that he had a "few too many" beers with the guys last night. On examination, he was found to have a temperature of 38.2°C (100.8°F) and in moderate distress secondary to the pain in his right toe. The right big toe was swollen, warm, red, and exquisitely tender. The remainder of the examination was normal. Synovial fluid was obtained and revealed rod- or needle-shaped crystals that were negatively birefringent under polarizing microscopy, consistent with gout.

- **What is the likely diagnosis?**
- **How would you make a definite diagnosis?**
- **What is the pathophysiology of this disorder?**

ANSWERS TO CASE 43: GOUT

Summary: A 47-year-old male presents with an acute onset of right toe pain in the middle of night after drinking alcohol and no history of trauma or any other joint pain.

- **Diagnosis:** Gouty arthritis.
- **Confirming diagnosis:** Demonstration of the presence of the monosodium urate crystals within the synovial leukocytes or in material derived from tophi under polarizing microscopy.
- **Pathophysiology:** Increased conversion of purine bases to uric acid or a decreased excretion of uric acid by the kidney. Elevated levels of the insoluble uric acid result in precipitation of urate crystals in the joints.

CLINICAL CORRELATION

Gout is a disorder that occurs when uric acid crystallizes in the joints of the body, usually the great toe or large joints. Hyperuricemia is a clinical condition characterized by elevated levels of uric acid. This leads to the formation of sodium urate crystals, which are found primarily in the joints of the extremities and in the renal interstitium. The presence of urate crystals is associated with extreme swelling and tenderness in the joints of the extremities. This condition is often referred to as gout or gouty arthritis. In this condition, elevated levels of uric acid are detectable in the blood and urine, and definitive diagnosis can be made by observing the presence of urate crystals in synovial fluid removed from affected joints. The preference of urate crystal formation in the joints of the extremities, such as the big toe, is thought to be associated with the decreased temperature of the extremities that aids in urate crystal formation when levels exceed solubility.

APPROACH TO URIC ACID CRYSTALLIZATION

Objectives

1. Be familiar with the uric acid pathway.
2. Know about purine base metabolism.
3. Be aware of treatment of gout with allopurinol and colchicine.

Definitions

Allopurinol: An inhibitor of the enzyme xanthine oxidase used to treat gout to decrease the amount of sodium urate in the blood and thus prevent its crystallization in the joints.

Colchicine: A tricyclic, water-soluble alkaloid isolated from the autumn crocus. Colchicine will inhibit microtubule formation and inhibit phagocytosis of urate crystals, thus preventing the inflammatory events associated with a gouty attack.

Gout: An inflammatory event triggered by the crystallization of sodium urate crystals in the joints as a result of increased levels of sodium urate in the blood.

HGPRT: Hypoxanthine-guanine phosphoribosyltransferase; the enzyme that catalyzes the synthesis of inosine monophosphate (IMP) and guanosine monophosphate (GMP) from hypoxanthine and guanine, respectively. It makes up part of the purine salvage pathway, a way of recycling purine bases back to the nucleotides.

Lesch-Nyhan syndrome: A genetic disease caused by a deficiency in HGPRT that is characterized by mental retardation and self-destructive behavior. Lesch-Nyhan patients have increased levels of uric acid and sodium urate that lead to gout and kidney stones.

Purine salvage pathway: The synthesis of purine nucleotides by the condensation of the purine bases with phosphoribosyl pyrophosphate. As the name suggests, it is a way in which purine bases can be recycled back to nucleotides. The purine salvage pathway consists of two enzymes, HGPRT and adenine phosphoribosyltransferase (APRT).

Uric acid: The final product in the degradation of purine nucleotides in human metabolism. The salt form of uric acid, sodium urate, is present at about saturation levels in the bloodstream. When sodium urate levels increase above this point, it can crystallize into sharp crystals, usually in the joints where the temperature is lower.

Xanthine oxidase: The enzyme that catalyzes the final steps in purine degradation to produce urate. This enzyme is inhibited by compounds such as allopurinol in treatment regimens designed to decrease sodium urate concentrations in the blood.

DISCUSSION

Purine bases are used in many important biological processes including the formation of nucleic acids (ribonucleic acid [RNA] and deoxyribonucleic acid [DNA]), **energy currency** (adenosine triphosphate [ATP]), **cofactors** (nicotinamide adenine dinucleotide [NAD], flavin adenine dinucleotide [FAD]), and **cellular signaling** (guanosine triphosphate [GTP], ATP, adenosine). Purines are both synthesized *de novo* and taken in through the diet. Their degradation is a ubiquitous process; however, increased levels of the enzymes that carry out the metabolism of purine bases suggest that purine catabolism is higher in the liver and the gastrointestinal tract. Abnormalities in purine biosynthesis and degradation are associated with numerous disorders suggesting that the regulation of purine levels is essential.

Degradation of purine nucleotides, nucleosides and bases follow a common pathway (Figure 43-1). During purine catabolism, the purine nucleotides

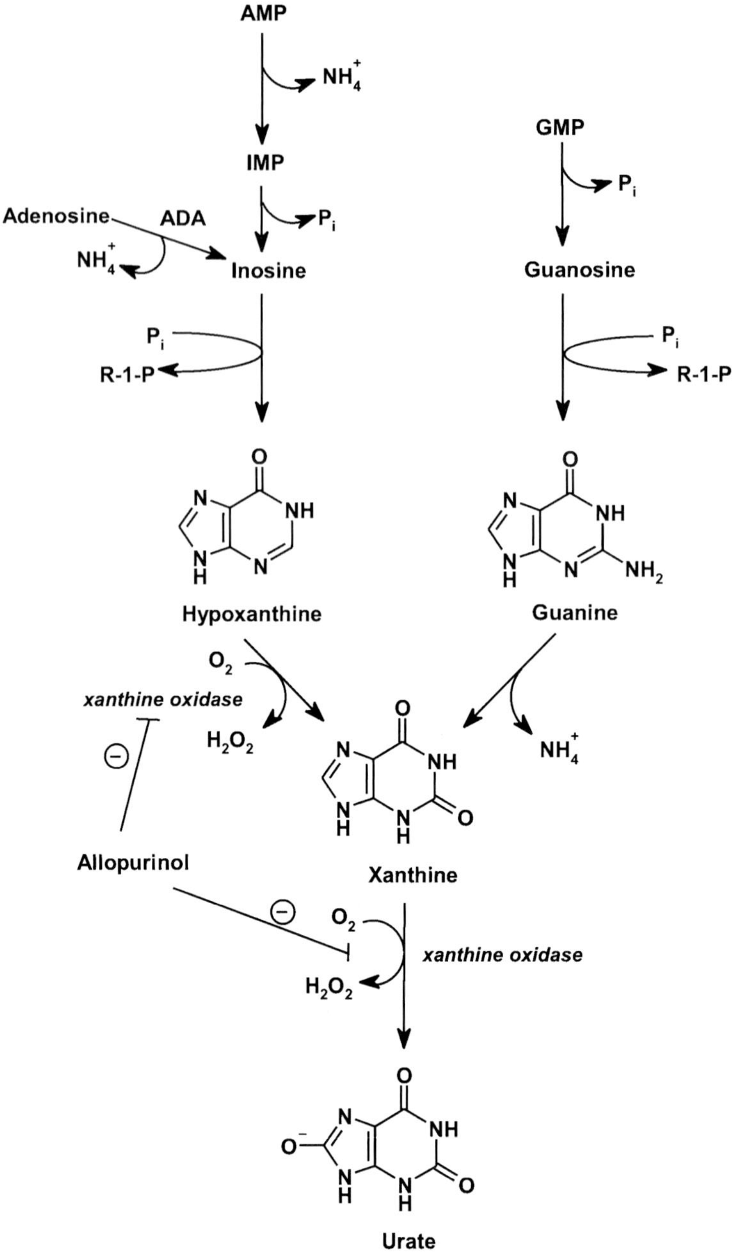

Figure 43-1. Purine catabolic pathway. The enzyme steps inhibited by allopurinol are indicated.

adenosine monophosphate (AMP) and GMP are generated from the dephosphorylation of ATP and GTP, respectively. AMP is then deaminated to IMP by AMP deaminase. Subsequently, GMP and IMP are dephosphorylated by specific nucleotidases to produce the nucleosides inosine and guanosine. Alternatively, AMP can be dephosphorylated to form adenosine, which is then deaminated by adenosine deaminase (ADA) to form inosine. Inosine and guanosine are further broken down by the cleavage of the purine base from the ribose sugar to yield ribose 1-phosphate and hypoxanthine and guanine, respectively. Similar reactions are carried out for the degradation of purine deoxyribonucleotides and deoxyribonucleosides. Guanine is deaminated to form xanthine, whereas hypoxanthine is oxidized to form xanthine by the enzyme xanthine oxidase. Xanthine is further oxidized, again by xanthine oxidase, to form uric acid, which is excreted in the urine. Uric acid has a pK_a of 5.4 and is in the ionized urate form at physiologic pH. Urate is not very soluble in an aqueous environment and the concentration of urate in human blood is very close to saturation. Therefore, conditions that lead to excessive degradation of purine bases can lead to the formation of urate crystals.

Metabolic abnormalities that lead to the overproduction of purine nucleotides through the *de novo* pathway lead to increased purine degradation and subsequent hyperuricemia. An example of this is an increase in the activity of 5-phosphoribosyl-1-pyrophosphate (PRPP) synthetase. This enzyme is responsible for the production of PRPP, which is an important precursor of both purine and pyrimidine de novo biosynthesis. Elevations in PRPP lead to increased purine nucleotide production that can in turn increase the rate of degradation and hence increased uric acid production. Hyperuricemia can also result from defects in the purine salvage pathway. The enzyme hypoxanthine-guanine phosphoribosyltransferase (HGPRT) is responsible for reforming IMP and GMP from hypoxanthine and guanine, respectively. In this manner purine bases are salvaged back into the purine nucleotide pool. Lesch-Nyhan syndrome results from an inherited deficiency in HGPRT. This syndrome is associated with mental retardation and self-destructive behavior, which may be associated with inadequate production of purine nucleotides through the salvage pathway in certain neuronal cells. In addition, Lesch-Nyhan patients have gout resulting from the inability to salvage purine bases, which leads to increased levels of uric acid. Hyperuricemia and gout can also arise from numerous undefined mechanisms that include dietary issues.

One approach for the treatment of gout is to decrease the production of uric acid to prevent the development of urate crystals. **Allopurinol is an inhibitor of xanthine oxidase enzymatic activity** (see Figure 43-1). The administration of allopurinol is an effective treatment of gout because it decreases the amount of uric acid produced, which in turn alleviates the amount of sodium urate crystals that are formed. Additional drugs used for the treatment of gout include alloxanthine, another inhibitor of xanthine oxidase, and **colchicine, which inhibits microtubule formation and prevents phagocytic cells from engulfing the urate crystals.** This prevents the urate crystals from rupturing the phagocytes and causing inflammation in the joints.

COMPREHENSION QUESTIONS

[43.1] A patient presents with extreme swelling and tenderness in the joints of the extremities. Examination of synovial fluid extracted from the big toe reveals the presence of urate crystals and confirms the diagnosis of gouty arthritis. The drug allopurinol is prescribed to inhibit which of the following enzymes?

A. Adenosine deaminase
B. AMP deaminase
C. Nucleoside phosphorylase
D. Uricase
E. Xanthine oxidase

[43.2] Hyperuricemia (gout) is a clinical condition characterized by elevated levels of uric acid that lead to the formation of sodium urate crystals that are found primarily in the joints of the extremities. Which of the following factors contributes most to the formation of urate crystals in the extremities?

A. Decreased blood flow
B. Decreased temperature
C. Exposure to sunlight
D. Increased blood flow
E. Increased mobility

[43.3] Inherited defects in components of purine catabolism and salvage are associated with various conditions and syndromes. The enzyme hypoxanthine-guanine phosphoribosyltransferase (HGPRT) is a key enzyme in the purine salvage pathway. It is responsible for reforming IMP and GMP from hypoxanthine and guanine, respectively. In this manner purine bases are salvaged back into the purine nucleotide pool. Genetic defects that lead to the loss of HGPRT activity are the primary cause for which of the following conditions?

A. Gout
B. Lesch-Nyhan syndrome
C. Orotic aciduria
D. Severe combined immunodeficiency syndrome
E. Tay-Sachs disease

Answers

[43.1] **E.** One approach for the treatment of gout is to decrease the levels of uric acid production to prevent the development of urate crystals. Allopurinol is an inhibitor of xanthine oxidase enzymatic activity. The administration of allopurinol is an effective treatment of gout because it decreases the amount of uric acid produced, which in turn alleviates the formation of sodium urate crystals.

[43.2] **B.** Uric acid has a pKa of 5.4 and is ionized in the body to form urate. Urate is not very soluble in an aqueous environment, and the quantity of urate in human blood is very close to the solubility range. Therefore, situations that lead to excessive degradation of purine bases can increase the urate concentration past the solubility point and lead to the formation of urate crystals. The decreased body temperature found in the joints contributes to the formation of urate crystals under these conditions.

[43.3] **B.** Lesch-Nyhan syndrome results from an inherited deficiency in HGPRT. This syndrome is associated with mental retardation and self-destructive behavior, which may be associated with inadequate production of purine nucleotides through the salvage pathway in certain neuronal cells. In addition, Lesch-Nyhan patients have gout resulting from the inability to salvage purine bases, which leads to increased levels of uric acid. However, most patients with gout do not have a defect in HGPRT but have hyperuricemia resulting from a number of factors, including diet.

BIOCHEMISTRY PEARLS

- Purine bases are used in many important biologic processes including the formation of nucleic acids.
- Because urate is not very soluble in an aqueous environment and the concentration of urate in human blood is very close to saturation, conditions that lead to excessive degradation of purine bases can lead to the formation of urate crystals.
- Allopurinol is an inhibitor of xanthine oxidase enzymatic activity.
- Colchicine inhibits microtubule formation and prevents phagocytic cells from engulfing the urate crystals.

REFERENCES

Becker MA. Hyperuricemia and gout. In: Scriver CR, Beaudet AL, Sly WS, et al., eds. The Metabolic and Molecular Basis of Inherited Disease, 8th ed. New York: McGraw-Hill, 2001:2513–35.

Marks DB, Marks AD, Smith CM, eds. Basic Medical Biochemistry. Baltimore, MD: Lippincott Williams & Wilkins, 1996:633–5.

❖ CASE 44

A 21-year-old healthy male college student went to celebrate his birthday with some friends at a bar. His friends convinced him to have his first beer since he just turned 21. After consuming the beer, he began to experience intense, worsening abdominal pain that was nonspecific in location and described as cramping. Nausea and vomiting then ensued and he was taken to the ER. Upon arrival to the ER, he was found to be very anxious with hallucinations. He was noted to be hypertensive, tachycardic, and diaphoretic. Peripheral neuropathy was also noticed on examination. Initial laboratory test revealed a normal CBC, drug screen, and EtOH level. Serum and urine aminolevulinic acid (ALA) and prophobilinogen (PBG) were both found to be elevated.

- **What is the likely diagnosis?**

- **What is the underlying biochemical problem?**

ANSWERS TO CASE 44: PORPHYRIA (ACUTE INTERMITTENT PORPHYRIA)

Summary: A 21-year-old healthy male patient with sudden onset abdominal pain, nausea and vomiting, hypertension, tachycardia, and peripheral neuropathy after consumption of first alcoholic beverage. Further testing revealed elevated levels of both serum and urine ALA and PBG.

- **Diagnosis:** Porphyria (likely acute intermittent porphyria, i.e., variegate)
- **Biochemical problem:** Enzymatic deficiency in heme biosynthetic pathway

CLINICAL CORRELATION

Porphyrias are inherited disorders in the heme biosynthetic pathway. Porphyrias are classified as either hepatic or erythropoetic depending on primary site of accumulation. Inheritance is usually autosomal dominate. Patients often are asymptomatic unless exposed to factors that increase production of porphyrias (drugs, alcohol, sunlight). Erythropoetic etiologies primarily present with photosensitivity. Hepatic porphyrias present with primarily neurovisceral symptoms such as: abdominal pain, nausea and vomiting, tachycardia and hypertension, peripheral neuropathy, and mental symptoms (hallucinations, anxiety, seizures). Diagnosis is confirmed with elevated levels of ALA and PBG in the urine and serum. Specific tests can be performed to detect which enzyme is deficient (i.e., variegate porphyria is caused by deficiency in the PPO enzyme). Treatment is supportive with avoidance of triggers in the future.

Objectives

1. Describe the biosynthesis of heme.
2. Explain why certain triggers (such as EtOH) cause an increase in ALA and PBG.
3. Explain why treatment with intravenous heme or hematin is effective.

Definitions

Aminolevulinic acid synthase (ALAS): Mitochondrial matrix enzyme that catalyzes the rate-limiting synthesis of ALA via condensation of succinyl-CoA and glycine.

ALA dehydratase (ALAD): Cytosolic enzyme that catalyzes the asymmetric condensation of two molecules of ALA to form PBG.

Autonomic neuropathy: Autonomic nervous disruption or deregulation affecting the cardiovascular, urogenital, gastrointestinal systems; symptoms include abdominal pain, nausea and vomiting, tachycardia and hypertension; (aka visceral neuropathy).

Coproporphyrinogen oxidase (CPO): Mitochondrial enzyme that catalyzes specifically the conversion of coproporphyrinogen III to protoporphyrinogen IX.

Ferrochelatase: Assists in the insertion of the ferrous iron into the protoporphyrin IX; the final step in heme synthesis.

Hematin: Compound similar in structure to heme, except that the iron is in the ferric state (tetracoordinated with the pyrrole nitrogens and a hydroxyl group); can restore negative regulation of ALAS.

Heme: An essential metalloorganic cofactor, consisting of one ferrous iron atom coordinated within a tetrapyrrole ring, protoporphyrin IX.

Porphobilinogen deaminase (PBGD): Cytosolic enzyme that processes six PBG molecules through a hexapyrrole adduct to catalyze the formation of a free linear tetrapyrrole, hydroxymethylbilane (aka **uroporphyrinogen I synthase**).

Porphyria: Any of a number of diseases characterized by derangement in porphyrin metabolism; many are caused by genetic defects in the biosynthetic enzymes.

Protoporphyrinogen oxidase (PPO): Catalyzes the oxidation of protoporphyrinogen IX to produce protoporphyrin IX.

Pyridoxal phosphate: Coenzyme active derivative of vitamin B_6.

Uroporphyrinogen (URO) III cosynthase: Cytosolic enzyme that catalyzes formation of the URO III isomer from hydroxymethylbilane.

URO decarboxylase (UROD): Cytosolic enzyme that catalyzes the removal of the carboxyl groups from the side chains of both URO isoforms converting them to their respective coproporphyrinogens (i.e., COPRO I and COPRO III).

DISCUSSION

The cofactor heme is required for numerous processes throughout the body. Most importantly, the heme iron facilitates systemic oxygen transfer via hemoglobin, participates in mitochondrial electron transport, and mediates oxidative drug metabolism in the liver through various cytochromes P450. All cellular tissues are capable of synthesizing heme, but expression of the pathway enzymes and levels of intermediates are greatest in erythropoietic and hepatic tissues due to high demand for heme incorporation into hemoglobin and cytochromes, respectively. Heme is a metalloorganic compound, consisting of one ferrous iron atom coordinated within a tetrapyrrole ring, protoporphyrin IX. Protoporphyrin IX is derived from eight molecules each of succinyl-CoA and glycine. Heme is a relatively planar molecule and highly stabilized by strong resonance throughout the tetrapyrrole ring system. The structure of heme is shown in Figure 44-1.

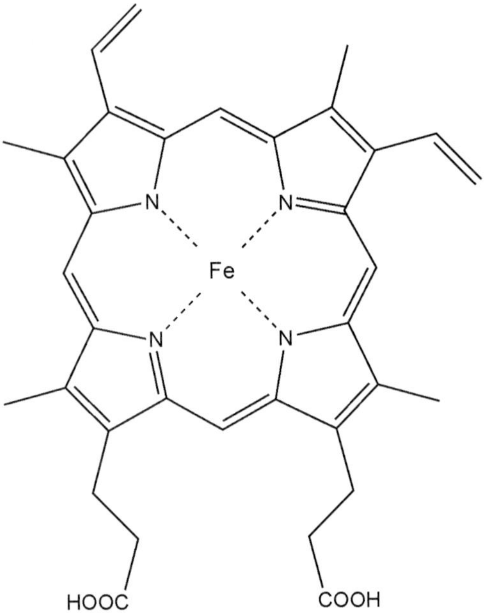

Figure 44-1. Structure of heme.

The biosynthetic pathway includes eight steps, shown in Figure 44-2. The first step is the rate-limiting condensation reaction between succinyl-CoA and glycine to form δ-aminolevulinic acid (ALA). This reaction is catalyzed by a mitochondrial matrix enzyme, **ALA synthase** (ALAS), and requires the cofactor **pyridoxal phosphate.** The ALAS protein is produced in the cytosol but remains unfolded or inactive until it is directed to the mitochondrial matrix where its *N*-terminal signaling sequence is cleaved. Additionally, this enzyme catalyzes the primary regulatory step in heme biosynthesis and is negatively regulated by any accumulation of free heme in the mitochondrial matrix. In the next step of the pathway, **ALA dehydratase** catalyzes the asymmetric condensation of two molecules of ALA to form porphobilinogen (PBG). Through the addition of water and the removal of four amino groups, **PBG deaminase** produces the linear tetrapyrrole intermediate, hydroxymethylbilane (HMB) from four molecules of PBG. At this point, HMB can close either in an enzyme-independent manner to form uroporphyrinogen (URO) I, or in an enzyme-dependent manner through **uroporphyrinogen III cosynthase** to form uroporphyrinogen III, the intermediate that will ultimately lead to heme formation. URO I and URO III differ in the order of the carboxymethyl and carboxyethyl substituents around the tetrapyrrole ring. In the last cytosolic

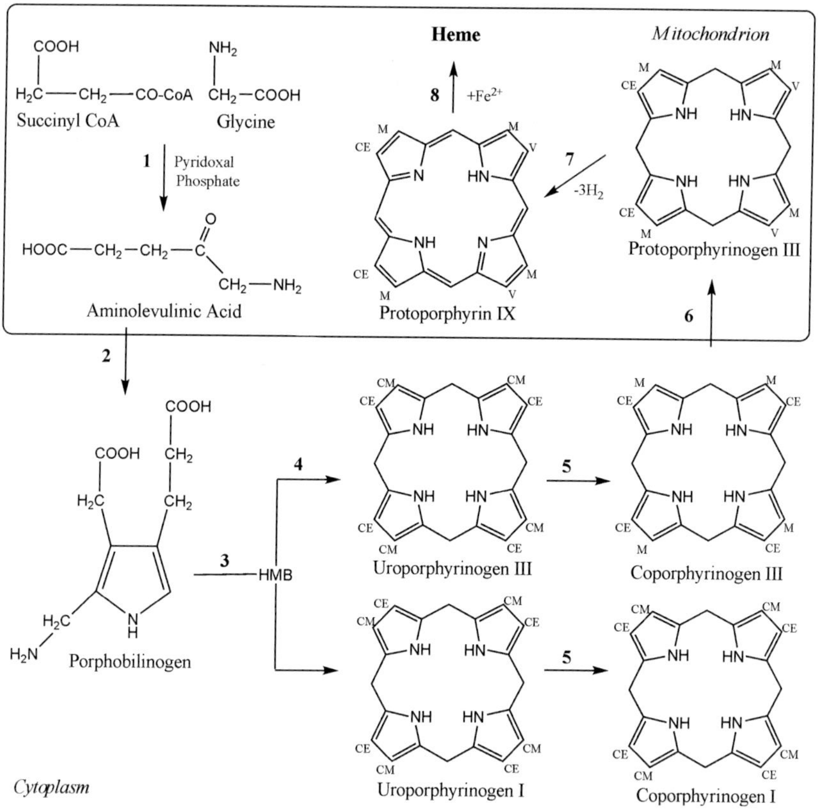

Figure 44-2. Heme biosynthetic pathway involving (1) ALAS, (2) ALAD, (3) PBGD, (4) UROS, (5) UROD, (6) CPO, (7) PPO, and (8) Ferrochelatase. HMB = hydroxymethylbilane; CM = carboxymethyl; CE = carboxyethyl; M = methyl; V = vinyl.

steps, **uroporphyrinogen decarboxylase** (UROD) recognizes either isomer URO I or III and removes specific carboxyl groups leaving coproporphyrinogen I or III, respectively. **Coproporphyrinogen oxidase** (CPO) acts exclusively on the type III isomer of coproporphyrinogen; after its substrate enters the mitochondrion CPO catalyzes the conversion of COPRO III to protoporphyrinogen III. This intermediate is oxidized by **protoporphyrinogen oxidase** (PPO) to form protoporphyrin IX. In the final step of heme synthesis, the enzyme **ferrochelatase** inserts a ferrous iron into water-insoluble protoporphyrin IX. The excess protoporphyrin IX not converted to heme is removed via biliary excretion into the intestine.

The porphyrias are a group of inherited disorders resulting from specific enzymatic defects of the heme biosynthetic pathway. Figure 44-3 shows the heme biosynthetic pathway and characteristics of porphyrias associated with each step. A derangement in any of the involved enzymes can result in an overproduction or backlog of heme precursors prior to the deficient enzymatic step. All of the intermediate compounds in this pathway are potentially toxic. The disorders are divided into two broad groups, erythropoietic and hepatic porphyrias, based on the primary source of precursor accumulation. Furthermore, the porphyrias are also categorized based on the appearance of distinctive acute or chronic symptoms. Pressure to overcome the metabolic "roadblock" contributes to the rapid accumulation of precursors in acute attacks. Early pathway intermediates, that is, ALA and PBG are associated with acute neurovisceral symptoms, whereas later intermediates (which undergo porphyrinogen-porphyrin conversions upon exposure to light) cause skin photosensitivity through free-radial damage. The porphyrins can be detected and identified by spectrofluorometry based on characteristic excitation and emission wavelengths for each porphyrin. For example, serum uroporphyrin levels can be determined by detection at 615 nm after excitation at

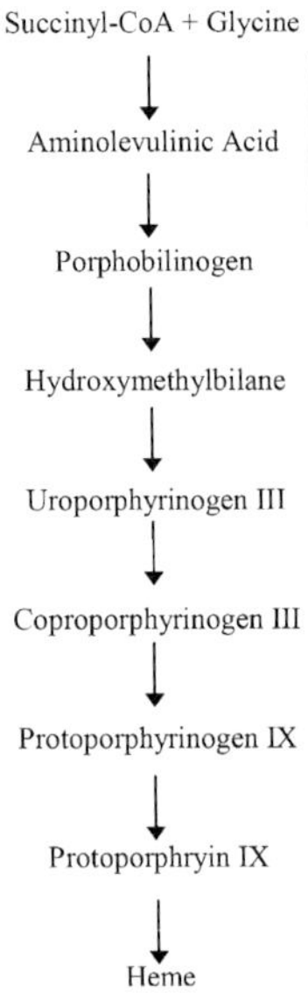

Disease	Inheritance	Enzyme	Type	Manifestation	Symptoms
ADP	Autosomal Recessive	2. ALA Dehydrase	Hepatic	Urinary ALA	Neurovisceral
AIP	Autosomal Dominant	3. PBG Deaminase	Hepatic	Urinary ALA & PBG	Neurovisceral
CEP	Autosomal Recessive	4. URO III Synthase	Erythropoietic	Urinary & RBC URO I and COPRO I	Photosensitivity
PCT, HEP	Variable, Autosomal Recessive	5. URO Decarboxylase	Hepatic/ Erythropoietic	Urinary URO- and Heptaporphyrin; Fecal isocoproporphyrin	Photosensitivity, Hemolytic anemia
HCP	Autosomal Dominant	6. COPRO Oxidase	Hepatic	Urinary ALA, PBG, coproporphryin	Neurovisceral, Photosensitivity
VP	Autosomal Dominant	7. PROTO Oxidase	Hepatic	Urinary ALA & PBG; Fecal protoporphyrin	Neurovisceral, Photosensitivity
EPP	Autosomal Dominant	8. Ferrochelatase	Erythropoietic	RBC protoporphyrin; Fecal protoporphyrin	Photosensitivity

Figure 44-3. Heme biosynthetic pathway and characteristics associated with specific enzyme-deficiency porphyrias. ADP = ALA dehydratase deficiency porphyria; AIP = acute intermittent porphyria; CEP = congenital erythropoietic porphyria; PCT = porphyria cutanea tarda; HEP = hepatoerythropoietic porphyria; HCP = hereditary coproporphyria; VP = variegate porphyria; EPP = erythropoietic protoporphyria.

395 nm to 398 nm, while protoporphyrin is detected at 626 nm. The excretion pathway of the porphyrins is determined by their water solubility. The first possible porphyrin byproduct of the heme biosynthetic pathway, uroporphyrin is by far the most water-soluble, while protoporphyrin the least soluble. Accordingly, uroporphyrin is excreted predominantly in urine, coproporphyrin in urine and in bile, and protoporphyrin exclusively in bile.

In most cases, these disorders are inherited in an autosomal dominant manner where the individual carries one normal allele and one loss-of-function allele. Under normal circumstances, the wild type allele allows for expression of enough functional enzymes to meet the individual's requirement for heme (symptom-free carriers). However certain environmental triggers for example, drugs, alcohol, steroids, fasting, trauma, and/or high stress can increase this demand beyond a level which can be compensated by the single functional allele. Upon exposure to the trigger(s), patients with acute hepatic porphyrias can shift from a compensated phase to a decompensated latent phase (increased precursor production and excretion without symptoms) or to a clinically manifest stage (marked by abdominal, peripheral neurologic, cardiovascular, and psychiatric symptoms).

Alcohol ingestion induces increased activity of ALAS significantly in the liver and moderately in peripheral tissues. This effect is mediated by alleviating the negative regulation of ALAS by free heme. Mitochondrial concentrations of free heme are diminished by increased heme utilization and/or reduced activity of downstream pathway enzymes. In the liver, the demand for heme is exacerbated by requirement for heme incorporation into alcohol-eliminating cytochromes P450. Upon increased activity of ALAS, a genetic defect in either ALAD or PBGD would cause the rapid accumulation of ALA or both ALA and PBG, resulting in **ALAD deficiency porphyria** (very rare) or **acute intermittent porphyria**, respectively. During an attack, excess ALA and PBG produced in the liver are secreted into systemic circulation and are later excreted in the urine. In circulation, these neurotoxic compounds have the greatest effect on the autonomic and peripheral nervous systems resulting in the peripheral neuropathy and neurovisceral symptoms, that is, abdominal pain, nausea and vomiting, tachycardia, and hypertension.

Alternatively, patients with cutaneous porphyrias, for example, **porphyria cutanea tarda** or **hereditary coproporphyria,** chronic symptoms develop as a result of sun exposure (400 nm radiation). The excess porphyrins accumulated in the skin can absorb light energy and transfer it to damaging chemical reactions, such as peroxidation of membrane lipids. This manifests as thickening of the dermal vessel walls leading to damage of the epidermal basement membrane, excessive fragility, blistering, and scarring. In **variegate porphyria,** patients can present with neurovisceral symptoms (acute symptoms are identical but often milder than those of AIP) and/or cutaneous symptoms.

Porphyrias are diagnosed after demonstration and biochemical identification of the increased porphyrin precursor(s). Acute porphyrias are often misdiagnosed, and attacks can be fatal. The intravenous administration of heme or hematin can aid in reestablishing the negative regulation of ALAS during acute attacks. With ALAS activity attenuated, the precursor backlog at the enzymatic deficient step can begin to return to manageable levels, and the patient's condition can shift back to the compensated, symptom-free phase. After diagnosis of acute porphyria is confirmed, strict avoidance of environmental triggers can prevent acute attacks. Most patients with acute porphyrias can lead a normal life; however, complications such as hypertension, chronic renal failure, and hepatoma can become problematic.

COMPREHENSION QUESTIONS

For Questions 44-1 and 44-2 refer to the following case:

A 30-year-old white woman enters the emergency room complaining of nausea, severe abdominal pain, and prolonged constipation. She appeared distraught and was sweating. She described beginning an extremely low calorie diet within the past 2 months in an attempt to lose weight. Physical examination determined rapid heart rate, moderate hypertension, and weakness in the extremities. Also noted were mild dermatitis/blistering on her hands and scarring on her face.

[44.1] You suspect porphyria. Which biochemical laboratory test(s) would be sufficient to determine the type of porphyria?

A. Urinary ALA and PBG.
B. Urinary and fecal PBG and porphyrins.
C. Porphyrin spectrofluorometry (plasma scan).
D. None of the above is sufficient alone.

[44.2] Laboratory tests revealed elevated urinary PBG and coproporphyrin, and plasma fluorescence emission at 626 nm. Which type of porphyria does the patient have and what is the most likely biochemical explanation?

A. Acute intermittent porphyria; PBGD heterozygous enzyme deficiency causing PBG backlog.
B. Porphyria cutanea tarda; UROD homozygous enzyme deficiency causing uroporphyrin backlog.
C. Variegate porphyria; PPO heterozygous deficiency causing protoporphyrin IX backlog.
D. Variegate porphyria; PPO homozygous deficiency causing protoporphyrin IX backlog.

[44.3] The second enzyme in the heme pathway, ALAD, is very sensitive to inhibition by heavy metals, for example, lead. Which of the following test results would distinguish lead-based poisoning from acute intermittent porphyria?

A. Decreased serum and urinary ALA and PBG
B. Increased serum and urinary ALA and PBG
C. Increased serum and urinary ALA, decreased serum and urinary PBG
D. Decreased serum and urinary ALA, increased serum and urinary PBG

Answers

[44.1] **C.** Given the patient's symptoms, neurologic *and* cutaneous, the main porphyrias to consider are hereditary coproporphyria and variegate porphyria. Both HCP and VP show elevated ALA and PBG during acute attacks, so urine and serum tests for these molecules would not distinguish. Similarly both disorders show coproporphyrin in stool. In this case, only a spectrofluorometric assay to distinguish between coproporphyrin and protoporphyrin in plasma would be sufficient.

[44.2] **C.** Variegate porphyria patients present with the same, but generally milder, symptoms as AIP. The emission wavelength at 626 nm is characteristic of protoporphyrin IX; the compound is responsible for the skin sensitivity not observed in AIP. PCT can also be ruled out because the characteristic fluorescence emission for uroporphyrin is 615 nm. Homozygous deletion of PPO is unlikely because the patient's symptoms were triggered by stress in the form of caloric restriction. Patients with homozygous deficiency in the heme synthesis enzymes show early onset and more severe symptoms.

[44.3] **C.** An increase in urine/serum ALA without concomitant increase in PBG indicates disrupted activity of ALA dehydratase. ALA alone can cause the same neurovisceral symptoms as ALA and PBG together cause in AIP. ALAD-deficiency porphyria is an extremely rare autosomal recessive disorder.

BIOCHEMISTRY PEARLS

- The porphyrias are a group of inherited disorders resulting from specific enzymatic defects of the heme biosynthetic pathway.
- The major types of porphyria are each caused by mutations in one of the genes required for heme production.
- The first step in heme formation is the rate-limiting condensation reaction between succinyl-CoA and glycine to form δ-aminolevulinic acid (ALA). This reaction is catalyzed by a mitochondrial matrix enzyme, **ALA synthase** (ALAS).
- Forms of porphyria include ALAS deficiency porphyria, acute intermittent porphyria, congenital erythropoietic porphyria, erythropoietic protoporphyria, hepatoerythropoietic porphyria, hereditary coproporphyria, porphyria cutanea tarda, and variegate porphyria.

REFERENCES

Awad W. Iron and Heme Metabolism, in Devlin TM, ed. Textbook of Biochemistry with Clinical Correlations, 5th ed. New York: Wiley-Liss, 2002:1053–80.

Doss MO, Kuhnel A, Gross U. Alcohol and porphyrin metabolism. Alcohol Alcohol, 2000;35(2):109–25.

Kauppinen R. Porphyrias. Lancet 2005; 365(9455):241–52.

Sassa S. Modern diagnosis and management of the porphyrias. Br J Haematol 2006;135(3):281–92.

CASE 45

A 65-year-old female presents to the clinic feeling tired and fatigued all the time. She has also noticed an increasing problem with constipation despite adequate fiber intake. She is frequently cold when others are hot. Her skin has become dry, and she has noticed a swelling sensation in her neck area. On examination she is afebrile with a pulse of 60 beats per minute. She is in no acute distress and appears in good health. She has an enlarged, nontender thyroid noted on her neck. Her reflexes are diminished, and her skin is dry to the touch.

- **What is the most likely diagnosis?**
- **What laboratory test would you need to confirm the diagnosis?**
- **What is the treatment of choice?**

ANSWERS TO CASE 45: HYPOTHYROIDISM

Summary: A 65-year-old female presents with weakness, fatigue, cold intolerance, constipation, dry skin, and goiter.

- **Diagnosis:** Hypothyroidism
- **Laboratory tests:** TSH and free T4
- **Treatment:** Thyroid hormone replacement with levothyroxine

CLINICAL CORRELATION

Hypothyroidism is quite common in older adults and may present with an indolent course, or it may induce dramatic mental changes such as coma or pericardial effusion with tamponade. The most common etiology is primary hypothyroidism, or failure of the thyroid gland to manufacture and release sufficient thyroid hormone. The diagnosis is established by an elevated thyroid-stimulating hormone (TSH). The treatment is by thyroxine replacement.

APPROACH TO THYROID GLAND

Objectives

1. Be familiar with thyroid hormone metabolism.
2. Know about the regulation of thyroid hormones.
3. Understand the role of iodine on synthesis of thyroid hormone.

Definitions

Graves disease: An autoimmune disorder in which antibodies overstimulate the production of thyroid hormones leading to a condition of hyperthyroidism, or elevated thyroid hormone synthesis and secretion.

Hashimoto thyroiditis: An autoimmune disorder in which the thyroid gland is destroyed by the action of antibodies leading to a condition of hypothyroidism, or decreased thyroid hormone synthesis and secretion.

Thyroid response elements: TRE; A domain on deoxyribonucleic acid (DNA) that will bind the complex formed by thyroid hormone binding to the thyroid hormone receptor. When the complex binds to the TRE, which are located in the promoter region of the DNA, it activates transcription of the gene. When thyroid hormone is not bound to the receptor the receptor acts as a transcription repressor.

Thyroxine: T_4; a thyroid hormone derived from the amino acid tyrosine that contains four iodine atoms per molecule.

TRH: Thyrotropin-releasing hormone; a tripeptide hormone that is released by the hypothalamus and that acts on the anterior pituitary to stimulate the release of thyroid-stimulating hormone.

Triiodothyronine: T_3; a thyroid hormone derived from the amino acid tyrosine that contains three iodine atoms per molecule.

TSH: Thyroid-stimulating hormone or thyrotropin; a glycoprotein hormone released from the anterior pituitary in response to increased levels of TRH. TSH binds to TSH receptors on the basal membrane of epithelial cells of the thyroid gland to stimulate the release of the thyroid hormones, T_3 and T_4.

DISCUSSION

The major circulating forms of thyroid hormone are thyroxine (T_4), containing four iodine atoms per molecule, and triiodothyronine (T_3), with three iodine atoms per molecule (Figure 45-1). Of these, T_3 is eightfold more active. These are synthesized in the thyroid gland after stimulation by thyroid-stimulating hormone (TSH). **TSH binds a G protein–coupled receptor to activate adenylate cyclase and trigger a signaling cascade leading to thyroid hormone biosynthesis.** TSH is released from the pituitary in response to negative feedback by circulating levels of thyroid hormone as well as regulation by circulating levels of thyrotropin-releasing hormone (TRH), a tripeptide synthesized in the hypothalamus.

Thyroid hormones are the only major biochemical species known to incorporate iodine. In fact, **in third-world countries, iodine deficiency is the major cause of hypothyroidism** (deficiency of thyroid hormones). Iodine deficiency is characterized by the development of a goiter, representing enlargement of the thyroid gland. In the **developed world, where iodine deficiency is rare because of the use of iodized salt, autoimmune disorders are a leading cause of thyroid disease.** These are characterized by the presence of antibodies in the blood that either stimulate or damage the thyroid gland. The most common examples are Graves disease, characterized by antibody overstimulation of thyroid hormone production, and Hashimoto thyroiditis, leading to autoimmune destruction of the thyroid gland. In addition, inherited human disorders resulting in mutations in the thyroid hormone receptor abolishing hormone binding have been reported. These individuals exhibit symptoms of hypothyroidism as well as a high incidence of attention-deficit disorder. This trait is genetically dominant, indicating that the mutant receptors act in a dominant negative manner.

Thyroid hormone biosynthesis (Figure 45-2) involves the concentrative uptake of iodide into thyroid cells where it is converted into iodine by **thyroid peroxidase** in the colloid space of the follicular lumen. Iodine is incorporated into tyrosine residues of thyroglobulin contained within the colloid space at the basal surface of the thyroid follicular cell. Tyrosine residues are iodinated

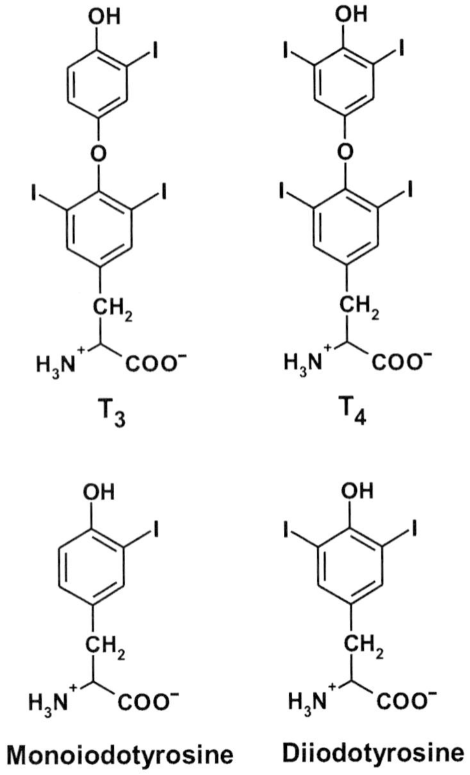

Figure 45-1. Structures of the thyroid hormones thyroxine (T_4) and triiodothyronine (T_3). Also shown are the intermediates monoiodotyrosine (MIT) and diiodotyrosine (DIT) that are also formed on thyroglobulin.

at either one or two sites, and then these residues are coupled to generate either T_3 or T_4 residues within thyroglobulin. The iodinated thyroglobulin is then taken up from the extracellular matrix into the cytoplasm of the thyroid cell where lysosomal proteases cleave T_3 and T_4 from thyroglobulin. The hormones are then carried in the blood bound primarily to thyroid-binding globulin. T_4 is converted to T_3 **in the liver** and to a lesser extent other tissues, accounting for 80 percent of circulating T_3.

Thyroid hormones stimulate protein synthesis in most cells of the body. They also **stimulate oxygen consumption by increasing the levels of the Na^+, K^+-ATPase ion transporter.** The generation of plasma membrane Na^+ and K^+gradients by the Na^+, K^+-ATPase is a major consumer of cellular adenosine triphosphate (ATP), leading to stimulation of ATP synthesis in the mitochondria

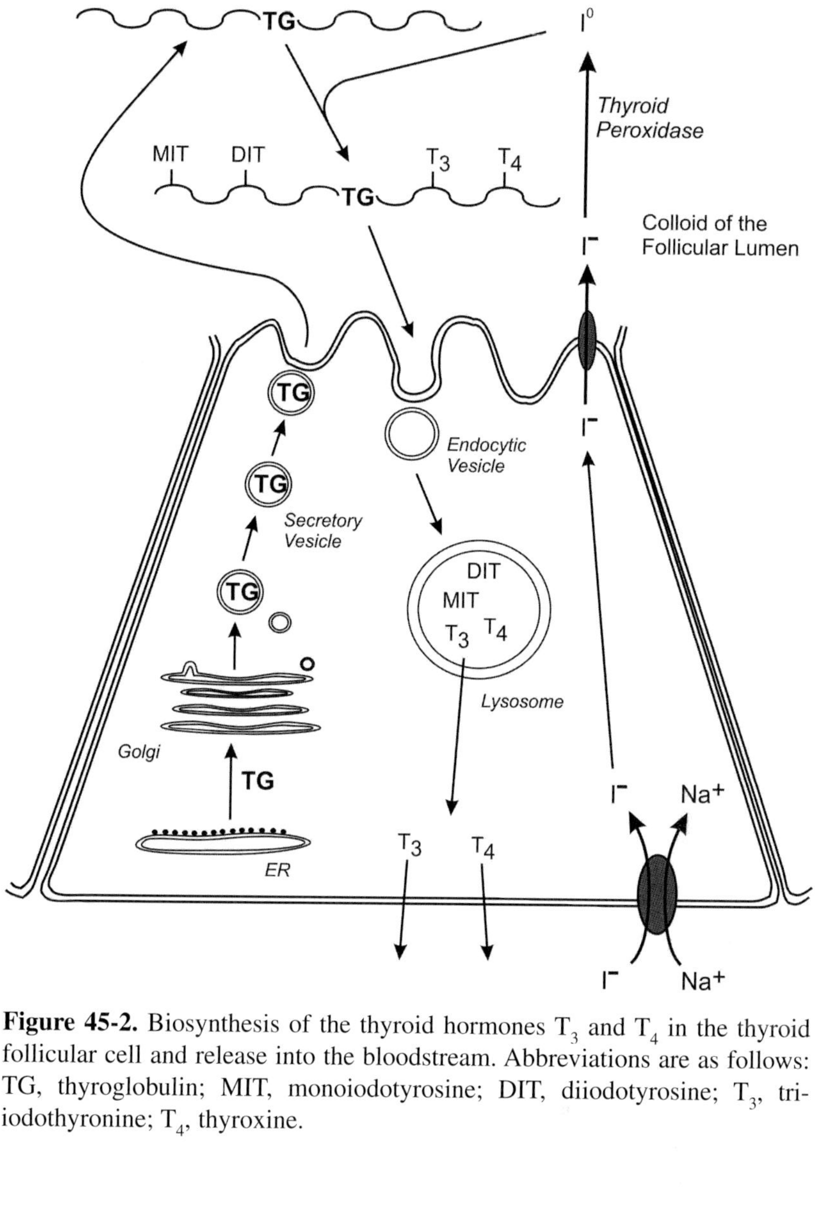

Figure 45-2. Biosynthesis of the thyroid hormones T_3 and T_4 in the thyroid follicular cell and release into the bloodstream. Abbreviations are as follows: TG, thyroglobulin; MIT, monoiodotyrosine; DIT, diiodotyrosine; T_3, triiodothyronine; T_4, thyroxine.

thus directly increasing mitochondrial energy metabolism. By this means, **thyroid hormones aid the conversion of food into energy and heat.** In all of its known actions, thyroid hormone exerts its effects by interaction with its receptor in the cell nucleus and activation of transcription of the target genes.

Thyroid hormone receptors are members of a large nuclear receptor superfamily that includes receptors for steroid hormones, vitamin D_3, and retinoic acid. Members of this receptor superfamily contain a **DNA-binding domain** responsible for binding to hormone response elements contained within the **promoters of target genes.** In addition, members of this superfamily contain a region responsible for specific binding to the hormone or biologically active agent. DNA-binding specificity is mediated by receptor sequence motifs known as *zinc fingers,* owing to their chelation of one zinc ion per loop, or *finger.* In the case of thyroid hormone, receptors bind thyroid response elements (TREs). Thyroid-hormone receptors can bind TREs as monomers, as homodimers, or as heterodimers with the retinoid X receptor, another member of this superfamily. The latter exhibits the highest DNA-binding affinity and is the major functional form of the receptor.

In essence, these **receptors serve as hormone-activated transcription factors that directly regulate transcription of messenger ribonucleic acid (mRNA) from target genes.** In contrast to other members of this superfamily, thyroid hormone receptors bind their sites on the promoter regions of DNA in the absence of bound hormone, usually resulting in transcriptional repression. **Binding of thyroid hormone triggers a conformational change in the receptor, converting it to a transcriptional activator.** In this state, it is competent to bind a group of coactivator proteins including histone transacetylase, an activity that serves to create a more open configuration on adjacent chromatin. Mammalian thyroid receptors are encoded by two different genes, each of which can be alternatively spliced, yielding four different receptor isoforms. These isoforms differ in their functional characteristics as well as their tissue-specific and developmental stage–dependent expression, underscoring the complexity of the multiple physiologic effects of thyroid hormones.

COMPREHENSION QUESTIONS

[45.1] A 25-year old female sought treatment for her constant fatigue, lethargy, and depression. She was small in stature and had previously been diagnosed with attention-deficit disorder. On physical examination she was found to have an enlarged thyroid gland (goiter). Blood tests revealed elevated levels of T_3, T_4, and TSH, yet she did not exhibit typical symptoms of hyperthyroidism. Which one of the following possibilities offers the most likely explanation of her symptoms?

A. Thyroid hormone overproduction because of a thyroid gland tumor
B. Hypersecretion of TSH because of a pituitary tumor
C. Genetic alteration in the thyroid receptor reducing its ability to bind thyroid hormone
D. Mutation in the TSH receptor in the thyroid gland reducing its ability to bind TSH
E. Iodide deficiency in the diet

[45.2] In individuals with iodide deficiency, which one of the following is most likely?

A. TSH levels are elevated and directly stimulate growth of the thyroid gland to a very large size.
B. Mono- and diiodinated thyroid hormone molecules are produced, and elevated levels of these derivatives compensate for the deficiency.
C. TSH levels are decreased, relieving their inhibitory effects on thyroid cell proliferation.
D. Synthesis of the Na^+, K^+-ATPase is increased.
E. Tissue utilization of oxygen is increased.

[45.3] In women taking thyroid hormone replacement pills, the dosage must be adjusted if they start taking birth control pills. Which one of the following best explains this situation?

A. Thyroid hormones block the action of estrogens, so the estrogen dose must be increased.
B. Estrogens block the action of thyroid hormones, so the dose of thyroid hormone must be increased.
C. Progestins block the action of thyroid hormone, so the dose of thyroid hormone must be increased.
D. Thyroid hormones stimulate the action of estrogens, so the estrogen dose must be decreased.
E. Estrogens stimulate the action of thyroid hormone, so the dose of thyroid hormone must be decreased.

Answers

[45.1] **C.** The patient exhibits symptoms of hypothyroidism including goiter, yet thyroid hormone levels are elevated. This pattern can only be explained by resistance of target cells to thyroid hormone, for example, a mutation of the receptor decreasing its binding affinity for hormone. Iodide deficiency would lead to goiter but not increased hormone levels.

[45.2] **A.** Elevation of TSH is the mechanism for goiter formation. Decreased thyroid hormone levels reduce feedback inhibition of TSH secretion by the pituitary. TSH secretion is therefore increased. TSH acts as a growth factor for the thyroid gland, increasing its mass and therefore its capacity to synthesize thyroid hormones.

[45.3] **B.** Estrogens partially block the action of thyroid hormones, making them less effective.

BIOCHEMISTRY PEARLS

- The major circulating forms of thyroid hormone are thyroxine (T_4), containing four iodine atoms per molecule, and triiodothyronine (T_3), with three iodine atoms per molecule.
- TSH binds a G protein–coupled receptor to activate adenylate cyclase and trigger a signaling cascade leading to thyroid hormone biosynthesis.
- In developed countries, where iodine deficiency is rare because of the use of iodized salt, **autoimmune disorders are a leading cause of thyroid disease.**
- Thyroid hormone receptors bind their sites on the promoter regions of DNA in the absence of bound hormone, usually resulting in transcriptional repression.
- **Binding of thyroid hormone triggers a conformational change in the receptor converting it to a transcriptional activator.**

REFERENCES

Barrett EJ. The thyroid gland. In: Boron WF, Boulpaep EL. Medical Physiology: A Cellular and Molecular Approach. Philadelphia, PA: W.B. Saunders, 2003.

Litwack G, Schmidt TJ. Biochemistry of hormones I: polypeptide hormones. In: Devlin TM, ed. Textbook of Biochemistry with Clinical Correlations, 5th ed. New York: Wiley-Liss, 2002.

Pathophysiology of the endocrine system. An online textbook from Colorado State University: http://www.vivo.colostate.edu/hbooks/pathphys/endocrine/

CASE 46

A 32-year-old male was seen in the emergency department yesterday after suffering a concussion and head trauma from a motor vehicle accident. The patient was stabilized in the emergency department and transferred to the intensive care unit (ICU) for observation. The patient had computed tomography (CT) scan of the head that revealed a small amount of cerebral edema but was otherwise normal. During the second day in the ICU, the nurse informs you that the patient has had a large amount of urine output in the last 24 hours. The nursing records report his urine output over the last 24 hours was 6400 cc. He has been given no diuretic medications. A urine osmolality was ordered and was found to be low. His physician remarks that the kidneys are not concentrating urine normally.

- **What is the most likely diagnosis for the increasing dilute urine output?**

- **What is the biochemical mediator that is responsible for this disorder?**

ANSWERS TO CASE 46: DIABETES INSIPIDUS

Summary: A 32-year-old male who had suffered head trauma is in stable condition in the ICU with increasing dilute urine output.

- **Diagnosis:** Diabetes insipidus.
- **Biochemical mechanism:** Absence of vasopressin leading to inability to retain free water.

CLINICAL CORRELATION

Vasopressin is a hormone produced in the hypothalamus and stored in the posterior pituitary gland. It is also called antidiuretic hormone, because its presence stimulates water resorption in the distal renal tubule. Excess antidiuretic hormone can lead to "water intoxication" and hyponatremia. In contrast, lack of vasopressin leads to excess loss of free water and hypernatremia. The clinical presentation is that of a patient who is excessively thirsty, having to drink large amounts of water, and urinating large amounts of dilute urine. Head trauma is one of the most common causes, particularly if the posterior pituitary stalk is disrupted. Excessive water (psychogenic water drinking) may present similarly, but these individuals will have normal decreased urine response to water restriction, and during sleep. In contrast, patients with diabetes insipidus (DI) will have excessive urinary loss even with water restriction and even during the night. The treatment is administration of desmopressin acetate (DDAVP), a synthetic analogue of AVP.

APPROACH TO VASOPRESSIN AND WATER BALANCE

Objectives

1. Understand the role of vasopressin and control of water metabolism.
2. Know the role of aldosterone in regulating salt and water balance.
3. Be aware of hormones that regulate salt and water balance (renin; angiotensin I, II, III).

Definitions

Aldosterone: A mineralocorticoid hormone that is synthesized from cholesterol in the adrenal cortex and released in response to angiotensin II or III. It increases the ability of the kidney to absorb Na^+, Cl^-, and water from the glomerular filtrate.

Angiotensin converting enzyme: An enzyme found primarily in the lung (as well as the vascular epithelium and other tissues) that removes two amino acids from angiotensin I to form angiotensin II.

Angiotensinogen: α_2-Globulin; a 14-amino acid peptide that circulates in the plasma. It is cleaved by the protease renin to give the inactive decapeptide **angiotensin I.** Angiotensin converting enzyme hydrolyzes two amino acids from the C-terminus of angiotensin I to give the active **angiotensin II,** which gives rise to **angiotensin III** by the action of an aminopeptidase.

Diabetes insipidus: The chronic excretion of large quantities of very pale urine of low specific gravity resulting in dehydration and extreme thirst.

Neurophysin: A protein that is secreted with oxytocin and vasopressin from the posterior pituitary gland. Neurophysin binds to these two hormones and stabilizes them.

Renin: A protease that is synthesized by the juxtaglomerular cells of the kidney and secreted into the bloodstream in response to conditions of hypovolemia and hyponatremia. It hydrolyzes circulating angiotensinogen to angiotensin I.

Vasopressin: Antidiuretic hormone; a nine-amino acid peptide that is synthesized by the hypothalamus and controls resorption of water by distal tubules of the kidney. It stimulates the insertion of water channels (aquaporins) into the apical membranes of kidney tubules.

DISCUSSION

Vasopressin (antidiuretic hormone) is a nonapeptide that controls resorption of water by distal tubules of the kidney to regulate the osmotic pressure of blood. It functions to conserve body water by reducing the output of urine, and thus it is known as an **antidiuretic. Vasopressin is synthesized in the supraoptic nucleus of the hypothalamus** where it is bound to a neurophysin protein carrier, packaged in granules, and delivered by intracellular transport to nerve terminals in the **posterior pituitary.** Vasopressin bound to neurophysin is released from the granules in response to increased extracellular osmolarity sensed by hypothalamic osmoreceptors, signaling by atrial stretch receptors or after a rise in angiotensin II levels. **Its secretion is increased by dehydration or stress and decreased after alcohol consumption.**

Vasopressin promotes increased resorption of water in the renal distal tubule by stimulating insertion of water channels or aquaporins into the apical membranes of kidney tubules. Water is resorbed across the renal epithelium into the blood leading to a decrease in plasma osmolarity and an increase in the osmolarity of urine. In DI, this process is impaired, leading to excessive urine production. In the absence of vasopressin, the kidney cannot resorb water and it flows out as urine. This condition can arise from a deficiency in vasopressin secretion from the posterior pituitary as a result of hypothalamic tumors, injury (as in the case of this patient) or infection. Alternately, the condition may result from mutations in the vasopressin receptor or aquaporin genes or other diseases impairing renal response to vasopressin. When injected in pharmacologic doses, vasopressin acts as a vasoconstrictor.

There are two major types of vasopressin receptors, V1 and V2. The V1 receptor occurs in vascular smooth muscle and is coupled via G_q to activation of the phosphoinositide cascade-signaling system and generation of the second messenger inositol trisphosphate (IP_3) and diacylglycerol. V2 receptors are found in kidney and are coupled via G_s to activation of adenylate cyclase and production of the second messenger cyclic AMP.

The steroid hormone aldosterone, synthesized in the zona glomerulosa of the adrenal cortex, also plays an important role in maintaining blood osmolarity. It binds its receptors in the cytoplasm of epithelial cells of the distal colon and the renal nephron, followed by translocation of the hormone-receptor complex to the nucleus and activation of transcription of ion transport genes to increase Na^+ reabsorption and K^+ secretion. Water follows Na^+ movement by osmosis. These transporters include the luminal amiloride-sensitive epithelial Na^+ channel, the luminal K^+ channel, the serosal Na^+, K^+-ATPase, the Na^+/H^+exchanger, and the Na^+/Cl^- cotransporter.

Vasopressin and aldosterone each act on the kidney to increase fluid retention and both are in turn regulated by angiotensin II. Renin is a proteolytic enzyme that is released from the kidney in response to sympathetic neuron stimulation, renal artery hypotension, or decreased Na^+ delivery to renal distal tubules. Renin cleaves circulating angiotensinogen to form the decapeptide angiotensin I. Then angiotensin converting enzyme (ACE) found primarily in lung removes two amino acids from angiotensin I to form the octapeptide angiotensin II. In addition to other important targets in regulating blood volume, arterial pressure and cardiac and vascular function, angiotensin II stimulates aldosterone release from the adrenal cortex and vasopressin release from the posterior pituitary. The actions of angiotensin II are mediated by plasma membrane seven-helix receptors coupled via G_q signaling by the phosphoinositide pathway.

COMPREHENSION QUESTIONS

For Questions 46.1 to 46.3 refer to the following case scenario:

Following brain surgery involving transsphenoidal removal of a pituitary adenoma, a patient experienced polyuria, polydipsia, and nocturia. These symptoms appeared shortly after the surgery and had never been observed previously. Osmolarity of the urine was below normal, even if liquid consumption was restricted. Administration of desmopressin alleviated these symptoms.

[46.1] Which of the following possibilities is the most likely hypothesis to explain these symptoms?

A. Damage to the thirst mechanism from surgical trauma, leading to excessive consumption of liquids
B. Damage to the pituitary or hypothalamus from surgical trauma, leading to decreased secretion of vasopressin
C. Onset of diabetes mellitus following surgery
D. Renal injury
E. Oversecretion of angiotensin II following surgery

[46.2] Desmopressin acts mainly by which of the following mechanisms?

A. Stimulating aldosterone secretion by the adrenal gland
B. Increasing synthesis of Na^+ transporters in kidney distal tubule
C. Increasing aquaporin insertion into renal distal tubule apical membranes
D. Acting as an insulin sensitizer
E. Stimulation of angiotensin II release

[46.3] Which of the following would be least likely to stimulate vasopressin release from the posterior pituitary?

A. Dehydration
B. Stress
C. Angiotensin II
D. Atrial stretch receptors
E. Aldosterone

Answers

[46.1] **B.** Desmopressin is a vasopressin analog. Symptoms are consistent with DI arising from surgical trauma to the pituitary or hypothalamus and impairment of vasopressin release. The other possibilities would not be alleviated by desmopressin.

[46.2] **C.** Vasopressin, and its analog desmopressin, acts by increasing insertion of aquaporin (water channels) into the renal distal tubule membrane, permitting increased resorption of water from the urinary filtrate.

[46.3] **E.** Aldosterone acts independent of vasopressin to increase water resorption by the kidney, by stimulating the insertion of ion transporters into the membrane of distal colon and kidney distal tubule.

BIOCHEMISTRY PEARLS

- Vasopressin (antidiuretic hormone) is a nonapeptide that controls resorption of water by distal tubules of the kidney to regulate the osmotic pressure of blood.
- **Vasopressin is synthesized in the hypothalamus and stored in the posterior pituitary.**
- Vasopressin is released in response to increased extracellular osmolarity sensed by hypothalamic osmoreceptors, signaling by atrial stretch receptors or after a rise in angiotensin II levels. **Its secretion is increased by dehydration or stress.**
- Vasopressin promotes increased resorption of water in the renal distal tubule by stimulating insertion of water channels or aquaporins into the apical membranes of kidney tubules.

REFERENCES

Booth RE, Johnson JP, Stockand JD. Aldosterone. Adv Physiol Educ 2002;26(1–4):8–20.

de Gasparo M, Catt KJ, Inagami T, et al. International union of pharmacology. XXIII. The angiotensin II receptors. Pharmacol Rev 2000;52(3):415–72.

Litwack G, Schmidt TJ. Biochemistry of hormones I: polypeptide hormones. In: Devlin TM, ed. Textbook of Biochemistry with Clinical Correlations, 5th ed. New York: Wiley-Liss, 2002.

Pathophysiology of the endocrine system. An online textbook from Colorado State University: http://www.vivo.colostate.edu/hbooks/pathphys/endocrine/

❖ CASE 47

A 36-year-old male comes into the physician's office because his hands and feet are "swelling," and his face has coarse features with oily skin. On examination, he is above the 95th percentile for height at his age. He is noted to have some coarse facial features including large nose, large tongue, and frontal bossing of his forehead. His hands are enlarged with soft tissue swelling, and his heel pad is thickened. He is noticed to have a slightly enlarged liver and spleen. The remainder of the examination was otherwise normal.

- **What is the most likely diagnosis?**
- **What is the biochemical mechanism of this disorder?**

ANSWERS TO CASE 47: ACROMEGALY

Summary: A 36-year-old male comes into the physician's office with height greater that the 95th percentile, coarse facial features (frontal bossing, macroglossia, large nose), oily skin, organomegaly, and thickened soft tissue of the hand and feet.

- **Diagnosis:** Acromegaly
- **Biochemical mechanism:** Excess and autonomous secretion of growth hormone

CLINICAL CORRELATION

Acromegaly is a disorder with excessive growth hormone (GH), usually as a result of autonomous secretion from a nonmalignant anterior pituitary tumor. When acromegaly affects individuals prior to bone growth plate closure, giantism may result; after the bone growth plates close, the patients usually develop coarse features and large hands and feet. The diagnosis is confirmed by demonstrating the failure of GH suppression within 1 to 2 hours of an oral glucose load (75 gm), because GH is usually decreased with glucose. As a consequence of the pulsatility of GH secretion, a single random GH level is not useful for diagnosis. Morbidity can ensue because of diabetes or hypertension. The most feared complications are cardiac in nature, which may affect up to 30 percent of patients. Cardiac arrhythmias, cardiomyopathy, left ventricular hypertrophy, hypertension, and coronary heart disease can be present. Arrhythmias can be associated with cardiomyopathy or coronary heart disease. The primary therapy of acromegaly is surgical, but there is often only partial response. Thereafter somatostatin analogues or dopamine agonists are used, and as a last line of therapy, radiation.

APPROACH TO GROWTH HORMONE ABNORMALITIES

Objectives

1. Be familiar with growth hormone regulation and function.
2. Know about the association with other hormonal problems (diabetes mellitus [DM], thyroidism, etc.).

Definitions

Ghrelin: A 28-amino acid peptide that is octanoylated on serine-3. Ghrelin is synthesized and secreted by endocrine cells in the stomach and will bind to somatotrophic cells in the anterior pituitary to promote the release of growth hormone (somatotropin).

GHRH: Growth hormone–releasing hormone; a peptide synthesized and secreted by the hypothalamus. GHRH stimulates the synthesis and secretion of GH.

IGF-1: Insulin-like growth factor-1, also called somatomedin C; a 70-amino acid peptide that shares structural homology with proinsulin. IGF-1 is synthesized and released by the liver in response to growth hormone binding to receptors on the hepatocyte.

JAK2: Janus kinase 2; an enzyme that will phosphorylate tyrosine residues on target proteins. JAK2 is bound to the growth hormone receptor but is inactive until the receptor binds the hormone. When hormone binds the receptor, it triggers dimerization and activates JAK2.

Somatotropin: Growth hormone; a 191-amino acid polypeptide hormone that is synthesized and secreted from the somatotroph cells of the anterior pituitary. It acts primarily on hepatocytes, although muscle and adipose cells also have receptors for growth hormone.

DISCUSSION

Growth hormone (somatotropin) is a polypeptide synthesized and secreted by somatotrophs in the anterior pituitary. Growth hormone synthesis and secretion is **stimulated by the hypothalamic peptide growth hormone–releasing hormone (GHRH). Somatostatin,** a polypeptide produced by several tissues including the hypothalamus, **opposes GHRH stimulation of growth hormone release.** High levels of insulin-like growth factor-1 (IGF-1) directly suppress growth hormone production in somatotrophs and also act indirectly by stimulating somatostatin release from the hypothalamus. Furthermore, growth hormone itself directly feedback inhibits its synthesis in the somatotroph and also inhibits GHRH production by the hypothalamus. A potent stimulator of growth hormone release is ghrelin, a polypeptide hormone secreted by the stomach that acts directly on somatotrophs. In general, many environmental factors including stress, exercise, nutrition, and sleep influence circulating levels of growth hormone. The overall effect is a pulsatile pattern of growth hormone release reaching a maximum in children and young adults shortly after the onset of deep sleep. Levels of growth hormone fluctuate dramatically and are suppressed in normal individuals after oral glucose administration.

Growth hormone plays a major role in cell proliferation and in regulating protein, lipid, and carbohydrate metabolism. It stimulates protein synthesis and the accompanying amino acid uptake in many tissues. In adipocytes, it increases fat utilization by stimulating triglyceride breakdown and oxidation. It **opposes effects of insulin** by suppressing its ability to stimulate glucose uptake and by stimulating gluconeogenesis. Interestingly, injection of growth hormone stimulates insulin secretion, leading to hyperinsulinemia.

In most of these examples, growth hormone acts at two levels. Growth hormone acts directly by binding its receptors in the plasma membrane of target cells to influence cell proliferation and metabolism. Binding of growth hormone to its receptor triggers receptor dimerization. In the absence of bound hormone, the receptor is a monomer and is bound on its cytoplasmic side to an inactive protein kinase (Janus kinase 2 [JAK2]). The JAK2 protein kinase is a different polypeptide from the receptor and the receptor itself lacks protein kinase activity. After hormone binding and dimerization, JAK2 is activated by cross-phosphorylation. Active JAK2 phosphorylates target proteins and itself on specific tyrosine residues to activate them, thus initiating an autocatalytic regulatory cascade.

Growth hormone also exerts important indirect effects by stimulating the liver and other tissues to secrete IGF-1. IGF-1 stimulates the proliferation of chondrocytes (cartilage cells) leading to bone growth. It also stimulates myoblast proliferation, leading to increased muscle mass. The IGF-1 receptor contains a tyrosine kinase activity in its cytoplasmic domain that is activated autocatalytically after IGF-1 binding, triggering activation of downstream signaling molecules. In this case, tyrosine kinase activity resides on the same polypeptide as hormone-binding activity.

Normal proliferation of somatic cells requires both thyroid hormone and growth hormone. Thyroid hormone stimulates growth hormone secretion, and many thyroid hormone actions on the insulin-like growth factor (IGF) system can be explained by this mechanism.

The profound physiologic role of growth hormone is revealed by conditions resulting from either its deficiency or excess. For example, **mutations in growth hormone or its receptor lead to dwarfism.** By contrast, **excessive secretion leads to giantism, if expressed before the growth plates have closed, or acromegaly, if overproduction is initiated in the adult.** Usually, growth hormone overproduction in the adult is the result of a noncancerous pituitary tumor. Overgrowth (thickening) of bones and connective tissues leads to the characteristic features of acromegaly, with accompanying enlargement of other tissues including the heart. In women, breast milk secretion may result. Left untreated, acromegaly may lead to DI (glucose intolerance), hypertension, heart failure, and sleep apnea.

COMPREHENSION QUESTIONS

[47.1] A 30-year-old female of normal weight was recently diagnosed with type II diabetes and hypertension. Menstrual cycles were irregular. In appearance she had unusually coarse features; a noticeable enlargement of the tongue, hands, and feet; and a deep voice. Although not pregnant or nursing, she unexpectedly began producing breast milk (galactorrhea). Which one of the following possibilities is most likely to explain all of these symptoms?

A. Hyperinsulinemia and insulin resistance
B. Pituitary tumor and growth hormone overproduction
C. Testosterone overproduction
D. Ovarian cysts
E. Transforming growth factor β overproduction

[47.2] A 7-year-old child who was very small for his age began receiving treatment with growth hormone. Which one of the following metabolic alterations is most likely to be observed after beginning this treatment?

A. Inhibition of cartilage formation
B. Inhibition of gluconeogenesis
C. Inhibition of triglyceride breakdown and oxidation in adipocytes
D. Stimulation of IGF-1 secretion
E. Stimulation of protein breakdown

[47.3] The following polypeptide hormones each interact with receptors in the plasma membrane of their target cells. Which one triggers a signaling pathway that is directly stimulated by treatment of the cell with an inhibitor of cyclic AMP phosphodiesterase?

A. ACTH
B. Epidermal growth factor
C. Growth hormone
D. Insulin
E. Nerve growth factor

Answers

[47.1] **B.** The symptoms are consistent with acromegaly, or GH overproduction, in the adult. This condition usually is caused by a pituitary tumor. In the female, breast milk secretion is sometimes observed, either as a result of GH overproduction or an accompanying overproduction of prolactin. Growth hormone opposes insulin action resulting in decreased glucose utilization and symptoms of diabetes mellitus. Growth hormone also increases IGF-1 production by the liver, leading to stimulation of cartilage synthesis and muscle mass. Excessive bone and tissue growth lead to the characteristic coarse facial features, enlarged tongue and heart, bone thickening, and other characteristics associated with this syndrome.

[47.2] **D.** Growth hormone stimulation of IGF-1 secretion is an important aspect of its action.

[47.3] **A.** ACTH acts by activation of adenylate cyclase and production of cAMP. Inhibition of cAMP breakdown synergistically increases the intracellular response to this hormone.

BIOCHEMISTRY PEARLS

- ❖ Growth hormone (somatotropin) is a polypeptide synthesized and secreted by somatotrophs in the anterior pituitary.
- ❖ Mutations in growth hormone or its receptor lead to dwarfism.
- ❖ Excessive GH secretion leads to giantism, if expressed before the growth plates have closed, or acromegaly, if overproduction is initiated in the adult.
- ❖ The most common cause of growth hormone overproduction in the adult is a noncancerous pituitary tumor.

REFERENCES

Litwack G, Schmidt TJ. Biochemistry of hormones I: polypeptide hormones. In: Devlin TM, ed. Textbook of Biochemistry with Clinical Correlations, 5th ed. New York: Wiley-Liss, 2002.

Pathophysiology of the endocrine system. An online textbook from Colorado State University: http://www.vivo.colostate.edu/hbooks/pathphys/endocrine/

❖ CASE 48

A 52-year-old female presents to your office with complaints of hot flushes, mood swings, irritability, and vaginal dryness and itching. Her last menstrual period was a little over a year ago. She denies any vaginal discharge. The patient is concerned about having thyroid problems because her friend has similar symptoms and was diagnosed with hyperthyroidism. On examination, the patient is in no acute distress with normal vital signs. Her physical is normal other than thin, atrophic vaginal mucosa. A thyroid-stimulating hormone (TSH) level is drawn and is normal. The follicle-stimulating hormone (FSH) level is drawn and is markedly elevated.

◆ **What is the organ that secretes the follicle-stimulating hormone (FSH)?**

◆ **What is the signal that stimulates the release of FSH?**

ANSWERS TO CASE 48: MENOPAUSE

Summary: A 52-year-old female presents with hot flushes, mood swings, irritability, vaginal dryness and atrophy, and last menstrual period over 1 year ago.

- **Organ secreting follicle-stimulating hormone (FSH):** Anterior pituitary gland.
- **Signal stimulating FSH release:** Gonadotropin releasing hormone from the hypothalamus binds to membrane receptors of the anterior pituitary cells, triggered phosphatidylinositol signaling to stimulate FSH production and release.

CLINICAL CORRELATION

This 52-year-old woman has symptoms of estrogen insufficiency, such as hot flushes, mood swings, and vaginal dryness. Her last menstrual period was 1 year ago, consistent with the menopause. The etiology of the hypoestrogen state is follicular atresia of the ovaries. Once in puberty, a woman usually has fairly regular menstrual cycles as dictated by the estrogen and progesterone secretion of the ovaries until about age 40 to 50 years. During a period of 2 to 4 years, some women may experience irregular menses because of irregular ovulation until finally there are no further menses. The diagnosis of the menopause is made by clinical criteria, although the gonadotropins, FSH, and luteinizing hormone (LH) are usually elevated.

APPROACH TO THE MENSTRUAL CYCLE

Objectives

1. Be familiar with the hormones that control reproduction.
2. Know about the regulation of the normal menstrual cycle.
3. Understand the hormone changes during menopause.

Definitions

Apoptosis: Programmed cell death that leads to the destruction of cells leaving membrane-bound particles that are shed or taken up by phagocytosis.

Estradiol: 17-β-Estradiol; an estrogen steroid hormone that is synthesized in the granulose cells of the ovary and secreted in response to binding of FSH released from the anterior pituitary.

FSH: Follicle-stimulating hormone; a polypeptide hormone that contains α and β subunits that is synthesized in and released from the anterior pituitary in response to binding of GnRH. When binding to receptors on the plasma membrane of the granulose cells of the ovarian follicle, it stimulates the synthesis and secretion of estradiol.

GnRH: Gonadotropin-releasing hormone; a decapeptide containing an *N*-terminal pyroglutaminyl residue and a C-terminal glycinamide residue. GnRH is synthesized and secreted from the hypothalamus and binds receptors on the anterior pituitary.

LH: Luteinizing hormone; a polypeptide hormone that is similar in structure to FSH and, like FSH, is synthesized and secreted from the anterior pituitary in response to binding of GnRH. It binds to receptors on the thecal cells of the ovarian follicle to increase synthesis of androgens. When binding to receptors on the corpus luteum, it increases the synthesis of progesterone.

Progesterone: A steroid hormone synthesized from cholesterol secreted from the corpus luteum in the luteal phase of the menstrual cycle.

DISCUSSION

With the onset of **puberty,** the normal menstrual (ovarian) cycle, as shown in Figure 48-1, is initiated by the pulsatile **release of gonadotropin-releasing hormone (GnRH)** from the hypothalamus. **GnRH binds to plasma membrane receptors** in its target cells in the **anterior pituitary** and triggers activation of the **phosphatidylinositol signaling** pathway. This in turn signals the pituitary to release both FSH and LH from the same cell. **FSH binds its plasma membrane receptor in the ovarian follicle to stimulate, via adenylate cyclase activation, cyclic AMP production and protein kinase A activation,** increased ovarian synthesis and secretion of 17-β-estradiol, the female sex hormone. This leads to maturation of the follicle and ovum. **Estradiol also induces progesterone receptors.** Estrogens circulate in the bloodstream to maintain female primary and secondary sex characteristics. The steroid hormones estradiol and progesterone act by activating their intracellular receptors (E_2R and PR, respectively) to bind their response elements in promoter regions of target genes. Together they promote the thickening and vascularization of the uterine endometrium in preparation for implantation of the fertilized ovum. Ovarian synthesis of inhibin, a negative feedback regulator of FSH (but not LH) production, is also stimulated. As **17-β-estradiol levels reach a maximum,** around **day 13** of the cycle, they stimulate a **massive release of LH** and to a lesser extent FSH, from the pituitary, known as the LH spike. The **LH spike,** together with other factors such as prostaglandin $F_2\alpha$, **triggers ovulation.**

After ovulation, estrogen biosynthesis by the follicle declines, leading to a drop in blood levels of estrogen. The Graafian follicle now differentiates into the corpus luteum, under the mediation of LH. LH binds plasma membrane receptors in the corpus luteum, acting by the adenylate cyclase/protein kinase A signaling pathway, to stimulate progesterone biosynthesis. The uterine endometrial wall becomes secretory in preparation for implantation of the fertilized egg. In the absence of fertilization, the corpus luteum dies, because of decreased levels of LH. This leads to decreased production of progesterone

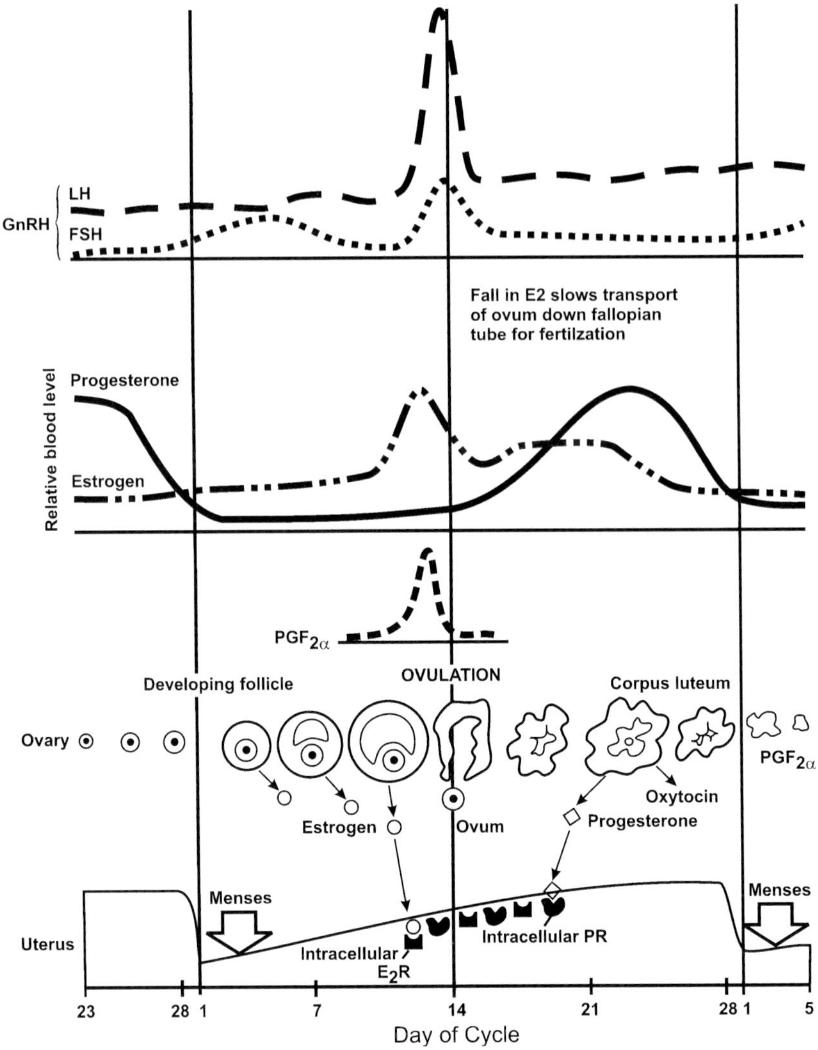

Figure 48-1. The ovarian cycle. *(Reproduced, with permission, from Devlin TM, ed.* Textbook of Biochemistry with Clinical Correlations, *5th ed. New York: Wiley-Liss, Copyright © 2002:935. This material is used by permission of John Wiley & Sons, Inc.)*

and estradiol, culminating with apoptosis of the uterine endometrial cells, and their shedding at day 28 in menstruation. A new follicle then begins to develop. The decline in blood levels of estradiol and progesterone relieves feedback inhibition of the gonadotropes and hypothalamus, leading to GnRH release and initiation of another ovarian cycle.

If fertilization occurs, the corpus luteum remains viable and begins secretion of human chorionic gonadotropin (HCG), a function eventually taken over by the placenta. This hormone **(HCG) is necessary for the maintenance of the endometrium during the first trimester of pregnancy.**

At menopause, beginning on an average at age 51, ovarian production of estrogen and progesterone gradually declines. The **resulting release of feedback inhibition on the pituitary leads to its greatly increased release of FSH and LH.** The adrenal glands continue to produce a minor amount of estrogen. Ovulation stops, and menstruation becomes less frequent and eventually ceases. The postmenopausal ovary and the adrenal gland continue to secrete androgens. The conversion of these androgens to estrogens mainly in fat cells and skin via the enzyme aromatase provides most of circulating estrogen in postmenopausal women.

COMPREHENSION QUESTIONS

[48.1] An obese 57-year-old woman did not yet exhibit symptoms of menopause but was diagnosed with polycystic ovary syndrome (PCOS) and insulin resistance. Her plasma levels of testosterone were above normal. Which one of the following is most likely in this case?

A. Hyperinsulinemia leading to androgen overproduction by the ovary and its conversion to estrogen in fat cells
B. Androgen overproduction by the adrenal gland and its conversion to estrogen in fat cells
C. Progesterone overproduction by the polycystic ovary leading to its conversion to estrogen
D. LH/FSH ratio = 1
E. Estrogen overproduction by the ovary and conversion to testosterone

[48.2] Which one of the following changes is most likely to be observed in a postmenopausal woman who is not taking hormone supplementation?

A. Cessation of androgen secretion
B. Increased levels of FSH and LH
C. Increased osteoblast activity
D. Decreased levels of gonadotropin-releasing hormone
E. Increased progesterone levels

[48.3] In a normal premenopausal woman, which one of the following is stimulated by progesterone?

A. Release of gonadotropin-releasing hormone by the pituitary
B. Ovulation
C. Development of the endometrium in preparation for possible pregnancy
D. Uterine contraction
E. Follicle development

Answers

[48.1] **A.** Hyperandrogenism is primarily ovarian in origin in PCOS women, although a minor contribution from the adrenal gland may occur. Hyperinsulinemia is the primary stimulus. Fat cells convert androgens to estrogens. This may explain the late menopause in this case. In normal postmenopausal women, ovarian production of testosterone continues and this provides the main source of circulating estrogen. Typically an LH/FSH ratio of at least 2.5 is associated with PCOS.

[48.2] **B.** Increased levels of FSH and LH result from decreased estrogen levels and release of feedback inhibition. Androgen secretion continues although diminished. Osteoblast activity decreases, eventually leading to risk of osteoporosis. Progesterone levels decrease, and GnRH levels increase.

[48.3] **C.** Progesterone is secreted by the corpus luteum under the influence of LH. Together with estrogen, it promotes the thickening and maintenance of the endometrium. Progesterone inhibits GnRH release, uterine contraction, and follicle development.

BIOCHEMISTRY PEARLS

- ❖ Puberty is initiated by the pulsatile release of gonadotropin-releasing hormone (GnRH) from the hypothalamus.
- ❖ GnRH binds to plasma membrane receptors in its target cells in the anterior pituitary and triggers activation of the phosphatidylinositol signaling pathway, stimulating the release of both follicle-stimulating hormone (FSH) and luteinizing hormone (LH) from the same cell.
- ❖ FSH binds its plasma membrane receptor in the ovarian follicle to stimulate, via adenylate cyclase activation, cyclic AMP production, and protein kinase, increasing the ovarian synthesis and secretion of 17-β-estradiol, the female sex hormone.
- ❖ Follicular atresia is the cause of hypoestrogenemia in the menopause, which is associated with elevated gonadotropin levels (FSH and LH).

REFERENCES

Litwack G, Schmidt TJ. Biochemistry of hormones I: polypeptide hormones. In: Devlin TM, ed. Textbook of Biochemistry with Clinical Correlations, 5th ed. New York: Wiley-Liss, 2002.

Pathophysiology of the endocrine system. An online textbook from Colorado State University: http://www.vivo.colostate.edu/hbooks/pathphys/endocrine/

❖ CASE 49

A 36-year-old male presents to the clinic with concerns about increasing weakness and fatigue. He reports that his symptoms are present at all times, but at times of stress, the weakness is much worse. He has had some nausea, vomiting, and nonspecific abdominal pain, resulting in involuntary weight loss. His skin coloration has changed similar to a tan and is all over his body. On exam, he is in no distress with normal vitals other than a slightly low blood pressure. His skin is bronze in color with darkening in elbows and creases of the hand, although he states that he avoids the sun. The remainder of the exam is normal.

◆ **What is the likely diagnosis?**

◆ **What is the underlying molecular disorder?**

ANSWERS TO CASE 49: ADDISON DISEASE

Summary: A 36-year-old male presents with fatigue, weakness, nonspecific GI symptoms, hypotension, and darkening skin coloration despite being away from sunlight.

- **Diagnosis:** Addison disease
- **Underlying problem:** Adrenal insufficiency with lack of mineralocorticoids and cortisol

CLINICAL CORRELATION

This 36-year-old male has the clinical stigmata of Addison disease, or adreno-corticoid deficiency. The weakness, fatigue, and hypotension are caused by decreased levels of cortisol as well as mineralocorticoids. The darkening of the skin is a result of increased melanocyte-stimulating hormone, which is metabolized from adrenocorticotropin hormone (ACTH), which is elevated because of the low levels of the adrenal hormones. This is a potentially fatal disorder and requires prompt clinical recognition. Serum electrolytes give a clue to the disorder, because the patient would likely have hyponatremia and hyperkalemia. An ACTH stimulation test revealing low levels of adrenal corticoid response is confirmatory. The treatment includes hydrocortisone (cortisol) to replace glucocorticoid deficiency and a mineralocorticoid supplementation. The most common cause of adrenal insufficiency on a chronic basis is autoimmune destruction of the adrenal gland. Other autoimmune diseases should be sought, such as diabetes mellitus and systemic lupus erythematosus. Adrenal insufficiency may occur acutely when the adrenal gland is not able to generate high levels of steroids in times of stress, such as surgery or infection; the most common reason in these cases is chronic corticosteroid therapy leading to relative adrenal suppression. These situations can be avoided by the administration of a "stress dose" hydrocortisone intravenously around the time of anticipated physiologic stress.

APPROACH TO ADRENAL HORMONES

Objectives

1. Be familiar with the clinical manifestations of mineralocorticoid deficiency.
2. Know about the regulation of ACTH and cortisol.

Definitions

ACTH: Adrenocorticotropin, adrenocorticotropic hormone; a polypeptide hormone released from the anterior pituitary into the bloodstream in response to the binding of CRH. ACTH binds to receptors in the adrenal cortex, causing the synthesis and release of cortisol.

CRH: Corticotropin-releasing hormone; a polypeptide hormone consisting of 41 amino acids that is released from the hypothalamus and travels to the adrenal cortex in a closed portal system. It binds to receptors in the adrenal cortex causing the release of ACTH and β-lipotropin.

Cortisol: Hydrocortisone; a glucocorticoid steroid hormone synthesized by the adrenal gland in response to binding of ACTH. Cortisol binds to cytosolic or nuclear receptors that act as transcription factors for glucocorticoid-responsive genes. In general, cortisol is a catabolic hormone that promotes the breakdown of proteins.

Corticosteroid: A class of steroid hormones that include the glucocorticoids (e.g., cortisol) and mineralocorticoids (e.g., aldosterone). Glucocorticoids are involved in the maintenance of normal blood glucose levels, whereas mineralocorticoids are involved in mineral balance.

Hypoadrenocorticism: A failure of the adrenal cortex to produce glucocorticoid (and in some cases mineralocorticoid) hormones.

α-MSH: Melanocyte-stimulating hormone; a polypeptide hormone derived from the breakdown of ACTH (the *N*-terminal 13-amino acid residues). Released from the anterior pituitary, it acts on skin cells to cause skin darkening by the dispersion of melanin.

Proopiomelanocortin: A precursor protein synthesized in the anterior pituitary from which are generated several polypeptide hormones. It gives rise to ACTH, β-lipotropin, γ-lipotropin, α-MSH, γ-MSH, CLIP, and β-endorphin (and potentially enkephalins and β-MSH).

DISCUSSION

Cortisol is a member of the glucocorticoid steroid hormone family. It acts on almost every organ and tissue in the body in carrying out its vital role in the body's response to stress. Among its crucial functions, it helps maintain blood pressure and cardiovascular function; acts as an antiinflammatory; modulates insulin effects on glucose utilization; and regulates metabolism of protein, carbohydrates, and lipids. In all of its actions, **cortisol interacts with intracellular receptors** to trigger their binding to specific response elements in the **promoters of target genes** to influence **transcription of their messenger ribonucleic acids (mRNAs).**

Cortisol is produced by the adrenal gland under the precise control of the hypothalamus and pituitary. The hypothalamus secretes corticotropin-releasing hormone (CRH) in response to stress. CRH acts on plasma membrane receptors in corticotrophic cells in the anterior pituitary to stimulate their release of adrenocorticotropin (ACTH). Secretion is also under circadian regulation. Vasopressin and angiotensin II augment this positive response but by themselves do not initiate it. ACTH is derived from the precursor polypeptide proopiomelanocortin after cleavage in the pituitary to release ACTH plus β-lipotropin, an endorphin precursor with melanocyte-stimulating activity. **ACTH binds to plasma membrane receptors in the adrenal gland to stimulate production of cortisol, a signaling response mediated by adenylate cyclase** (Figure 49-1). Increased blood cortisol levels exert feedback inhibition of ACTH secretion, a feedback loop acting at multiple levels including the hypothalamus, the pituitary and the central nervous system. Under separate positive control by norepinephrine, the intermediate pituitary converts ACTH to melanocyte-stimulating hormone (β-MSH) plus CLIP (corticotropin-like intermediary peptide).

Most cases of Addison disease are a result of idiopathic atrophy of the adrenal cortex induced by autoimmune responses, although a number of other causes of adrenal cortex destruction have been described. Hypoadrenocorticism results in decreased production of cortisol and, in some

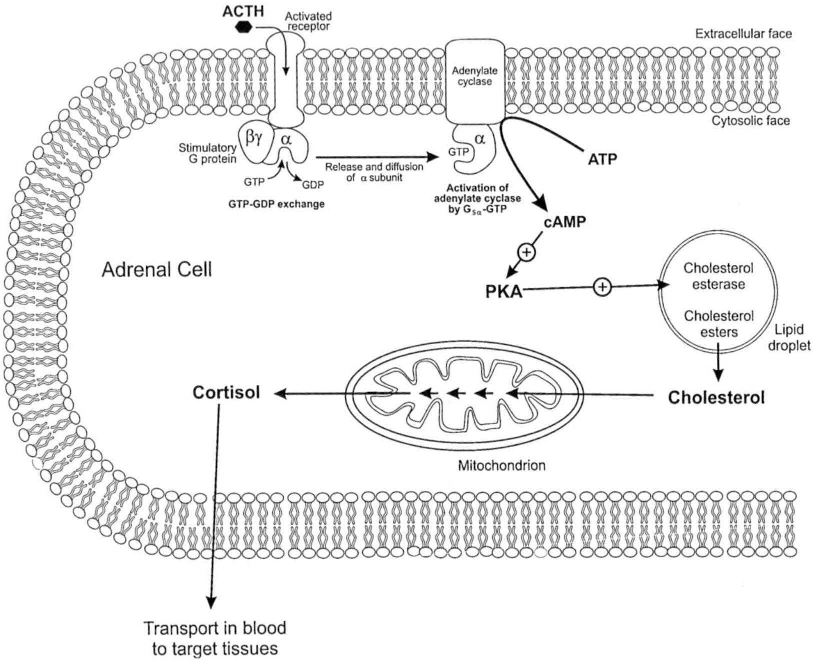

Figure 49-1. The sequence of events leading to the release of cortisol from the adrenal gland. Binding of ACTH to cell surface receptors activates adenylate cyclase to produce cAMP, which in turn activates protein kinase A. This causes phosphorylation events to occur that cause cholesterol to be released from cholesterol ester droplets in the cell. This initiates the conversion of cholesterol to cortisol, which is then released into the bloodstream.

cases, also results in decreased production of aldosterone, the other principal steroid hormone produced by this gland. If aldosterone levels are insufficient, characteristic electrolyte abnormalities are evident owing to **increased excretion of Na^+ and decreased excretion of K^+** chiefly in urine but also in sweat, saliva, and the GI tract. This condition leads to isotonic urine and **decreased blood levels of Na^+ and Cl^- with increased levels of K^+.** Left untreated, **aldosterone insufficiency** produces **severe dehydration, plasma hypertonicity, acidosis, decreased circulatory volume, hypotension, and circulatory collapse.**

Cortisol deficiency impacts carbohydrate, fat, and protein metabolism and produces severe insulin sensitivity. Gluconeogenesis and liver glycogen formation are impaired, and hypoglycemia results. As a consequence, hypotension, muscle weakness, fatigue, vulnerability to infection, and stress are early symptoms. A characteristic hyperpigmentation on both exposed and unexposed parts of the body is evident. Decreased blood levels of cortisol in Addison disease lead to increased blood levels of both ACTH and β-lipotropin reflecting decreased feedback inhibition by cortisol of their synthesis. β-Lipotropin has melanocyte-stimulating activity accounting for the increased pigmentation. The disease is progressive, with the risk of a potentially fatal adrenal crisis triggered by infection or other trauma.

Secondary adrenal insufficiency may result from lesions in the hypothalamus or pituitary, leading to impaired release of ACTH. This condition does not exhibit hyperpigmentation because release of feedback inhibition of ACTH production by low cortisol levels cannot overcome the primary defect in ACTH production. Usually, aldosterone secretion is normal.

By contrast, Cushing syndrome results from cortisol overproduction and most commonly is caused by a tumor in either the adrenal gland or pituitary.

COMPREHENSION QUESTIONS

[49.1] A 29-year-old female patient exhibited a rounded face, hirsutism, upper body obesity, easily bruised skin, severe fatigue, muscle weakness, and anxiety. She also complained of irregular periods. A long-term asthma sufferer, she had been prescribed prednisone for the past 2 years. Findings on examination revealed high fasting blood glucose levels and high blood pressure. Cortisol levels were below normal.

Which one of the following is the most likely explanation to account for the patient's symptoms?

A. Decreased levels of insulin
B. Increased levels of testosterone
C. Decreased secretion of ACTH
D. Excess exogenous glucocorticoid hormone
E. Increased hepatic metabolism of steroid hormones

[49.2] A patient suffering from weakness, fatigue, nausea, and vomiting was found to have low blood concentrations of Na^+ and Cl^- and high levels of serum K^+. Physical examination revealed a deep tanning of both exposed and unexposed parts of the body and dark pigmentation inside the mouth. The hyperpigmentation in this patient most likely resulted from which of the following?

A. Increased secretion of ACTH
B. Prolonged exposure of the patient to ultraviolet radiation
C. Excessive ingestion of β-carotene–containing foods
D. Activation of melanocytes caused by medication side effects
E. Inhibition of plasma membrane Na^+, K^+-ATPase

[49.3] Metyrapone is used to block the mitochondrial 11β-hydroxylase in the corticosteroid synthetic pathway and is administered to evaluate hypothalamic-pituitary-adrenal axis function. Which of the following results is most likely from this overnight diagnostic test in a healthy individual?

A. Feedback inhibition of cortisol biosynthesis
B. Increase in the levels of cortisol precursors
C. Decrease in ACTH levels
D. Inhibition of adenylate cyclase activity in adrenal cortical cells
E. Inhibition of pituitary function

Answers

[49.1] **D.** Prednisone acts as a glucocorticoid hormone analog, giving rise to Cushing syndrome symptoms after prolonged administration.

[49.2] **A.** Hyperpigmentation is a feature of Addison disease, the diagnosis in this case. Decreased plasma cortisol because of adrenal insufficiency releases feedback inhibition of ACTH secretion by the pituitary, resulting in elevation of ACTH biosynthesis. The ACTH precursor peptide is cleaved to also yield melanocyte-stimulating hormone, the factor responsible for hyperpigmentation even in areas not exposed to sunlight.

[49.3] **B.** Inhibition of this step in cortisol biosynthesis relieves feedback inhibition of its biosynthetic enzymes, leading to accumulation of cortisol precursors, particularly 11-deoxycortisol. Normal patients will also have a compensatory increase in the secretion of ACTH.

BIOCHEMISTRY PEARLS

- Most cases of Addison disease are because of idiopathic atrophy of the adrenal cortex induced by autoimmune responses.
- Cortisol interacts with intracellular receptors to trigger their binding to specific response elements in the promoters of target genes to influence transcription of their mRNAs.
- Under separate positive control by norepinephrine, the intermediate pituitary converts ACTH to melanocyte-stimulating hormone, which accounts for the "well tanned" appearance.
- Aldosterone insufficiency produces severe dehydration, plasma hypertonicity, acidosis, decreased circulatory volume, hypotension, and circulatory collapse.

REFERENCES

Litwack G, Schmidt TJ. Biochemistry of hormones I: polypeptide hormones. In: Devlin TM, ed. Textbook of Biochemistry with Clinical Correlations, 5th ed. New York: Wiley-Liss, 2002.

Pathophysiology of the endocrine system. An online textbook from Colorado State University: http://www.vivo.colostate.edu/hbooks/pathphys/endocrine/

❖ CASE 50

A 32-year-old woman presents to her obstetrician/gynecologist with complaints of irregular periods, hirsutism, and mood swings. She also reports weight gain and easy bruising. On examination, she is found to have truncal obesity, a round "moon" face, hypertension, ecchymoses, and abdominal striae. The patient is given a dexamethasone suppression test which reveals an elevated level of cortisol.

- **What is the likely diagnosis?**

- **What would an elevated adrenocorticotropic hormone (ACTH) level indicate?**

ANSWERS TO CASE 50: CUSHING SYNDROME

Summary: A 32-year-old female with irregular menses, hirsutism, mood swings, weight gain, truncal obesity, hypertension, abdominal striae, ecchymoses, and elevated cortisol levels.

- **Diagnosis:** Cushing Syndrome
- **Elevated ACTH level:** Likely cause of adrenal hyperplasia from ACTH-producing tumor

CLINICAL CORRELATION

This patient presents with many of the classic findings of Cushing syndrome. Adrenal hyperplasia can be caused by excessive stimulation from ACTH (pituitary or ectopic production) or from a primary adrenal problem such as adenomas/carcinomas. In addition to above symptoms, patients with Cushing syndrome are also at risk for osteoporosis and diabetes mellitus (DM). The diagnosis is confirmed with elevated cortisol levels after a dexamethasone suppression test. Treatment depends on the underlying etiology and is often surgical.

APPROACH TO CUSHING SYNDROME

Objectives

1. Describe the biosynthesis of steroids in the adrenal gland.
2. Explain from a biochemical standpoint why hypertension is a common consequence of Cushing syndrome.

Definitions

Abdominal striae: Stretch marks of the abdominal region.

ACTH (adrenocorticotropic hormone or corticotropin): Hormone produced in the anterior pituitary, which stimulates adrenal production of cortisol.

Adenoma: Any benign tumor of glandular origin; typically found in the adrenal, pituitary, and thyroid glands (note: once an adenoma has progressed to malignancy, it is referred to as an adenocarcinoma).

CRH (corticotropin-releasing hormone): Hormone produced in the hypothalamus, which stimulates release of ACTH from the anterior pituitary.

Cushing disease: A specific form of Cushing syndrome, which is caused by an ACTH-secreting pituitary adenoma; represents approximately 66 percent of all cases of Cushing syndrome. Because of structural similarities with melanocyte-stimulating hormone (MSH), excess ACTH from pituitary adenomas can induce dermal hyperpigmentation.

Dexamethasone suppression test: An overnight test used to screen patients for Cushing syndrome by administering dexamethasone to a patient. Positive results for this test are indicated by a patient's inability to reduce cortisol levels after dexamethasone treatment—usually because the patient's feedback loop mechanism is ineffective at inhibiting cortisol release.

Ecchymosis: Bruise or contusion; normally comes from damage to the capillaries at the site of injury, allowing blood to seep out into the surrounding tissue, presenting initially as a blue or purple color.

Ectopic ACTH syndrome: Form of Cushing syndrome in which benign or malignant tumors arise in places other than the pituitary, leading to excessive release of ACTH and subsequently, cortisol into the bloodstream; represents approximately 10 to 15 percent of Cushing syndrome cases.

Hirsutism: Increased presence of hair in women on body regions where hair does not normally grow.

Hypercortisolism: A condition in which the body is exposed to an excess of cortisol for an extended period of time.

Iatrogenic Cushing syndrome: Condition in which all symptoms of Cushing syndrome are brought on by administration of synthetic forms of cortisol, such as prednisone and dexamethasone. "Iatrogenic" originates from Greek and literally translates to mean "born from" the "healer."

Pseudo-Cushing syndrome: Condition in which alcohol induces symptoms of Cushing syndrome without the tumor that leads to increased cortisol levels.

DISCUSSION

Cortisol is a stress hormone released in response to trauma—physical and emotional—that leads to several physiologic changes aimed at reducing the stress associated with this trauma. This process is helpful to the body because the activity of cortisol can limit the harmful effects of stress. However, if too much cortisol is secreted (hypercortisolism) symptoms of Cushing syndrome may appear.

Cortisol is secreted from the adrenal glands of the kidneys ("adrenal" literally means near or at the kidney). All steroid hormones are synthesized from cholesterol, with the rate-limiting step in steroid biosynthesis being the cleavage of the cholesterol side chain. This is done by several enzymes that make up the cytochrome P450 side-chain cleavage complex. High levels of Ca^{2+} and protein phosphorylation—due to increased cAMP in the cytosol—increase the rate of cholesterol side-chain cleavage in the mitochondria. First, cholesterol is mobilized into the mitochondria of the adrenal cortex cells, where its side chain is cleaved by the cytochrome P450 cleavage complex (*CYP11A1*) to yield pregnenolone (Figure 50-1). Pregnenolone is oxidized by 3β-hydroxysteroid dehydrogenase to form progesterone, which is then converted

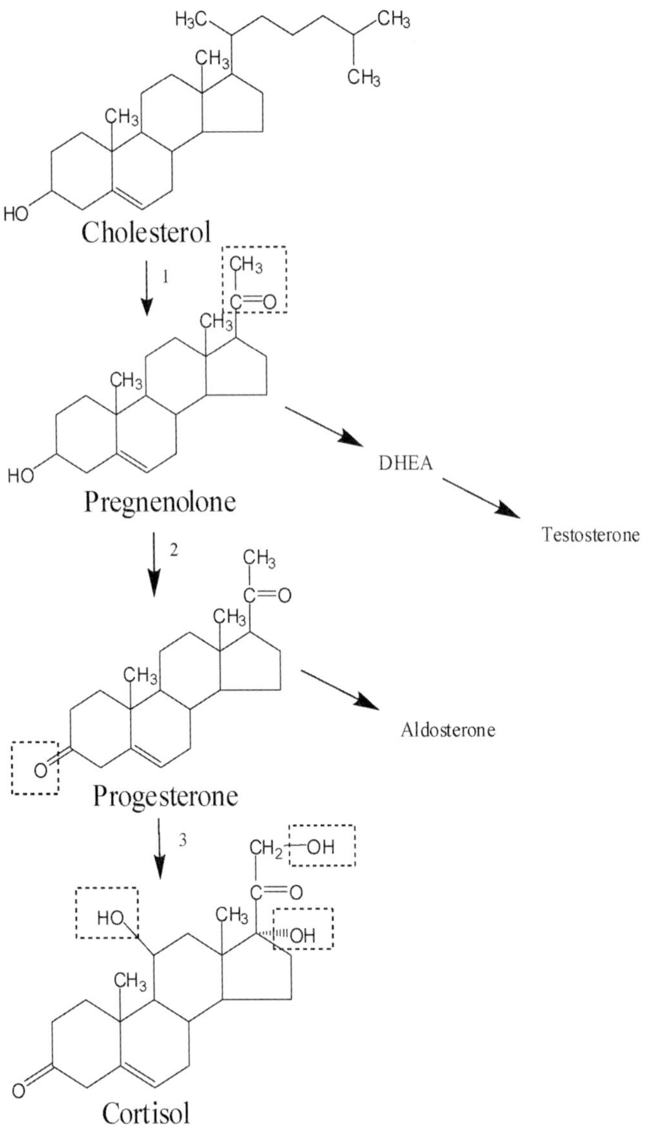

Figure 50-1. Biosynthesis of cortisol. Cholesterol is the starting material for all steroid synthesis. Cortisol synthesis involves a series of oxidation reactions catalyzed by cytochrome P450 enzymes (1) The cleavage of cholesterol's six-carbon side chain is catalyzed by *CYP11A1*—the rate-limiting step in steroid synthesis. (2) This is followed by dehydrogenation of the hydroxyl group in pregnenolone, which is catalyzed by 3β-hydroxysteroid dehydrogenase. (3) This molecule is then oxygenated at carbons 17, 21, and 11 by *CYP17*, *CYP21A2*, and *CYP11B1*, respectively, to yield cortisol. All structure modifications are indicated by dashed lines.

into cortisol by the action of three cytochrome P450 enzymes, *CYP17, CYP21A2,* and *CYP11B1*. After synthesis is complete, cortisol is released from the zona fasciculata in the adrenal cortex via free diffusion into the blood stream for distribution to its target organs, such as liver and kidney.

The adrenal gland produces cortisol in response to intermediate hormones, called **adrenocorticotropin hormone (ACTH)** and **corticotropin-releasing hormone (CRH)** via the humoral stress pathway, which extends from the brain to the adrenals (Figure 50-2). Once the body has received an environmental stress signal, it is detected by neurons in the cerebral cortex and transmitted to the hypothalamus. The hypothalamus then releases CRH via the classic secretory pathway into the anterior pituitary. CRH stimulates release of

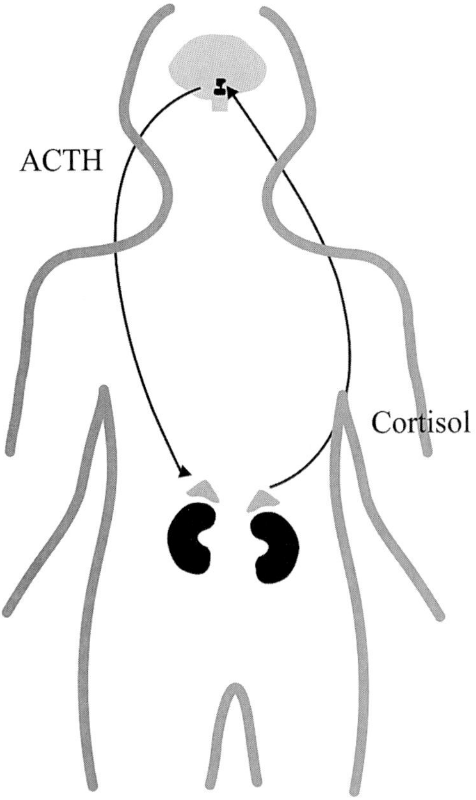

Figure 50-2. Induction of the humoral stress pathway. Once an environmental stress is detected by the CNS, this signals the hypothalamus to release CRH into the anterior pituitary. ACTH is then released from the pituitary into the blood and eventually binds membrane receptors in cells of the adrenal cortex. Cortisol is synthesized and leaves the cell via free diffusion into the blood and is eventually distributed to the hypothalamus and pituitary to inhibit release of CRH and ACTH, respectively, through a negative feedback loop.

ACTH, which then moves into the blood and down to the adrenal glands. Once in the zona fasciculata of the adrenal cortex, ACTH binds membrane receptors stimulating transfer of cholesterol to the mitochondria for cleavage events leading to synthesis and release of cortisol. After diffusing out of the cell into the blood stream, cortisol binds corticosteroid-binding globulin (CBG), also known as transcortin. Cortisol then moves to receptors back in the pituitary to signal reduction of ACTH and CRH production, thus regulating cortisol synthesis by a negative feedback mechanism.

As the major stress hormone, cortisol has many functions. For example, in trabecular bone, cortisol inhibits synthesis of new bone by osteoblasts and decreases absorption of Ca^{2+} in the GI tract, leading to osteopenia. However, the two principal influences of cortisol are on metabolism and the immune system.

Metabolism: Cortisol is catabolic and carries out lipolysis and muscle tissue degradation. Muscle catabolism provides a source of amino acids used by the liver to fuel gluconeogenesis and increases blood glucose levels. Proteolysis of collagen can lead to skin fragility, easy bruising, and striae. Lipolysis, or lipid degradation, generates free fatty acids in the blood, which when degraded by β-oxidation in the liver provide an alternative energy source, decreasing the demand for glucose. Increased lipolysis in Cushing syndrome is also thought to cause the redirection of fat deposition away from the limbs toward the trunk, leading to symptoms, such as "buffalo hump" and "moon face." In addition to increasing blood glucose levels, too much cortisol can inhibit insulin activity and exacerbate diabetic complications. Cortisol and cAMP induce two enzymes phosphoenolpyruvate carboxykinase (PEPCK) and glucose-6-phosphatase (G6Pase), both of which lead to increased glucose levels. The effect of cortisol is to induce PEPCK and G6Pase gene expression through a glucocorticoid response element (GRE) upstream of each gene. PEPCK and G6Pase both increase the rate of gluconeogenesis and antagonize insulin response activity, leading to increased blood glucose.

Immune System: Cortisol also has immunosuppressive effects and can reduce inflammation. For example, other synthetic forms of cortisol, such as hydrocortisone, are used medicinally to reduce inflammation. Cortisol's immunosuppressive effects are partly a result of its ability to sequester lymphocytes in the spleen, thymus, and bone marrow. Most other immunosuppressive effects come from cortisol's ability to modulate gene transcription. Cortisol diffuses into the cell and binds glucocorticoid receptors, which disaggregate into single, or monomer, proteins. Now that the activated receptor, or transcription factor, has a free DNA binding domain it will translocate to the nucleus to alter gene expression. This activity induces transcription of several immunosuppressive genes that inhibit expression of target genes, such as IL-2.

Cushing syndrome is a rare condition in which elevated levels of cortisol are present in the patient for an extended period of time (hypercortisolism). Typically, it affects those between the ages of 20 and 50 years. On rare occasions, this disease may result from an inherited condition, which leads to growth of adenomas in endocrine glands, such as the adrenal, parathyroid, pancreas or pituitary glands.

Benign tumors in the pituitary or adrenal glands can lead to excess release of ACTH or cortisol into the blood, causing the symptoms of Cushing syndrome. Typically, excess levels of endogenous cortisol—not synthetic forms of cortisol—can cause hypertension, most likely because of its weak binding capacity with mineralocorticoid receptors (i.e., aldosterone receptors).

Typically, hypercortisolism comes from: (1) A pituitary adenoma, which secretes excess ACTH, thus causing release of cortisol from the adrenals; (2) an adrenal adenoma, which secretes excess cortisol; (3) a lung tumor, which secretes excess ACTH; or (4) administration of synthetic forms of cortisol (i.e., dexamethasone, prednisone, etc.) due to previously diagnosed diseases, such as rheumatoid arthritis. In the case of a pituitary adenoma, the location of the adenoma makes the pituitary gland insensitive to the negative feedback mechanism brought on by excess cortisol in the blood. Sixty-six percent of all cases of Cushing syndrome are derived from pituitary adenomas. When hypercortisolism is due to exogenous administration of steroids it is referred to as **iatrogenic Cushing syndrome.**

Also, because cortisol is a stress hormone, people who suffer a great deal of stress, such as athletes, alcoholics, and pregnant women may have high blood cortisol levels and exhibit symptoms of Cushing syndrome (also known as **pseudo-Cushing syndrome**).

Treatment of Cushing syndrome is intended to return cortisol levels back to normal and usually occurs through surgery. In some cases medications, such as mitotane, which lower blood and urine cortisol levels, can be used alone or in combination with radiation therapy.

COMPREHENSION QUESTIONS

A 45-year-old female patient presents with hirsutism, striae, bruising, acne, and hyperpigmentation of the skin. After a thorough physical examination the physician notes that she also suffers from hypertension and shows signs of a "buffalo hump" on her back between the shoulders. Cushing syndrome is suspected and after laboratory tests show elevated blood cortisol levels she is given a dexamethasone suppression test. Her results are positive.

[50.1] Following administration of dexamethasone, this patient exhibits elevated cortisol levels (a positive result) because of which of the following?

A. The humoral stress pathway can no longer regulate cortisol levels via a negative feedback loop.
B. There is a deficiency in the enzyme that breaks down dexamethasone, leading to excess amounts of glucocorticoid in the blood
C. The anterior pituitary is nonresponsive to excess cortisol and is aberrantly producing excess CRH
D. A CRH-secreting tumor of the adrenal glands is stimulating cortisol synthesis and is no longer responding to the negative feedback loop

[50.2] The hirsutism observed in the patient above is best explained by which of the following?

A. ACTH stimulating synthesis of adrenal androgens
B. CRH stimulating synthesis of adrenal androgens
C. Cortisol activating aldosterone receptors
D. Cortisol stimulating expression of adrenal androgen biosynthetic enzymes

[50.3] The characteristic accumulation of adipose tissue in the facial, truncal, and cervical regions of the body in patients with Cushing syndrome is best explained by which of the following?

A. Excess adrenal androgens due to adrenal tumor
B. Excess cortisol over long periods of time
C. Cross stimulation of mineralocorticoid receptors by cortisol
D. Increased production of MSH
E. Increased proteolysis

[50.4] The most common cause of hypercortisolism is which of the following?

A. Adrenal tumor that secretes excess cortisol and mineralocorticoid hormones.
B. Lung tumor that secretes excess ACTH that leads to excess cortisol in the blood.
C. Administration of synthetic cortisol by physician
D. Pituitary adenoma that secretes excess ACTH, leading to excess cortisol in the blood.

Answers

[50.1] **A.** The body's humoral stress pathway can no longer regulate cortisol levels via a negative feedback loop. By administering a synthetic form of cortisol (dexamethasone), the physician is testing the body's natural ability to reduce cortisol production. However, a patient with Cushing syndrome produces excess cortisol from one of several major sources, such as an ACTH-secreting pituitary adenoma, an ACTH-secreting lung tumor, or a cortisol-secreting adrenal tumor. Either one of these sources will be unaffected by the dexamethasone suppression test, thus exhibiting no change in blood cortisol levels. If the stress humoral pathway is intact, the test will show a drop in cortisol levels.

[50.2] **A.** Androgenic phenotypes are commonly seen in Cushing syndrome because ACTH, frequently in excess, stimulates synthesis of adrenal androgens in addition to cortisol.

[50.3] **B.** While the exact mechanism is not known, excess cortisol over long periods of time mobilizes lipids and redirects fat deposition away from the peripheral regions to the truncal region creating a "buffalo hump" appearance between the shoulders and a "moon face."

[50.4] **D.** While all of the choices can lead to hypercortisolism, about two-thirds of all cases are due to a pituitary adenoma that secretes excess ACTH, leading to excess cortisol in the blood.

BIOCHEMISTRY PEARLS

- ❖ All steroid hormones are synthesized from cholesterol, with the rate-limiting step in steroid biosynthesis being the cleavage of the cholesterol side chain.
- ❖ Cortisol is catabolic and carries out lipolysis and muscle tissue degradation.
- ❖ Cushing syndrome occurs from hypercortisolism over a long period of time.
- ❖ Cortisol has numerous effects on the body, and in excess, can cause hypertension, lipolysis, "buffalo hump," "moon faces," easy bruising of the skin, and truncal obesity.

REFERENCES

Boron WF, Boulpaep EL. Medical Physiology: A Cellular and Molecular Approach. Philadelphia, PA: Elsevier Science, 2003; 1319.

Lin B, Morris DW, Chou JY. Hepatocyte nuclear factor 1alpha is an accessory factor required for activation of glucose-6-phosphatase gene transcription by glucocorticoids. DNA Cell Biol 1998;17:967–4.

Mahmoud-Ahmed AS, Suh JH. Radiation therapy for Cushing's disease: a review. Pituitary 2002;5:175–80.

Muller OA, von Werder K. Ectopic production of ACTH and corticotropin-releasing hormone (CRH). J Steroid Biochem Mol Biol 1992;43:403–8.

Petersen DD, Magnuson MA, Granner DK. Location and characterization of two widely separated glucocorticoid response elements in the phosphoenolpyruvate carboxykinase gene. Mol Cell Biol 1988;8:96–104.

Phillips PJ, Weightman W. Skin and Cushing syndrome. Aust Fam Physician 2007;36:545–7.

❖ CASE 51

A 42-year-old female presents to the clinic with complaints of vague abdominal discomfort, weakness and fatigue, and bone pain. The patient gives no personal or family history of medical problems. The patient did remember that she suffer from frequent urinary tract infections and has had several episodes of kidney stones. Her physical examination was within normal limits. The patient had a normal complete blood count (CBC), and electrolytes revealed a significantly elevated calcium level and low phosphorus level.

- **What is the most likely diagnosis?**

- **What is the biochemical secondary messenger activated in this disorder?**

ANSWERS TO CASE 51: HYPERPARATHYROIDISM

Summary: A 42-year-old female, who has a history of frequent urinary tract infections and kidney stones, is found to have vague abdominal pain and weakness, with hypercalcemia and decreased phosphorus levels.

- **Diagnosis:** Hyperparathyroidism, leading to hypercalcemia and hyperphosphatemia.
- **Biochemical mechanism:** Elevated parathyroid hormone level acts by binding its 7-helix plasma membrane receptor to activate the adenylate cyclase/protein kinase A signaling system.

CLINICAL CORRELATION

This patient presents with kidney stones, which causes severe flank pain. The most common causes of hypercalcemia include malignancies or hyperparathyroidism. Other causes include granulomatous disorders such as sarcoid and tuberculosis, and less commonly, hypercalcemia may be the presentation of intoxication with vitamins A or D, or calcium-containing antacids, or occur as a side effect of drug therapies like lithium or thiazide diuretics. Genetic conditions like familial hypocalciuric hypercalcemia and hyperparathyroidism as part of a multiple endocrine neoplasia syndrome are also uncommon. **Primary hyperparathyroidism,** usually because of a **solitary parathyroid adenoma,** is the most likely cause when hypercalcemia is discovered in an otherwise **asymptomatic** patient on routine laboratory screening. Most patients have no symptoms with mild hypercalcemia below 12.0 mg/dL, except perhaps some polyuria and dehydration. With levels above 13 mg/dL, patients begin developing increasingly severe symptoms. These include central nervous system (CNS) symptoms (lethargy, stupor, coma, mental status changes, psychosis), gastrointestinal symptoms (anorexia, nausea, constipation, peptic ulcer disease), kidney problems (polyuria, nephrolithiasis), and musculoskeletal complaints (arthralgias, myalgias, weakness.) The **symptoms of hyperparathyroidism** can be remembered as: **stones** (kidney), **moans** (abdominal pain), **groans** (myalgias), **bones** (bone pain), and **psychiatric overtones.** Diagnosis can be established by finding hypercalcemia and hypophosphatemia, with inappropriately elevated PTH levels. Symptomatic patients can be treated with parathyroidectomy.

APPROACH TO CALCIUM METABOLISM

Objectives

1. Be familiar with calcium metabolism.
2. Know about regulation of serum calcium including the roles of parathyroid hormone and calcitonin.

Definitions

Calcitonin: A polypeptide hormone of 32-amino acid residues that is synthesized in the parafollicular cells (C cells) of the thyroid gland. Calcitonin is secreted in response to elevated blood Ca^{2+} levels.

IP_3: Inositol 1,4,5-trisphosphate (inositol trisphosphate); a second messenger released by the action of phospholipase C on phosphatidylinositol 4,5-bisphosphate (PIP_2). IP_3 will bind to receptors on the endoplasmic reticulum (ER) to cause the rapid efflux of Ca^{2+} from the ER into the cytoplasm.

Phosphoinositide cascade: The sequence of events that follow the binding of a hormone to a receptor that acts via a Gq-protein. Hormone binding to the Gq-coupled receptor activates phospholipase C, which cleaves PIP_2 to IP_3 and diacylglycerol, both of which are second messengers.

PTH: Parathyroid hormone; an 84-amino acid polypeptide hormone that is synthesized in the parathyroid gland and is secreted in response to low blood Ca^{2+} levels. PTH acts to increase the Ca^{2+} concentration in the blood by stimulating osteoclast formation and activity, thus releasing bone calcium and phosphate into the blood.

Vitamin D: Vitamin D_3, a secosteroid (a steroid in which one of the rings has been opened) formed by the action of UV light on 7-dehydrocholesterol. The active form of vitamin D is the hormone 1,25-dihydroxycholecalciferol (calcitriol), formed in the kidney in response to elevated PTH levels. It binds to nuclear receptors in intestine, bone, and kidney to activate the expression of calcium-binding proteins.

DISCUSSION

Owing to the critical importance of calcium ion in a wide range of physiologic processes, blood calcium ion concentrations are tightly regulated. **Hypocalcemia rapidly leads to muscle spasm, tetany, cardiac dysfunction, and numerous other symptoms.** About half of the calcium ions in blood are bound to protein, and half are in the unbound state. Blood calcium ion concentrations are close to 1 mM, a value 10,000 times higher than free calcium ion concentrations in the cytoplasm. Normal calcium and phosphate concentrations in blood are near their solubility limit. Thus elevation in these levels leads to precipitate formation and organ damage.

Intracellular calcium ion is largely sequestered within mitochondria and the lumen of the endoplasmic reticulum. Inositol trisphosphate (IP_3)—a second messenger of hormones that acts via stimulation of the phosphoinositide cascade (Figure 51-1)—binds its receptor on the endoplasmic reticulum membrane, triggering a rapid efflux of calcium ion from this intracellular store into the cytoplasm. The **calcium-binding protein calmodulin** senses fluctuations in intracellular calcium ion concentration by altering its conformation, thus influencing activities of its numerous enzyme ligands,

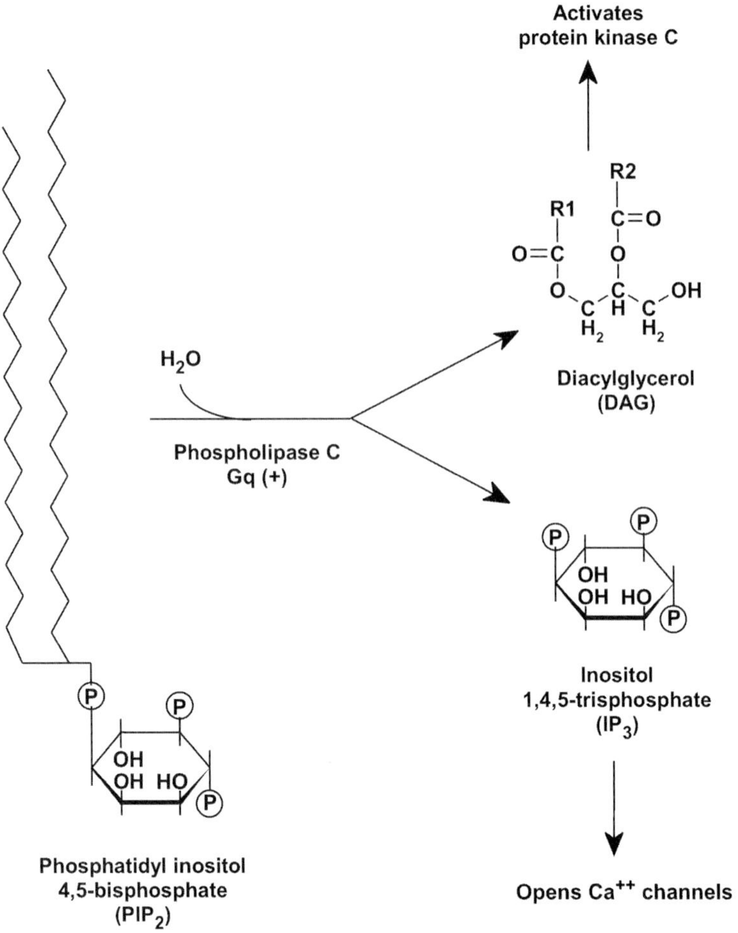

Figure 51-1. The phosphoinositide cascade. Binding of hormone to the Gq-coupled receptor leads to the activation of phospholipase C, which hydrolyzes PIP_2 in the membrane to release the two second messengers, DAG and IP_3.

including the **calcium ATPase.** This spike in intracellular calcium concentration is restored to the resting level by the activity of calcium transporters including the calcium ATPase and the sodium/calcium exchanger.

Parathyroid hormone (PTH) plays a crucial role in regulating concentrations of calcium and phosphorus in the extracellular fluid. The major signal for parathyroid hormone release is low extracellular levels of free calcium ion. PTH acts on 3 major targets—small intestine, kidney, and bone—to restore calcium ion concentrations in the extracellular fluid to the normal range if they fall too low.

PTH acts by binding its 7-helix plasma membrane receptor to activate the adenylate cyclase/protein kinase A signaling system. In some tissues receptor binding is also coupled to the phosphoinositide signaling system to activate protein kinase C.

PTH is synthesized as a 115-amino acid preprohormone in the chief cells of the parathyroid gland, where it is processed by proteolysis to an active 84-amino acid polypeptide and packaged in secretory vesicles. When **blood calcium ion levels fall** below normal, the active **hormone is secreted by exocytosis** into the blood. The parathyroid cell monitors calcium ion levels by means of a calcium-sensing receptor in the plasma membrane. Phosphate ion is a much less significant regulator of PTH secretion and does so by an indirect mechanism. Within seconds of calcium binding its receptor on the chief cells of the parathyroid gland, PTH secretion is decreased. Within a matter of hours, transcription of the preprohormone messenger ribonucleic acid (mRNA) diminishes. If hypocalcemia persists over days and months, the parathyroid gland enlarges in an effort to increase PTH production.

Bone serves as a vast reservoir of calcium in the body. Approximately 1 percent of calcium in bone can rapidly exchange with extracellular calcium ion. **PTH stimulates demineralization of bone and release of calcium and phosphate into the blood by stimulating osteoclast** formation and activity. This process is synergistically enhanced by vitamin D.

PTH also acts to **increase absorption of calcium ion by the small intestine.** It does this indirectly by promoting the formation of active vitamin D in the kidney. PTH acts on the final, rate-limiting step in vitamin D synthesis, the formation of 1,25-dihydroxycholecalciferol in the kidney. If PTH is low, formation of the inactive derivative, 24,25-dihydroxycholecalciferol, is stimulated instead. Vitamin D acts on intracellular receptors in the small intestine to increase transcription of genes encoding calcium uptake systems, to up-regulate their expression.

The kidney plays a critical role in calcium homeostasis. **PTH acts directly on the kidney to suppress calcium ion excretion in the urine** by maximizing tubular calcium reabsorption. It increases phosphate ion excretion in the kidney (phosphaturic effect) to prevent excessive accumulation of this anion released during bone demineralization.

Hyperparathyroidism results from oversecretion of PTH. This condition leads to excessive bone turnover and demineralization and must be treated by removal of the parathyroid gland. The disorder is classified into primary, secondary, and tertiary hyperparathyroidism. Sporadic primary hyperparathyroidism is the third most common endocrine disorder, after diabetes and hyperthyroidism. It is most common in females older than 55 years of age and the leading cause is a single adenoma, which secretes the hormone constitutively, without regulation. Symptoms can include osteopenia and bone fractures, renal stones resulting from hypercalciuria, peptic ulcer disease, and pancreatitis. In milder cases, patients are asymptomatic or suffer only muscle weakness, fatigue, and/or depression.

Secondary hyperparathyroidism arises from chronic hypocalcemia. This condition can result from renal failure leading to poor reabsorption of calcium from the urinary filtrate. It can also arise from poor nutrition or malabsorption of vitamin D by the intestine. In response, parathyroid glands increase their secretion of PTH. This condition also leads to decalcification of bone. Tertiary hyperparathyroidism is often seen after renal transplantation. In these patients, the parathyroid gland secretes the hormone independently of blood calcium levels.

Paradoxically, although **chronic exposure to high PTH levels leads to bone decalcification,** administration in **pulses, as a once-a-day injection, stimulates an increase in bone mass.** This treatment is now used as an effective therapy for osteoporosis. **Calcitonin,** secreted by the parafollicular C cells of the thyroid gland, **opposes effects of PTH.** Calcitonin is secreted when blood levels of calcium are too high, and it acts to suppress reabsorption of calcium in the kidney and inhibit bone demineralization. However, in humans, it plays a minor role in regulating blood levels of calcium ion.

COMPREHENSION QUESTIONS

[51.1] A 54-year-old patient complained of muscle weakness, fatigue, and depression. She had a recent episode of renal stones and a bone scan revealed osteopenia. She had not yet entered menopause. She has taken a daily multivitamin tablet plus an additional 500 mg of calcium citrate for the past 20 years. Results from blood chemistry analysis indicated elevated levels of serum calcium ion. Urinalysis indicated phosphaturia. The patient's symptoms are most likely caused by which of the following?

A. Excessive intake of vitamin D
B. Excess synthesis of parathyroid hormone
C. Excessive intake of calcium
D. Excess synthesis of calcitonin
E. Excess intake of phosphate

[51.2] Before the introduction of vitamin D-fortified milk, children who spent most of their time indoors often developed rickets. In these vitamin D-deficient children, the most likely explanation for their bone malformations is which of the following?

A. Excessive renal excretion of calcium
B. Excessive renal excretion of phosphate ion
C. Inadequate uptake of calcium in the intestine
D. Lack of weight-bearing exercise
E. Excessive renal tubular reabsorption of calcium

[51.3] Why can excessive ingestion of phosphate-containing soft drink in otherwise well-nourished individuals lead to decreased bone density?

A. Increased levels of blood phosphate ion interact with sensors on the membrane of parathyroid cells to stimulate parathyroid hormone release.
B. Phosphate ion binds the active site on calcium transporters in the intestine, inhibiting their ability to transport calcium.
C. Phosphate ion depletes cellular levels of ATP resulting in inhibition of Ca^{2+}–ATPase calcium transporters.
D. Phosphate ion spontaneously forms an insoluble precipitate with calcium ion, decreasing its absorption in the intestine.
E. Phosphate ion is excessively incorporated into bone, weakening its structure.

Answers

[51.1] **B.** Hyperparathyroidism is the likely cause of all of the patient's symptoms. Increased parathyroid hormone leads to bone demineralization, increased calcium uptake from the intestine, increased blood levels of calcium, decreased calcium ion excretion by the kidney, and increased phosphate excretion in the urine. Increased blood calcium levels caused renal stones, while bone demineralization progressed to osteopenia. The patient's intake of calcium and vitamin D are not excessive. Calcitonin acts to decrease bone demineralization. Muscle weakness and depression reflect the widespread role of calcium ion in many physiologic processes.

[51.2] **C.** The major action of vitamin D is to increase absorption of calcium from the small intestine. Deficiency of the vitamin leads to low blood calcium levels, stimulation of parathyroid hormone secretion, and acting synergistically, promotion of bone demineralization. Renal excretion of calcium is decreased by hypocalcemia but elevated parathyroid hormone levels promote renal excretion of phosphate, to prevent excessive accumulation of this product of bone demineralization. Although lack of exercise decreases bone density, it does not lead to rickets if vitamin D is sufficient.

[51.3] **D.** Precipitate formation effectively decreases available calcium ion. Because of widespread consumption of soft drinks rather than water by school age children aided by school vending machines, this consequence is of concern.

BIOCHEMISTRY PEARLS

- ❖ Intracellular calcium ion is largely sequestered within mitochondria and the lumen of the endoplasmic reticulum.
- ❖ PTH acts on three major targets—small intestine, kidney, and bone—to restore calcium ion concentrations in the extracellular fluid to the normal range if they fall too low.
- ❖ Hyperparathyroidism causes elevated levels of calcium and phosphate.
- ❖ PTH acts by binding its 7-helix plasma membrane receptor to activate the adenylate cyclase/protein kinase A signaling system.
- ❖ PTH stimulates demineralization of bone and release of calcium and phosphate into the blood by stimulating osteoclast formation and activity, increases small bowel absorption of calcium ion, and acts directly on the kidney to suppress calcium ion excretion in the urine.

REFERENCES

Chaney SG. Principles of nutrition II: micronutrients. In: Devlin TM, ed. Textbook of Biochemistry with Clinical Correlations, 5th ed. New York: Wiley-Liss, 2002.

Litwack G, Schmidt TJ. Biochemistry of hormones I: polypeptide hormones. In: Devlin TM, ed. Textbook of Biochemistry with Clinical Correlations, 5th ed. New York: Wiley-Liss, 2002.

Pathophysiology of the endocrine system. An online textbook from Colorado State University: http://www.vivo.colostate.edu/hbooks/pathphys/endocrine/

SECTION III

Listing of Cases

LISTING BY CASE NUMBER

LISTING BY DISORDER (ALPHABETICAL)

INDEX

Note: Page numbers followed by *f* or *t* indicate figures or tables, respectively.

D

E

G

I